LASERS IN ATOMIC, MOLECULAR AND NUCLEAR PHYSICS

LASERS IN ATOMIC, MOLECULAR AND NUCLEAR PHYSICS

Edited by V. S. Letokhov

Institute of Spectroscopy
Academy of Sciences USSR, Moscow

Proceedings of the Third International School on Laser Applications in Atomic, Molecular and Nuclear Physics

*August 27–September 4, 1984
Vilnius, USSR*

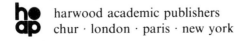

harwood academic publishers
chur · london · paris · new york

© 1986 by Harwood Academic Publishers GmbH, Poststrasse 22, 7000 Chur, Switzerland. All rights reserved.

Post Office Box 197
London WC2E 9PX
England

58, rue Lhomond
75005 Paris
France

Post Office Box 786
Cooper Station
New York, New York 10276
United States of America

Library of Congress Cataloging-in-Publication Data
International School on Laser Applications in Atomic,
  Molecular and Nuclear Physics (3rd: 1984:
  Vilnius, Lithuania)
  Lasers in atomic, molecular and nuclear physics.

Includes index.
1. Lasers—Congresses.   2. Nuclear physics—
Congresses.   I. Letokhov, V. S.   II. Title.
QC685.I58   1984   535.5′8   86-264
ISBN 3-7186-0348-9

ISBN: 3-7186-0348-9. No part of this book may be reproduced or utilized in any form or by any means, electronic or mechanical, including photocopying and recording, or by any information storage or retrieval system, without permission in writing from the publishers. Printed in Great Britain by Bell and Bain Ltd., Glasgow.

The following articles in this book were first published in the journal *Laser Chemistry*, Volume 6 (1986).

State-selected photodissociation dynamics of formaldehyde
By C. B. Moore and D. J. Bamford   Issue no. 2, p. 93

Inverse electronic relaxation at IR multiple-photon excitation of molecules
By A. A. Puretzky   Issue no. 2, p. 103

Laser induced chemical reaction in combustion and industrial processes
By J. Wolfrum   Issue no. 2, p. 125

Lasers and reactive collisions: the Cs^*-H_2 reaction
By R. Vetter   Issue no. 2, p. 149

Space-time holography of picosecond pulsed fields in highly-selective photochromic media
By R. Kaarli, J. Kikas, A Rebane and P. Saari   Issue no. 3, p. 165

Photochemical and photophysical hole burning in electronic spectra of complex organix molecules
By R. I. Personov and B. M. Kharlamov   Issue no. 3, p. 181

Nonlinear spectroscopy of highly-excited molecules and condensed media
By N. I. Koroteev   Issue no. 3, p. 203

Use of lasers in photophysical research of photosynthesis
By G. Laczkó, P. Maróti and L. Szalay   Issue no. 3, p. 219

Primary processes of photosynthesis studied by fluorescence spectroscopy methods
By A. Freiberg   Issue no. 4, p. 233

Influence of structural heterogeneity on energy migration in photosynthesis
By L. Valkunas   Issue no. 4, p. 253

Picosecond spectroscopy of photoreceptor molecules
By F. R. Aussenegg, M. E. Lippitsch and M. Riegler   Issue no. 4, p. 269

Picosecond processes of photosynthesis in laser adsorption studies
By A. Yu. Borisov, A. P. Razjivin, R. V. Danielius and R. J. Rotomskis
Issue no. 5, p. 291

The structure, function and assembly of the light-harvesting antenna of photosynthetic purple bacteria
By R. Van Grondelle   Issue no. 5, p. 307

Picosecond relaxation processes in monomeric and aggregated dyes under solvent influence
By S. K. Rentsch   Issue no. 5, p. 319

Picosecond spectroscopy in study of the photoinduced isomerizations of hexafluorobenzene
By J. L. Suijker, A. H. Huizer and G. A. G. O. Varma   Issue no. 5, p. 333

Contents

Preface   xi

1. Atoms and Nucleii

1.1. Laser detection of very rare isotopes
  By V. S. Letokhov   1
1.2. Collinear laser-fast beam spectroscopy on unstable nuclides
  By R. Neugart   23
1.3. Photoionization laser spectroscopy of short-lived isotopes
  By V. I. Mishin   47
1.4. Laser cooling and localization of atoms
  By V. G. Minogin   61
1.5. New trends in the structure and classification of the states of atoms and ions
  By Z. B. Rudzikas and J. M. Kaniauskas   77
1.6. Spectroscopy of highly-excited states of rare earth elements
  By E. Vidolova-Angelova   93
1.7. The atomic collisions in laser fields
  By A. M. Bonch-Bruevich, S. G. Prjibelsky and V. V. Kchromov   111

2. Molecules and Laser Induced Processes

2.1. State-selected photodissociation dynamics of formaldehyde
  By C. B. Moore and D. J. Bamford   125
2.2. Inverse electronic relaxation at IR multiple-photon excitation of molecules
  By A. A. Puretzky   135

2.3. Laser induced chemical reactions in combustion and industrial processes
By J. Wolfrum — 157

2.4. Lasers and reactive collisions: the Cs*-H_2 reaction
By R. Vetter — 181

3. Laser Study of Condensed Matter

3.1. Space-time holography of picosecond pulsed fields in highly-selective photochromic media
By R. Kaarli, J. Kikas, A. Rebane and P. Saari — 197

3.2. Picosecond processes in semiconductors and their application
By V. Brückner — 213

3.3. Optical bistability of the light induced orientation effects in the liquid crystals
By S. M. Arakelian, A. S. Karaian and Ju. S. Chilingarian — 229

3.4. Identification of relaxation mechanisms in the nonlinear spectroscopy of semiconductors
By V. M. Petnikova, S. A. Pleshanov and V. V. Shuvalov — 245

3.5. Photochemical and photophysical hole burning in electronic spectra of complex organic molecules
By R. I. Personov and B. M. Kharlamov — 259

3.6. Spectroscopy of nonlinear optical activity in crystals
By N. Zheludev — 281

3.7. Nonlinear spectroscopy of highly-excited molecules and condensed media
By N. I. Koroteev — 291

4. Laser Study of Photosynthesis

4.1. Use of lasers in photophysical research of photosynthesis
By G. Laczkó, P. Maróti, and L. Szalay — 307

4.2. Primary processes of photosynthesis studied by fluorescence spectroscopy methods
By A. Freiberg — 321

4.3. Influence of structural heterogeneity on energy migration in photosynthesis
By L. Valkunas — 341

4.4. Picosecond spectroscopy of photoreceptor molecules
By F. R. Aussenegg, M. E. Lippitsch and M. Riegler   357
4.5. Electrolyte control of photosynthetic electron transport in cyanobacteria
By G. C. Papageorgiou   379
4.6. Picosecond processes of photosynthesis in laser absorption studies
By A. Yu. Borisov, A. P. Razjivin, R. V. Danielius and R. J. Rotomskis   397
4.7. The structure, function and assembly of the light-harvesting antenna of photosynthetic purple bacteria
By R. Van Grondelle   413

5. Ultrafast Processes and Technique

5.1. Generation, propagation and compression of femtosecond light pulses
By B. Wilhelmi   425
5.2. Short pulses in optical fibers
By H. P. Weber, W. Hodel and B. Valk   457
5.3. Picosecond relaxation processes in monomeric and aggregated dyes under solvent influence
By S. K. Rentsch   473
5.4. Picosecond spectroscopy in study of the photoinduced isomerization of hexafluorobenzene
By J. L. Suijker, A. H. Huizer and G. A. G. O. Varma   487
5.5 The picosecond system in Jyväskylä
By J. Korppi-Tommola   503

6. New Laser Methods and Sources

6.1. Classical, semiclassical and quantum echo
By V. P. Chebotayev and B. Ya. Dubetsky   509
6.2. Fiber optic sensors
By H. P. Weber and Q. Munir   527
6.3. Horizons of detection of gravitational waves on the basis of frequency stable lasers
By S. N. Bagayev and E. V. Baklanov   539

6.4. Coherent vacuum UV sources using the methods of non-linear optics
By C. R. Vidal 553

6.5. Recent advances in tunable solid-state lasers
By G. S. Kruglik, G. A. Scripko and A. P. Shkadarevich 563

6.6. Line competition in optically pumped lasers
By R. Salomaa, M. A. Dupertuis and M. R. Siegrist 575

6.7. Optical pumping and coherence effects in Doppler-free laser spectroscopy
By W. Gawlik 585

Index of Contributors 603

Subject Index 605

Preface

Laser methods find use in many different fields of scientific research—nuclear physics, astrophysics, atomic physics, molecular and chemical physics, photochemisty and photobiology, and others. These fields of active invasion of laser methods have been the subject of many special conferences with many different names, but all of them sure to include the popular key word "lasers".

In 1960, at the time the laser was invented, I had to begin my work as a student in laser physics and its associated problems at the P. N. Lebedev Physical Institute in Moscow. And now, 25 years later, I still marvel at how quickly laser methods became established at research laboratories of different types. Many methods and ideas, however, turn out to be in principle rather similar to those in other fields which otherwise seem to have little in common. So it occurred to me, of course, that it is necessary sometimes to hold conferences and schools where scientists of different fields of science united by the use of the laser, this new instrument of research, could meet. I think that the international conference "Tunable Lasers and their Applications" held in 1976 in the small town of Loen in the heart of Nordfiord, Norway, is an example of a successful conference. The proceedings of that conference titled like the conference and edited by Drs. A. Mooradian, T. Jaeger and P. Stokseth, were published by Springer Verlag. This example inspired me to organize such a meeting in the U.S.S.R. and Prof. R. Khokhlov supported this idea.

As a result, the Institute of Spectroscopy, USSR Academy of Sciences, jointly with the Institute of Physics, Academy of Sciences of Lithuanian SSR (Prof. Yu. Vischakas), and Vilnius State University (Prof. A. Piskarskas) began to hold a new series of two-week international conferences-schools called "Lasers in Atomic, Molecular and Nuclear Physics". The first meeting took place in August 1978, the second in July 1981 and, finally, the third in September 1984.

The proceedings of the first two meetings were published in a joint English–Russian edition by the USSR Academy of Sciences.

This book is an English edition of papers at the last meeting, which was as international in character as the earlier ones. As the section titles show, the meeting dealt with a wide range of problems in nuclear, atomic and molecular physics, with a particular emphasis on the study of ultrafast processes in condensed media and biomolecules. About half of the authors are from different laboratories in the USSR. Therefore, the proceedings will be useful for many specialists from other countries who wish to gain a better idea of the studies being carried out in the USSR.

I thank Dr. A. Puretzky and V. Burimov for their help in preparing these lectures for publication.

V. S. Letokhov

1. ATOMS AND NUCLEII

1.1 Laser Detection of Very Rare Isotopes

V. S. LETOKHOV

Institute of Spectroscopy, USSR Acad. Sci., 142092, Troitzk, Moscow Region, USSR

Laser spectroscopy techniques have made it possible to solve the cardinal problems of optical spectroscopy: (1) the spectral resolution of the Doppler-free nonlinear spectroscopy techniques has already reached a value of $R = \nu/\Delta\nu \simeq 10^{11}$ at a spectral resonance width of $\Delta\nu \simeq 10^3$ Hz, the development of methods for further reduction of $\Delta\nu$ being under way[1]; (2) where subpicosecond mode-locked tunable lasers are used, the time resolution amounts to a few tens of femtoseconds, i.e. only a few tens of light oscillation periods[2]; (3) the sensitivity of some techniques, atomic and molecular photoionization spectroscopy in particular[3], reaches ultimate values - single atoms and molecules. There is one more, perhaps the last, problem of optical spectroscopy that is still to be solved; (4) high-selectivity detection of trace atoms and molecules in a real environment, particularly the detection of trace rare isotope atoms in the presence of an abundant isotope or the detection of trace molecules of a certain species in a molecular mixture. Subject to intensive development are now being various approaches that can combine a maximum possible sensitivity with an exceptionally high detection selectivity. The present lecture treats in short of some possible ways to solve the first of these problems - to attain a high selectivity of optical detection of very rare isotopes. This problem was introduced in Ref.[4] and discussed in Refs.[5,6].

1. RARE ISOTOPES AND EXISTING METHODS FOR THEIR DETECTION

There are a fairly large number of rare isotopes of cosmic origin, particularly those formed in the upper atmosphere as a result of nuclear reactions under the effect of cosmic rays. They include such isotopes as ^{10}Be resulting from interaction of galactic cosmic rays with the N and O nuclei, ^{14}C formed in the reaction between secondary neutrons and N, and ^{26}Al produced as a result of splitting of the Ar nucleus. These isotopes form in the upper atmosphere, precipitate, and accumulate on the earth's surface and ocean bottom. The rate of their precipitation in the ocean can be considered to remain constant during a long period of time exceeding their half-life $T_{1/2}$. Table I lists some rare isotopes of cosmic origin, along with their half-lives and concentrations relative to the content of their main stable isotopes.

TABLE I   Some rare cosmogenic isotopes

Isotope	$T_{1/2}$, years	Relative concentration
^{10}Be	$1.5 \cdot 10^6$	10^{-10}
^{14}C	$5.7 \cdot 10^3$	$10^{-12} - 10^{-16}$
^{26}Al	$7.4 \cdot 10^5$	10^{-14}
^{36}Cl	$3.1 \cdot 10^5$	10^{-17}
^{41}Ca	$8.0 \cdot 10^4$	10^{-21}

The best known radioisotope among these is radiocarbon, ^{14}C, which is used for estimating the age of objects of organic origin[7]. Radiocarbon, which is formed in the upper atmosphere in a concentration of ^{14}C/^{12}C $= 10^{-12}$, is involved in the Earth's biochemical

ATOMS AND NUCLEII 3

life cycle. After an organism has ceased to be living and participating in the carbon cycle, its content of ^{14}C decreases exponentially in accordance with the 5730 - years' half-life of this radioisotope. To date organic archaeological objects or events of 50 000 years, for example, it is necessary to detect ^{14}C in relative concentrations as low as 10^{-15}. By utilizing other, longer-lived isotopes, radioisotope dating can be extended to cover millions of years in the past.

At present, there exist two universal methods for detecting cosmogenic radioactive isotopes in low concentrations. The first, most popular one consists in measuring the specific radioactivity of the sample under analysis and comparing it with that of a specimen of zero age. This standard quantity for ^{14}C, for example, is well known and amounts to 15.3 beta-decay events per minute per gram of the natural mixture of carbon isotopes. To realize this method requires fairly large samples (around 5 g) and a long observation time (approximately 1 day). Therefore, a major part of a very valuable sample has frequently to be sacrificed in order that its age can be determind. Serious measures should also be taken to ensure proper protection against background activity. The nuclear method of detecting rare isotopes is disadvantageous in that it depends for its operation on radioactive transformations of the isotopes, which occur extremely seldom. Therefore, to have a reasonable observation time (a few days), the sample must, in principle, contain a large number of the rare isotope atoms of interest.

The second method for detecting rare isotopes consists in utilizing a linear^8 or cyclic^9 accelerator as a high-resolution mass spectrometer. The principal difficulty in implementing this method is the need to suppress background noise due to abundant isotopes and isobaric atoms such as ^{14}N in the case of ^{14}C. The

4   LASERS IN ATOMIC, MOLECULAR AND NUCLEAR PHYSICS

method handles much smaller samples (down to 5 mg), but the sample is completely lost as a result of measurements. The cost of the equipment required by the method is rather high.

It is clear that the shortcomings of the both generally accepted methods open up a wide field of application for laser techniques, for these are, in principle, capable of tackling the very difficult task of detecting a few rare isotope atoms against the background of 10^{10} to 10^{20} atoms of the most abundant isotope of the same atomic species. Taking as an example the detection of ^{10}Be in polar ice, Table II gives an idea of the parameters of the known methods and the requirements of the potential laser methods for detecting rare isotopes.

TABELE II   Sensitivity of various methods in detecting ^{10}Be ($T_{1/2} = 1.5 \cdot 10^6$ years, $^{10}Be/^9Be = 10^{-10}$) in polar ice

Method	Sample weight, kg	Number of ^{10}Be atoms	Experiment
1. Ultraweak radioactivity measurement technique	10^6	10^{13}	McCorkell et al., 1967[10]
2. Ultrasensitive accelerator mass spectrometry	1	10^7	Oeschgerr, Wölfli, 1983[11]
3. Potential laser technique for detecting single rare isotope atoms	10^{-3}	10^4	?

2. POSSIBLE LASER TECHNIQUES. LIMITATIONS AND WAYS TO OVERCOME THEM

All the existing laser spectroscopy methods whose ultimate sensitivity lies at a level of single atoms (see reviews[5,12]) can, in principle, be employed to effect a highly selective detection of

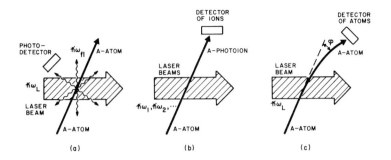

FIGURE 1   Various mechanisms of laser-atom interaction which can be used for atomic detection: (a) fluorescence; (b) photoionization; (c) deflection.

rare isotope atoms. Figure 1 presents simplified schemes of the three main laser techniques for detecting single atoms. These techniques take advantage of the effects due to resonant interaction of the atom with photons: (1) spontaneous reradiation of many photons absorbed from the laser beam; (2) photoionization of the atom as a result of absorption of a few photons; (3) changes of the atomic coordinate and velocity consequent upon reradiation of a large number of photons.

The main question relating to the applicability of these methods to detecting rare isotopes is their ultimate selectivity S, i.e. the ability to detect a small number (N_B) of the rare isotope atoms B in the presence of a much greater number (N_A) of the main isotope atoms A:

$$S = N_A/N_B \qquad (1)$$

The selectivity of these techniques stems from the presence of a small isotope shift $\Delta = \omega_A - \omega_B$ of the spectral line of one or several consecutive resonant transitions of the atom from its ground state to an excited state (Figure 2). The fact that the width of any spectral line is finite naturally limits the selectivity because of the overlapping of the wings of the close spec-

6 LASERS IN ATOMIC, MOLECULAR AND NUCLEAR PHYSICS

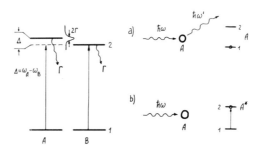

FIGURE 2   Limitation of detection selectivity if rare B atoms in presence of abundant A atoms with close spectral line: (a) method of fluorescent detection; (b) method of excitation detection.

tral lines of the atoms A and B, but the character of the limitation largely depends on the particular technique used.

With the fluorescence technique, the fluorescence excitation line usually has a Lorentzian shape, i.e. the probability that a photon will be scattered by the atoms A when the laser frequency coincides with the center of the spectral line of the rare atoms B is defined (in excitation conditions far from saturation) by the expression

$$W_{scat}^{A} = \sigma_o I \mathcal{L}(\Delta/\Gamma) \simeq (1/2)(\mu_{12} \mathcal{E}/\hbar \Delta)^2 \Gamma \quad (2)$$

where I is the laser radiation intensity (in photons/cm^{2}s), σ_o the cross section of the radiative transition $1 \rightarrow 2$ at a maximum, Γ the natural half-width or the rate of spontaneous decay of the atom into its initial ground state, μ_{12} the dipole moment of the transition $1 \rightarrow 2$, $\mathcal{E}$ the electric field strength of the light wave, and $\mathcal{L}(x) = 1/(1 + x^2)$ the Lorentzian function. By virtue of Eq. (2), the selectivity of the fluorescence detection of rare atoms is limited to the level of

$$S_{fl} = W_{scat}^{B}/W_{scat}^{A} = (\Delta/\Gamma)^2 \quad (\Delta \gg \Gamma) \quad (3)$$

For typical isotope shift and radiative linewidth values, $S_{fl} \simeq$

$10^4 - 10^6$, i.e. it is much lower than the required values indicated in Table I.

The photoionization technique detects excited atoms by their subsequent transition into an ionized state. Its selectivity is therefore governed by the probability of the atomic excitation and not by that of photon reradiation. The probability of the atom A being excited on the wing of its spectral line is determined by the probability of absorption of two photons with the frequency $\omega = \omega_B$ and concurrent spontaneous reradiation of a photon with the shifted frequency $\omega_{fl} = 2\omega - \omega_A = \omega + \Delta$ [13]:

$$W^A_{exc} \simeq (\mu_{12} \mathcal{E}/\hbar\Delta)^4 \Gamma \backsim \Delta^{-4} \qquad (4)$$

This expression differs from Eq. (2), which is the one commonly used for estimation purposes, by a stronger dependece of the excitation rate on the frequency shift Δ, the difference being substantial, but the selectivity of the method depends materially on the type of subsequent ionization of the excited atoms A. In the case of nonresonant simultaneous ionization where the difference between the energy of the first absorbed photon and the atomic excitation energy, $\hbar\Delta$, can be compensated for by the second absorbed photon, the selectivity is reduced to the former level difined by Eq. (3). Where the atomic ionization is a resonance process or delays from excitation pulse, a higher selectivity can be attained (see [13]).

It follows from the above simple estimates that none of the laser techniques in its simplest version can provide for the very high detection selectivity required. However, each of the techniques can be modified so as to ensure a substantial increase in selectivity. The methods that are possible here may be divided into two groups:

(1) Methods based on a <u>repeated</u> resonance interaction of the atom with the laser light, wherein the atom reradiates a large

number of photons. In this case, there occurs what may be called a "selectivity accumulation" on account of an effect which is possible only where a single atom reradiates a large number of photons following its repeated resonance excitation. This effect is made use of in the method of "fluorescent bursts"^{14,15} and that of laser deceleration and cooling of atoms^{16}.

(2) Methods based on a <u>multistep</u> resonance excitation of the atom in a multiple-frequency laser field, wherein use is made of the isotope shift on several consecutive resonant transitions. As a result of such a multistep resonance excitation, the selectivities S_i attained at each excitation and ionization stage are multiplied^{17}.

Let us consider briefly the capabilities of these methods as far as the detection of rare isotopes is concerned.

3. FLUORESCENCE TECHNIQUE

The fluorescence detection technique is, in principle, applicable to atoms and molecules alike. For isotope-selective detection of atoms, use can be made of the method of "fluorescent bursts", and molecules can be detected by the method of multistep fluorescence excitation. Consider both these possibilities.

<u>Atoms. Method of "fluorescent bursts"</u>. The high sensitivity of this method is due to the fact that a single atom in a laser beam is capable of reradiating a large number of photons which can be reliably registered. Let us consider a two-level atom in a resonant laser field of intensity I, whose frequency is tuned exactly to resonance with the transition frequency ω_{21} of the rare isotope to be detected. As a result of stimulated transition, the atom rises to the upper level during the time $T_{exc} = 1/\sigma_o I$, there σ_o is the transition cross section at exact resonance, and then spontaneously returns to the initial state during the time $\tau_{sp} = 1/A_{21}$ on the average, and so forth. The average duration of a "stimulat-

ed transition - spontaneous radiation" cycle, t_{cycle}, is equal to the sum $T_{exc} + \tau_{sp}$. As the laser intensity is increased ($I \gtrsim I_{sat} = 1/\sigma_o \tau_{sp}$), t_{cycle} tends to τ_{sp}. The time it takes the atom to make a transit across the laser beam is $\tau_{tr} = a/v_o$, a being the beam diameter and v_o the mean atomic velocity, and during this time the atom reradiates a large number of photons, given by

$$n_{ph} = \tau_{tr}/t_{cycle} = \tau_{tr}/(T_{exc} + \tau_{sp}) \simeq \tau_{tr}/\tau_{sp} \gg 1 \quad (5)$$

Even where the photon collection efficiency η_{coll} and the quantum efficiency of the photodetector, η_{det}, are moderate, the number of photoelectrons produced as a result of the fluorescent burst due to the atom crossing the laser beam, $K_{ph.el} = n_{ph} \eta_{det} \eta_{coll}$, is much greater than unity.

In conditions of such fluorescent bursts, the photons scattered in a resonance fashion reach the photodetector predominantly in groups, whereas those scattered in a nonresonance manner, randomly. In other words, the Poisson distribution for the photoelectrons due to the atoms crossing the laser beam is shifted towards greater K values at $\overline{K} \gg 1$. On the contrary, the Poisson distribution for background photoelectrons is shifted towards smaller K, as shown in Figure 3. By setting a certain discrimination threshold for the number of photoelectrons coming from the photodetector during the time τ_{tr}, e.g. $K_{thr} > 2$, one can distinguish between the photon burst due to the rare atoms B crossing the laser beam (which produce the useful multiphotoelectron signal) and random photons of stray light that give rise to single-photoelectron pulses on the detector[14,15].

Inasmuch as the main isotope atoms A are not in exact resonance with the laser field, the quantum transition cross section for them decreases by a factor of $(\Delta/\Gamma)^2$ far on the wing of the Lorentzian contour. Accordingly, the time t_{cycle} for them is increased and the average number of photoelectrons produced by them

10 LASERS IN ATOMIC, MOLECULAR AND NUCLEAR PHYSICS

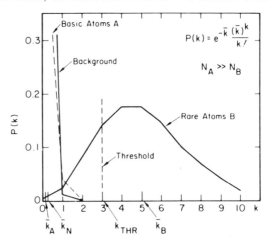

FIGURE 3 Multiphotoelectronic discrimination of the fluorescence singal from single resonant B atoms against background signals of nonresonant scattered light and fluorescence of A atoms. The discrimination is based on different distributions of photoelectrons from B atoms $W(K,\bar{K}_B)$, A atoms $W(K,\bar{K}_A)$, and background $W(K,\bar{K}_N)$.

during the transit time τ_{tr} reduced, so that $\bar{K}_A = \bar{K}_B (\Gamma/\Delta)^2 \ll 1$. The Poisson distributuion $W(K,\bar{K}_A)$ is correspondingly shifted towards smaller K. Obviously the transit of the rare resonance atoms B across the laser beam can be substantially discriminated from that of the abundant off-resonance atoms A when registering only multiphotoelectron events, single-photoelectron events being disregarded. Selectivity estimated[18] show that where photoelectron pulses contain three or more photoelectrons, the detection selectivity can, in principle, reach the values of $S_{mult.ph.el} \gtrsim 10^{20}$ at $\Delta = 10^3 \Gamma$.

In actual experiments, the detection selectivity is limited by the low intensity of the atomic beam, which is a must because the method requires that only a few atoms A should be in the laser beam region. As a result, there is a limitation on the number N_A of the atoms among which a search could be made for the rare isotope atoms B during the observation time τ_{obs}. As a matter of fact,

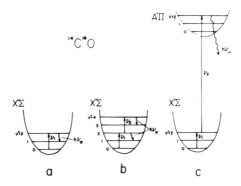

FIGURE 4 Methods of multistep highly-selective excitation of vibrational (a,b) and electronic (c) states of ^{14}CO molecules and the detection of IR (a,b) and UV (c) fluorescence.

it is for this reason that the detection selectivity is limited to the level of

$$S \simeq \tau_{obs}/\tau_{tr} \simeq 10^8\text{-}10^{10} \qquad (6)$$

Unfortunately, the absence of cw UW lasers as yet precludes the application of this method to the rare isotopes listed in Table I. Therefore, it would be interesting to analyze the molecular method as well.

Molecules. Stepwise excitation method. The rich rotational-vibrational structure of molecules makes it impossible to devise an effective scheme for their cyclic interaction with the laser beam, and so the method of fluorescent bursts is inapplicable to them. On the other hand, molecules can remain in the irradiated volume for a long time, the isotopic shift for vibrational frequencies is very large, and, finally, to enhance selectivity, use can be made of the multistep ixcitation method.

The capabilities of the molecular-fluorescent method were analyzed for the case of ^{14}C detection through isotope-selective excitation of the $^{14}C^{16}O$ molecules[19]. As shown in Figure 4, the following three detection schemes are possible:

12   LASERS IN ATOMIC, MOLECULAR AND NUCLEAR PHYSICS

(a) Single-step excitation of the vibrational level v" = 2 on account of absorption of IR radiation on the first overtone of the $^{14}C^{16}O$ molecule, followed by registration of IR fluorescence on the main band.

(b) Two-step excitation of the vibrational level v" = 4 by a two-frequency IR radiation, followed by registration of IR fluorescence on the main band.

(c) Two-step excitation of the $A^1\Pi$ electronic state from the vibrational level v" = 2 of the ground electronic state $X^1\Sigma^+$, followed by registration of UV fluorescence.

Estimates show that a ^{14}C detection selectivity of around 10^{13} can be reached with the scheme (a). The selectivity is enhanced with the two-step excitation schemes (b) and (c). In particular, the two-step IR-UV excitation according to the scheme (c) makes it possible to attain a ^{14}C detection selectivity of about 10^{16}. The amount of carbon necessary for the analysis does not exceed $5 \cdot 10^{-4}$ g. Note that with all the above three schemes, the thermal population of the vibrational level v" = 2 to which excitation is effected at the first stage must be less than 1/S. This condition is satisfied for CO for v" = 1 at a temperature of 77 K. So, an effective laser detection of ^{14}CO is really feasible and will be realized once tunable IR and UV lasers have progressed adequately.

4. METHOD OF RESONANCE COOLING OF ATOMS

When atoms interact with a laser beam in a cyclic manner, so that there occurs reradiation of a large number of photons, the resonance light pressure causes considerable changes in the velocities and coordinates of those atoms which are in resonance with the laser field[20]. For instance, a substantial monochromatization of atomic velocities (narrowing of the longitudinal atomic velocity distribution)[21] and formation of an intensive beam of cold atoms

ATOMS AND NUCLEII 13

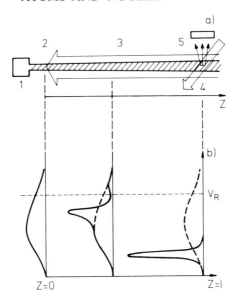

FIGURE 5 Scheme of fluorescent detection of single atoms through deceleration and velocity monochromatization of atoms by counter-running laser radiation: 1 - atomic beam source, 2 - atomic beam, 3 - decelerating laser-radiation, 4 - detecting laser beam, 5 - photodetector. A general view of the atomic-beam-velocity distribution deformation. The laser radiation is at resonance with atoms having velocity V_R.

with velocities distributed in the vicinity of the zero velocity[22] were observed when a beam of Na atoms was made to interact with a counter-propagating laser wave. The effect of resonance deceleration of atoms in a beam can also be used to improve the selectivity of fluorescence detection of rare isotopes[16].

Figure 5 is illustrative of the idea of such method. The frequency of the laser beam is tuned close to the absorption line maximum of the rare isotope atom B. Because of the deceleration and cooling of the resonant atoms B, their velocity distribution becomes deformed so that a peak forms in it in the vicinity of zero velocities. At the same time, the number of interaction cycles for the nonresonant atoms A, as well as the number of photons

14 LASERS IN ATOMIC, MOLECULAR AND NUCLEAR PHYSICS

reradiated by them, is smaller by a factor of $(\Delta/\Gamma)^2$, and so the velocity distribution $W_A(v)$ remains unchanged. At the end of the deceleration path the velocities of the atoms A and B may differ by a factor of 100. Accordingly, the interaction time of the slow atoms B with a probe laser beam is greatly increased, which allows them to be reliably discriminated from the faster atoms A, e.g. by the multiphotoelectron registration method. Detailed estimates of the limitations of the method of resonance cooling of atoms show that a detection selectivity of $S \gtrsim 10^{13}$ can, in principle, be reached. The method, however, suffers from the same shortcomings as the atomic fluorescent method considered above. In particular, it is difficult to realize today for the most interesting cosmogenic isotopes.

5. MULTISTEP PHOTOIONIZATION METHOD

The selectivity of two-step resonance ionization is limited to the level defined by Eq. (3) and therefore special methods to improve selectivity are necessary.

<u>Multiple repetition of the process</u>. Obviously there is a possibility of multiplying the limited selectivity by repeating again and again the process of resonance ionization. Such a detection scheme for rare isotopes was discussed in[12], and is shown in Figure 6a. Let sample 1 contain a mixture of isotopes A and B. By evaporating them and then subjecting the gas obtained to resonance ionization, one can increase their concentration ratio in the photoion stream by a factor of S_{opt} and then implant the ions produced into foil 2. The selectivity of such a single-shot process is equal to the ratio of the net ionization efficiencies Y of the atoms B and A:

$$S_{opt} = ([B]/[A])_2 / ([B]/[A])_1 = Y(B^+)/Y(A^+) \qquad (7)$$

Repeating this process m times can, in pronciple, provide for the

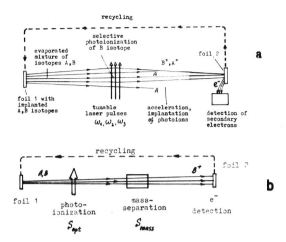

FIGURE 6 Enhancement of ionization selectivity detection of rare atoms by means of multiple repetition of ionization process; (a) selective ionization of rare B atoms by laser radiations; (b) nonselective ionization with mass-separation.

selectivity

$$S = (S_{opt})^m \qquad (8)$$

Unfortunately, if the net ionization effeciency $Y(B^+)$ largely differs from unity, such a repetition inevitably leads to a loss of sensitivity. In particular, the loss of sensitivity relative to its limit (a single atom) is equal to $[Y(B^+)]^m$. For example, repeating the process three times with $Y(B^+) = 0.1$ results in the loss of 99.9% of the rare isotope atoms B. This method is feasible only in schemes with exceptionally high efficiencies of collecting the rare isotope atoms B throughout the entire pathway from the sample to the foil.

Selectivity can be improved by placing a mass separator with a separation selectivity of S_{mass} in the way of photoions (Figure 6b). In this case, the total selectivity of a single-shot process is $S = S_{opt} S_{mass}$, and repeating the process m times will

16 LASERS IN ATOMIC, MOLECULAR AND NUCLEAR PHYSICS

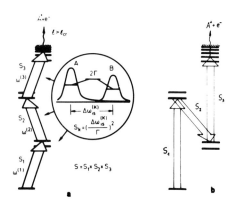

FIGURE 7 Multiplication of selectivity in case of multistep excitation with isotopically-selective excitation for each step of excitation: (a) consecutive excitation up; (b) consecutive excitation and deexcitation by laser radiation.

raise it accordingly:

$$S_m = (S_{opt} S_{mass})^m \qquad (9)$$

This method is also applicable in cases where the atomic ionization is nonselective ($S_{opt} = 1$). The entire selectivity is then accumulated as a result of repetition of the mass-separation process, the only function remaining for the laser radiation being highly effective ionization[23,24]. This method may prove useful where the isotope shift and, correspondingly, the isotopic selectivity are very small. Specifically, it can be used for detecting rare isotopes of noble gases.

<u>Multistep selectivity</u>. The selectivity of multistep resonance photoionization can be considerably increased by taking advantage of isotope shifts in several resonance excitation stages[17] as shown in Figure 7. For example, in the case of a three-step excitation where each stage is isotope-selective, the total selectivity is

$$S_{opt} = S_1 S_2 S_3 \qquad (10)$$

At moderate values of $S_i \simeq 10^4$-10^6 the total selectivity of ionnization may reach exceptionally high values of $S_{opt} = 10^{12}$-10^{18}. If the given energy level diagram does not allow one to realize ceveral consecutive excitation steps up to the ionization threshold, one can, in principle, stimulate with laser radiation transitions from excited states down (N-type multistep excitation) as shown in Figure 7b. This method can help to realize schemes with three or more excitation steps when using UV and visible laser light.

The idea of multiplication of isotopic selectivity at each excitation stage is difficult to realize for the most interesting cosmogenic isotopes (Table I) because it is hard to find for them a series of consecutive upward transitions with noticeable isotope shifts, such shifts being characteristic of the ground state only.

A universal way to overcome this difficulty and make the method of multistep resonance photoionization really applicable to the detection of rare isotopes was suggested in[25]. The idea of the method is based on collinear stepwise photoionization of a beam of accelerated atoms. It is common knowledge that acceleration of atoms in the form of ions under a given potential difference U and subsequent neutralization of the ions into atoms lead to the bunching of the longitudinal ionic velocities, hence to the narrowing of they Doppler width of all the spectral lines of the given atomic species (if viewed in a collinear fashion) as compared with the initial Doppler width $\delta \nu_D(0)$ at an ion source temperature of T (Figure 8)[26]:

$$\delta \nu_D(V) / \delta \nu_D(0) = (1/2)(kT/eU)^{1/2} \qquad (11)$$

At $U = 10^4$ V the narrowing factor reaches 10^3. What is important is that the atoms in this case group in a smaller volume of the phase space and their Doppler-free spectroscopy is effected with-

18 LASERS IN ATOMIC, MOLECULAR AND NUCLEAR PHYSICS

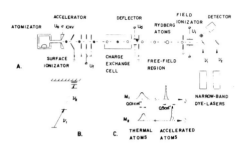

FIGURE 8 Laser detection of rare atoms by multistep ionization of accelerated atoms: (a) general scheme of set-up; (b) scheme of quantum transition to the Rydberg states; (c) formation of mass-isotopoc shift in spectra of accelerated atoms.

out any loss of sensitivity.

Along with the narrowing of the Doppler width, there also occurs the Doppler shift of all the spectral lines of the accelerated atoms, which depends on the mass of the ion. As a result, there occurs an artificial "mass" shift on any spectral transition of the atom:

$$\Delta \nu_{sh} / \nu_o = (1/c)(\sqrt{2eU})(1/\sqrt{M_1} - 1/\sqrt{M_2}) \qquad (12)$$

This shift at U = 10 kV is approximately by an order of magnitude greater than the ordinary isotope shift $\Delta \nu_{is} \simeq m_e (\Delta M/M^2) \nu_o$. What is most important is that such a kinematic isotope shift occurs on any spectral transition of the atom, this making it possible to realize in a natural way the idea of selectivity multiplication in multistep isotope-selective excitation.

Simple estimates for ^{26}Al show that a two-step resonance excitation of accelerated Al atoms can provide for a selectivity of $S_{opt} = S_1 S_2 \simeq 10^{14}$. Using a simple mass filter with a selectivity of $S_{mass} \simeq 10^2$, one can attain a sure detection of the rare isotope ^{26}Al against the background of the main stable isotope ^{27}Al, even without resorting to isotopic selectivity at the third excitation stage (on the transition to the autoionization state

ATOMS AND NUCLEII   19

$S_3 = 1$). Of course, to practically realize a high selectivity, it is necessary to devise a technique for obtaining sufficiently intensive ion currents. For instance, at an ion current of $I = 10^{-4}$ A in the above example it is necessary to detect by this method $6 \cdot 10^{3}$ ^{26}Al ions from a total of $6 \cdot 10^{17}$ Al ions during an observation time of 10^{3} s. Even at a total ionization efficiency of 1% there will be produced 60 ^{26}Al ions against the background of 0.6 ^{27}Al ions at a net ionization selectivity of $S \simeq 10^{16}$. Thus, it seems quite possible to develop a laser detection technique for very rare isotope atoms, based on the multistep isotope-selective photoionization of collinearly accelerated atoms.

Note that the multistep isotope-selective ionization method is applicable to isotopic molecules as well. For example, the scheme of Figure 4c can be used to effect a highly selective ionization of ^{14}CO molecules. To realize such a scheme, however, requires tunable IR and VUV lasers.

6. METHODS FOR LASER ISOTOPE ENRICHMENT

To use a preliminary isotope enrichment is of exceptional value to any detection method for rare isotopes. The following two approaches are possible here: (1) enrichment of the starting isotope mixture with the rare isotope B and (2) enrichment of the initial quantum state with the isotope B.

Enrichment of isotope mixture. For this purpose, use can be made of one of several effective laser isotope separation schemes based on the isotope-selective photoionization of atoms and photodissociation of molecules[27]. The first successful experiments on the enrichment of an isotope mixture with ^{14}C were performed on the basis of photopredissociation of H_2CO[28]. A preliminary 400-fold enrichment of a $^{14}C/^{12}C$ mixture makes it possible to shift the zero time reference 50 thousand years back and date samples as old as 80 thousand years. It is quite possible to use other laser

20 LASERS IN ATOMIC, MOLECULAR AND NUCLEAR PHYSICS

isotope separation methods to enrich the starting mixture with other rare isotopes.

Enrichment of metastable qunatum state. The idea of isotope enrichment is also applicable to a metastable quantum state which can be depleted in an isotope-selective manner with a laser radiation[29,30]. The method presupposes that the atom (or ion) has a suitable energy level structure. The atoms of all isotopes in a beam are first nonselectively excited by some method to the metastable state "1". A cw laser radiation is tuned to resonance with the transition $1 \rightarrow 2$ in the abundant isotope. Owing to the other spontaneous decay channel, $2 \rightarrow 0$ ("0" meaning the ground state or a metastable state below the state "1"), the relative concentration of the abundant isotope "1" is reduced exponentially during the entire time it takes the atom to make a transit across the laser beam. The extent of radiative de-excitation of the abundant isotope is governed by the reverse radiative process (Raman scattering with the $0 \rightarrow 1$ atomic transition):

$$N_A/N_A^{(o)} \simeq (\Gamma/2\Delta_{20})^2 \qquad (13)$$

where $\Gamma = \Gamma_{21} + \Gamma_{20}$ is the net rate of the spontaneous decay of the level 2 via the two channels and $\hbar\Delta_{20}$ the difference in energy between the initial and final states of the two-step Λ-type transition $1 \rightarrow 2 \rightarrow 0$.

Because of the isotope shift on the transiton $1 \rightarrow 2$, the atoms of the rare isotope B interact with the laser radiation much weaker than those of the abundant isotope A. As a matter of fact, the rare isotope atoms are not involved at all in the process of de-excitation if the isotope shift on the transition $1 \rightarrow 2$ is by an order of magnitude greater than the radiative width of this transition.

After its isotope-selective depletion, the state "1" has a much higher concentration of the rare isotope B, the detection of whose atoms now requires a correspondingly lower selectivity. To detect the excited atoms in the state "1" rich in the isotope B, use can be made of the above-considered fluorescence and photoionization methods.

By and large the laser techniques considered here pave the way for the development in the next few years of efficient laser ultralow radioactivity detectors independent of the radioactive decay phenomenon.

In conclusion, I wish to express my sincere gratitude to my colleagues - Drs. V.I. Balykin, Yu.A. Kudryavtsev, A.A. Makarov, and V.G. Minogin - for their co-operation.

REFERENCES

1. V. S. LETOKHOV, V. P. CHEBOTAYEV, Nonlinear Laser Spectroscopy, Vol. 4, Springer Series in Optical Sciences (Springer-Verlag, Berlin, Heidelberg, New York, 1977).
2. D. H. AUSTON, K. B. EISENTHAL, eds. Ultrafast Phenomena IV, Vol. 38, Springer Series in Chemical Physics (Springer-Verlag, Berlin, Heidelberg, New York, Tokyo, 1984)
3. V. S. LETOKHOV, Laser Photoionization Spectroscopy (Academic Press (in preparation)); Optica Acta (in press).
4. V. S. LETOKHOV, in Tunable Lasers and Applications, edited by A. Mooradian, T. Jaeger, P. Stokseth, Springer Series in Optical Sciences (Springer-Verlag, Berlin, Heidelberg, New York, 1976), Vol. 3, p. 122.
5. V. S. LETOKHOV, in Chemical and Biochemical Applications of Lasers, edited by C. B. Moore (Academic Press, New York, London etc., 1980), Vol. 5, p. 1.
6. V. S. LETOKHOV, Comm. Atom. Molec. Phys., 10, 257 (1981)
7. W. F. LIBBY, Collected Papers, Vol. 1, Tritium and Radiocarbon, edited by R. Berger and L. M. Libby (Santa Monica: Geo Science Analytical, 1980)
8. D. E. NELSON, R. G. KORTELING, W. R. SCOTT, Science, 198, 507 (1977)
9. R. A. Muller, Science, 196, 489 (1977)
10. R. MCCORKELL, E. L. FIREMAN, C. C. LANGWAY, Science, 158, 1690 (1967).
11. J. BEER, M. ANDREE, H. OESCHGER, B. STAUFFER, R. BALZER, G. BONANI, G. STOLLER, M. SUTER, W. WOLFLI, R. C. FINKEL, Radiocarbon, 25, 269 (1983)

12. V. I. BALYKIN, G. I. BEKOV, V. S. LETOKHOV, V. I. MISHIN, Uspekhi Fiz. Nauk (Russ), 132, 293 (1980) [Sov. Phys. Usp., 23, 651 (1980)]
13. A. A. MAKAROV, Zh. Eksp. Teor. Fiz. (Russ), 85, 1192 (1983)
14. G. W. GREENLESS, D. L. CLARK, S. L. KAUFMANN, D. A. LEWIS, J. F. TONN, J. H. BROADHURST, Opt. Comm., 23, 236 (1977)
15. V. I. BALYKIN, V. S. LETOKHOV, V. I. MISHIN, V. A. SEMICHISHEN, Pis'ma Zh. Eksp. Teor. Fiz. (Russ), 26, 492 (1977); Zh. Eksp. Teor. Fiz. (Russ), 77, 2221 (1979)
16. V. I. BALYKIN, V. S. LETOKHOV, V. G. MINOGIN, Appl. Phys., B33, 247 (1984)
17. V. S. LETOKHOV, V. I. MISHIN, Opt. Comm., 29, 168 (1979)
18. V. I. BALYKIN, V. S. LETOKHOV, V. I. MISHIN, Appl. Phys., 22, 245 (1980)
19. Yu. A. KUDRIAVTZEV, V. S. LETOKHOV, E. V. MOSKOVETZ, Possibilities of Laser Detection of Carbon-14 Stepwise Isotopically-Selective Excitation of CO Molecules, Preprint N°1 of Institute of Spectroscopy, 1983, pp. 42.
20. V. S. LETOKHOV, V. G. MINOGIN, Phys. Reports, 73, 1 (1981)
21. S. V. ANDREEV, V. I. BALYKIN, V. S. LETOKHOV, V. G. MINOGIN, Pis'ma Zh. Eksp. Teor. Fiz. (Russ), 34, 463 (1981); Sh. Eksp. Teor. Fiz. (Russ), 82, 1429 (1982)
22. V. I. BALYKIN, V. S. LETOKHOV, A. I. SIDOROV, Opt. Comm., 49, 248 (1984)
23. C. H. CHEN, G. S. HURST, M. G. PAYNE, Chem. Phys. Lett., 75, 473 (1980)
24. C. H. CHEN, S. D. KRAMER, S. L. ALLMAN, G. S. HURST, Appl. Phys. Lett., 44, 640 (1984)
25. Yu. A. KUDRIAVTZEV, V. S. LETOKHOV, Appl. Phys., B29, 219 (1982)
26. K. R. ANTON, S. L. KAUFMANN, W. KLEMPT, G. MORUZZI, R. HEUGART, E. W. OTTEN, B. SCHINZLER, Phys. Rev. Lett., 40, 642 (1978)
27. V. S. LETOKHOV, Nature, 277, 605 (1979)
28. C. B. MOORE, Nature, 276, 255 (1978)
29. A. A. MAKAROV, Appl. Phys., B29, 287 (1982)
30. A. A. MAKAROV, Kvantovaya Elektronika (Russ), 10, 1127 (1983)
31. V. I. BALYKIN, Yu. A. KUDRIAVTZEV, V. S. LETOKHOV, A. A. MAKAROV, V. G. MINOGIN, in Laser Spectroscopy VI, edited by H. P. Weber and W. Lüthy, Springer Series in Optical Sciences (Springer-Verlag, Berlin, Heidelberg, New York, Tokyo, 1983), Vol. 40, p. 103.

1.2 Collinear Laser-Fast Beam Spectroscopy on Unstable Nuclides

R. NEUGART

Institut für Physick, Universität Mainz, Mainz, Federal Republic of Germany

1. INTRODUCTION

These lectures will demonstrate the uniquely wide application of collinear laser fast (atom or ion) beam spectroscopy to unstable isotopes, as they are available at on-line isotope separators connected to accelerators or reactors. The outstanding example of such a facility is ISOLDE at CERN[1]. Using 600 MeV-proton-induced spallation, fission or fragmentation reactions, and various target-ion source combinations, it provides intense beams of unstable isotopes of about 40 elements.

The nuclear properties studied in atomic spectroscopy include spins, magnetic dipole and electric quadrupole moments as well as the variation in the mean square charge radius within a sequence of isotopes. These properties manifest themselves in the hyperfine structures and isotope shifts. The hyperfine energies of the different F states within a hyperfine multiplet $|J - I| \leqslant F \leqslant J + I$, given by the well-known formula

$$W_F = \frac{1}{2} KA + \frac{(3/4) K (K+1) - I (I+1) J (J+1)}{2 I (2I-1) J (2J-1)} B$$

with

$$K = F (F+1) - I (I+1) - J (J+1),$$

24 LASERS IN ATOMIC, MOLECULAR AND NUCLEAR PHYSICS

are determined by the nuclear spin I, the magnetic dipole interaction constant

$$A = \mu_I H_e(0)/IJ$$

and the electric quadrupole interaction constant

$$B = eQ_S \varphi_{JJ}(0)$$

The nuclear moments μ_I and Q_S can be extracted from A and B using empirical or theoretical values for the magnetic hyperfine field $H_e(0)$ and the electric field gradient $\varphi_{JJ}(0)$ at the nucleus. Similarly, the isotope shift $\delta\nu^{AA'}$ of an optical transition is related to the change in the nuclear mean square charge radius $\delta\langle r^2\rangle^{AA'}$ between the isotopes A and A' by

$$\delta\nu^{AA'} = F\delta\langle r^2\rangle^{AA'} + M\frac{A'-A}{A'A}$$

where the electronic factor F is proportional to the change of the electronic density at the nucleus in the optical transition. The second term represents the mass shift due to the change of nuclear recoil energy. In simple (s → p or s^2 → sp) transitions of heavy atoms, where F is large and (A' - A)/A'A is small, one can usually neglect the effects of electron momentum correlation and assume M = ν/1836 for the transition frequency ν.

For many years these observables of atomic spectroscopy have been an invaluable source of information about the nuclear ground-state structure. Various efforts have now been made to extend the scope of this work, beyond the limits of stable and long-lived redioactive nuclides, far into the regions of β-instability. The technical problems encountered were mainly due to the minute amounts of radioactive material available and to the short half-lives. The experimental approaches are briefly reviewed by Otten[2] in a lecture at the preceding 2nd ISLA, 1981. More detailed accounts are given in the volume "Lasers in Nuclear Physics"(Proceedings of the 1982 conference at Oak Ridge) by Kluge[3] for the experiments using optical cell techniques, Thibault and Touchard[4] for the atomic beam optical pumping experiments on alkali isotopes, and by Rebel and Schatz[5] for the off-line atomic beam fluorescence experiments. The recent progress in laser multistep photo-ionization at the Gatchina on-line isotope separator[6] is treated in the lecture by Mishin.

2. REMARKS ON THE COLLINEAR-BEAM TECHNIQUE

ISOLDE provides a variety of unstable nuclides in the form of 60 keV beams of singly-charged ions. As an example, Figure 1 shows the isotopic distribution of the radium yield from spallation in a uranium target. The spectroscopy with a laser beam along the ion-beam axis can profit from the narrow longitudinal velocity spread related to the spread in kinetic energy by

$$\delta E = mv \delta v$$

where $v = (2eU/m)^{1/2}$ is given by the acceleration voltage U and the atomic mass m. This leads to a Doppler width $\delta \nu_D = \nu \delta v/c$, or

$$\delta \nu_D = \nu \frac{\delta E}{(2eUmc^2)^{1/2}}$$

which is comparable with the natural line width in strong optical transitions of a few Megahertz[7]. The observation of nearly Doppler-free resonances involves no expense of sensitivity, because all atoms in the beam are excited simulataneously. A technical advantage for computer-controlled on-line accelerator experiments is the Doppler-tuning achieved by speeding up or slowing down the ion beam. A tuning voltage in the range ±10 kV offers an easy, fast and precise control of the effective laser frequency over typically ± 50 GHz.

The versatility of the method is not restricted by specific processes of sample preparation which depend on the chemical properties of the elements. In selecting a convenient optical transi-

26 LASERS IN ATOMIC, MOLECULAR AND NUCLEAR PHYSICS

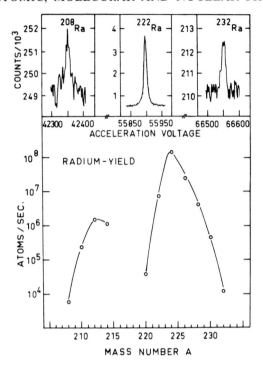

FIGURE 1   Radium yield from ISOLDE measured by laser spectroscopy. The 600 MeV protons hit a 15 g/cm² target of ^{238}U (in the form of UC_2 heated to 2000°C and connected to a tungsten orifice at 2400°C for surface ionization). The upper part of the figure shows the resonances of the transition $7s^2\ ^1S_0 - 7s7p\ ^1P_1$ in Ra I as an illustration of the sensitivity (cf. Section 4).

tion, one has the choice between singly-charged ions and neutral atoms produced by sharge exchange according to

$$B^+ + A \rightarrow B + A^+ + \Delta E$$

where B^+ and A represent an ion in the fast beam and an alkali atom in a vapour target, respectively. Neutral atoms are usually more convenient as their resonance lines can be excited by the visible light from cw dye lasers. Moreover, the charge-exchange process predominantly populates states whose binding energies are close to the ground-state energy of the reaction partner, i.e.

ATOMS AND NUCLEII 27

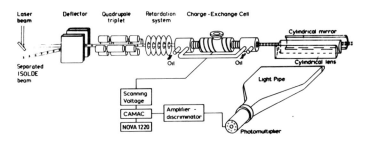

FIGURE 2 Basic experimental set-up for the collinear laser-neutral atomic beam technique.

the charge-exchange cross-section is large, if the energy defect ΔE is small. This offers an efficient mechanism of producing beams of metastable atoms, thus giving access to elements with low-lying ground-states and resonance lines in the UV. The advantages of this scheme have been first exploited for ytterbium[8] and used later for mercury[9] and the rare gases (cf. also Section 5).

A schematic diagram of the standard set-up at ISOLDE is shown in Figure 2. Its essential features are:

i) the electrostatic deflection of the ion beam into the laser-beam axis;

ii) a charge-exchange cell at a variable potential to convert the ion beam into a fast atomic beam and to Doppler-tune the effective laser frequency (For spectroscopy on ions, the Doppler-tuning potential is applied to the electrically insulated observation chamber.);

iii) a detection chamber to observe fluorescence photons from laser-excited atoms;

iv) precision high-voltage scanning and measuring devices for controlling the Doppler shifts that correspond to the resonance frequencies to be determined.

Further details can be found in[7].

28 LASERS IN ATOMIC, MOLECULAR AND NUCLEAR PHYSICS

3. EXPERIMENTS

About 130 isotopes of 8 elements produced at ISOLDE have been investigated during the past three years. The interest has been concentrated on the rare-earth region, to study the neutron-shell closure effect at N = 82 and the transition from spherical to strongly-deformed nuclear shapes around N = 90. On the other hand, there has been an effort to explore a similar and rather unknown region around and above N = 126. Here, the initial experiments led to the first observation of optical isotope shifts and hyperfine structures in radium.

3.1. Rare-Earth Region

The first manifestation of a transition from spherical to deformed nuclear shapes was found in optical isotope shifts and hyperfine structures. In 1949, Brix and Kopferman[10] interpreted the large isotope shifts - corresponding to big shanges in the nuclear mean square charge radius - between ^{150,152}Sm and ^{148,150}Nd, as well as the strong increase in the quadrupole moment between ^{151,153}Eu by the onset of a collective non-spherical charge distribution between the neutron numbers N = 88 and 90. These arguments found a firm proof in the discovery of rotational spectra. For a long time, the information on this sudden onset of deformation had been restricted to a few elements with proton numbers between Z = 60 and 64. Below and above, the relevant neutron numbers shift rapidly into the region of very neutron-rich and neutron-deficient nuclides, respectively.

The present work on the neutron-deficient isotopes of the heavier rare-earth elements was stimulated by the observation of a smooth transition from spherical to deformed shapes for the barium isotopes above the neutron-shell closure N = 82, up to N = 90[7]. Selected features of the experiments and results have been reported earlier[11,12], with particular emphasis on the systematic Z-dependence in the development of deformation. Up to now, the measurements cover the even-Z elements ytterbium[8,12], erbium, dysprosium, and gadolinium as well as the odd-Z element europium[13].

As an example for the resonance transitions in the rare-earth elements, Figure 3 shows the hyperfine structure pattern of ^{151}Dy in the 4212 Å line, with respect to the single components of the even reference isotopes ^{156}Dy and ^{158}Dy. The corresponding transition is essentially of the type $s^2 \rightarrow sp$ and can be described in a scheme in which the electron configurations of the open 4f shell and the valence shell are coupled:

$$4f^{10}(^5I_8)6s^2(^1S_0)^5I_8 \rightarrow 4f^{10}(^5I_8)6s6p(^1P_1), \; J = 9 \; (4212 \text{ Å})$$

Similar $6s^2 \rightarrow 6s6p$ transitions have been chosen for the other elements, mainly by considering the detection sensitivity which depends on the population of the initial state and the line strength. In Yb the filled 4f shell leaves a clean two-valence-electron spectrum of which the transitions $6s^{2\;1}S_0 \rightarrow 6s6p\;^3P_1$ (5556 Å) and $6s6p\;^3P_2 \rightarrow 6s7s\;^3S_1$ have been studied.

The neutron deficient isotopes of all heavier rare-earth elements are simultaneously produced by 600 MeV-proton-induced spallation in a Ta-foil target, heated to 2200°C. The ions are formed by surface ionization in a tungsten tube at 2400°C, and the mass-separated 60 keV beam is neutralized in sodium vapour at a pressure of about 10^{-3} Torr.

Measurements on the even-Z elements have been performed in the ranges $^{142-154}$Gd, $^{146-164}$Dy, $^{150-170}$Er, and $^{156-176}$Yb. The differences of the mean square nuclear charge radii within the isotopic sequences have been deduced from the isotope shifts, and the spins I, magnetic dipole and electric quadrupole moments μ_I and Q_S of the odd isotopes have been extracted from the hyperfine structures. The hyperfine structure analysis[14] in terms of the

30 LASERS IN ATOMIC, MOLECULAR AND NUCLEAR PHYSICS

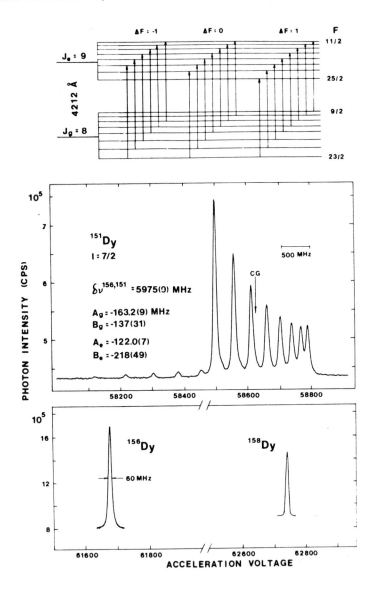

FIGURE 3   Example of a measured hyperfine structure pattern for ^{151}Dy, with the stable doubly-even reference isotopes ^{156}Dy and ^{158}Dy. The level scheme of the odd isotope is given at the top. The number of hyperfine components (of which only the strong $\Delta F = 1$ and the weak $\Delta F = 0$ transitions are seen) is determined by the nuclear spin.

magnetic dipole and electric quadrupole interaction parameters A and B uses hyperfine fields and electric field gradients deduced from moment measurements in the stable isotopes. Precise μ_I values are usually available from atomic beam or nuclear magnetic resonance experiments, whereas the spectroscopic quadrupole moments in the rare-earth region have recently been determined from the hyperfine structures in muonic atoms. These latter experiments, reported by Steffen[15], remove the considerable uncertainties of the theoretical or semi-empirical calculations of the electric field gradients including the Sternheimer correction. The analysis of the isotope shifts in terms of $\delta\langle r^2\rangle$ has been performed by the established semi-empirical procedures[16].

Figure 4 shows the differential changes of the mean square charge radii $\delta\langle r^2\rangle^{N-2,N}$ between the even isotopes with neutron number N-2 and N for the elements Ba, Dy, Er, and Yb. Between N=82 and 90 the curve of Dy represents essentially what is known from the stable isotopes of Nd and Sm and the preliminary results of the recent Gd experiment. For the sake of clarity, these data have been omitted in the figure. The peak value $\delta\langle r^2\rangle^{88,90}$, corresponding to the sudden onset of strong static deformation, disappears gradually in the sequence Dy, Er, Yb, as well as for Ba which has 4 protons less than Nd - the lightest element for which the effect has been observed. It is thus localized to a relatively narrow range of proton numbers around, say, Z=64.

Apart from the regular increase with neutron number described, for example, by the droplet model, the mean square radii are believed to be mainly determined by deformation effects, either dynamic or static. Although the droplet model has been refined recently to account also for the free quadrupole and haxadecapole deformation parameters β_2 $(=\beta)$ and β_4[17], we can approximately use the geometrical picture of a nuclear shape described by

32   LASERS IN ATOMIC, MOLECULAR AND NUCLEAR PHYSICS

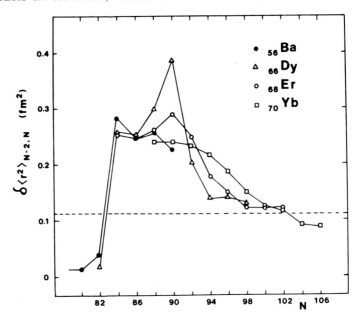

FIGURE 4   Differences in the nuclear mean square charge radii between even isotopes of neutron numbers $N-2$, N (Brix-Kopfermann diagramm). The horizontal dashed line corresponds to the spherical droplet-model prediction.

$R(\Theta) = R_o(1 + \beta Y_{20}(\Theta))$. According to

$$\langle r^2 \rangle = \langle r^2 \rangle_o + \frac{5}{4\pi} \langle r^2 \rangle_o \langle \beta^2 \rangle$$

the deviations from the spherical $\langle r^2 \rangle_o$ can be ascribed to the mean square deformation $\langle \beta^2 \rangle$. The latter decreases with neutron number below N=82 and increases above, leading to the dramatic change of $\delta \langle r^2 \rangle$ at the shell closure. Around N=90 the plot suggests a smooth transition into strong deformation for Yb, whereas for Dy, the onset of a strong static deformation occurs essentially in one big step at N=90. Er plays the role of an intermediate case.

The overall trends are better seen in an integral plot, shown in Figure 5, where the differences of $\langle r^2 \rangle$ to the isotopes with a

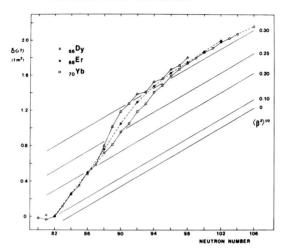

FIGURE 5 Development of the mean square charge radii in the Dy, Er, and Yb isotopes above the N=82 neutron-shell closure. The slope of the equi-deformation lines is given by the droplet model.

closed neutron shell N=82 are displayed. Also shown is the slope of the droplet-model line, with equi-deformation lines according to the equation above. With the assumption $\langle\beta^2\rangle^{1/2} = 0.1$ for N=82 nuclides, as suggested by B(E2) values, the strongly deformed nuclides end up at deformations slightly larger than 0.3 which is again consistent with the B(E2) values and with the measured quadrupole moments.

The nuclear spins and moments have been consistently interpreted[12] within the particle-rotor approach by Larsson et al.[18] which is based on the Nilsson model. This approach describes well the transition from the rather pure $f_{7/2}$ shell model states close to N=82 to the strongly coupled and almost pure Nilsson states in the region of strong deformation. Here, we shall not go into further details.

The key to a general interpretation of the nuclear ground-state behaviour in this region seems to be the proton-subshell closure Z=64 and its effect of stabilizing spherical nuclear

34 LASERS IN ATOMIC, MOLECULAR AND NUCLEAR PHYSICS

shapes. A naive picture would suggest the steepest increase of deformation in the middle between the closed proton shells at Z=50 and 82, i.e. around Z=66. This trend is counteracted by the subshell closure at Z=64, leading to equi-deformation curves almost parallel to the Z axis close to N=82, and thus explaining the similarity in the behaviour of $\langle r^2 \rangle$ for $82 \leqslant N \leqslant 88$ within the full sequence of elements. Since Z=64 loses its significance as a semi-magic proton number for large deformation, the bistability of the nuclear shape around N=88 to 90 leads to a situation similar to the light Hg isotopes. For $N \geqslant 90$, the deformation maximum is found around Gd and Dy, as expected. These observations are in agreement with the results of a recent analysis of the energies of 2^+ and 4^+ states in the ground-state vibrational/rotational bands[19].

As a prominent odd-Z candidate we have investigated europium (Z=63)[13], because of its single-proton hole in the $2d_{5/2}$ subshell and its relatively simple electronic structure. Of the three strong resonance transitions $4f^7 6s^2\ {}^8S_{7/2} \rightarrow 4f^7 6s6p\ {}^8P_{5/2,7/2,9/2}$ (4662 Å, 4627 Å, and 4594 Å) the two latter have the best resolved hyperfine structures including strong quadrupole interaction.

Measurements in these two transitions have been performed for the sequence $^{140-153}$Eu. All these isotopes ixhibit a stable $5/2^+$ proton configuration for which the magnetic moments and spectroscopic quadrupole moments can be followed in the odd-A isotopes over a wide range of neutron numbers. In Figure 6 these moments are plotted against neutron numbers. The μ_I values show a decrease on either side of the shell closure (N=82). The maximum at ^{145}Eu reaches 83% of the Schmidt value. This reflects the minimization of core polarization reducing the single-particle moments. The Q_s values increase on either side of N=82, with a big jump at the shape transition between N=88 and 90. A roughly estimated single-particle value of $Q_{sp} = -\langle r^2 \rangle (2j-1)/(2j+1) = 0.13b$ accounts for

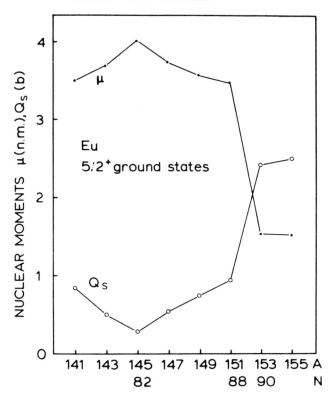

FIGURE 6 Magnetic moments μ_I and spectroscopic quadrupole moments Q_s of the odd-A Eu isotopes with $5/2^+$ ground-states ($^{141-155}$Eu). Except for the strongly deformed ^{153}Eu, they are well described by the $2d_{5/2}$ shell model configuration.

50% of the observed value at the shell-closure minimum, $Q_s(^{145}$Eu$) =$ = 0.28b. The small residual core polarization is similar to analogous cases of single-particle or hole states in a doubly-magic core (e.g. ^{209}Bi). The quadrupole moments and $\delta \langle r^2 \rangle$ give a consistent picture of the nuclear deformation between the nearly spherical ^{145}Eu and the strongly-deformed ^{153}Eu, with $\beta(^{153}$Eu$) =$ = 0.32. The step in deformation between ^{151}Eu and ^{153}Eu (N=88-90) is clearly more pronounced than in the neighbouring even-Z elements.

In parallel to the present work there has been a remarkable interest in measuring the nuclear moments and isotope shifts of the radioactive Eu isotopes. Alkhasov et al.[20,6] have measured the isotope shift with unresolved hyperfine sturcture in the 5765 Å line of Eu I for $^{145-150}$Eu, using the technique of laser multistep photoionization, and Dörschel et al.[21] have published high-resolution data for ^{147}Eu, ^{149}Eu, and $^{151-156}$Eu, obtained by collinear laser-ion beam spectroscopy in the 6049 Å line of Eu II. These latter data, mainly on the neutron-rich isotopes, are complementary to our results. On the other hand, the nuclear orientation technique has been used to determine the magnetic moments of $^{145-149}$Eu, by van den Berg et al.[22] and Kracikova et al.[23] with contradictory results of which the latter are in qualitative agreement with ours.

Apart from the global view given here, the present results allow a rather detailed examination of the nuclear structure. One example is the interpretation of the spin sequences and the corresponding moments of the odd-neutron isotopes in terms of shell-model or Nilsson-model wave functions for spherical or well-deformed nuclei, or within the particle-rotor approach in the transitional regions (see above). Moreover, the so-called odd-even staggering of the isotope shift shows interesting irregularities. They may contribute to a quantitative understanding of this effect in the future.

3.2 Radium

We shall now focus on the studies of optical isotope shifts and hyperfine structures in the radioactive element radium. In the past, classical spectroscopy had been performed only in samples of the long-lived doubly-even isotope ^{226}Ra($T_{1/2}$ = 1600 y), yielding a comprehensive atomic energy-level scheme[24]. Hyperfine spectroscopy on the shorter-lived odd isotopes can contribute considerably to a better knowledge, not only of the structure of the heaviest nuclei, but also of the atomic structure in the heaviest element with a simple two-valence-electron spectrum.

The production of radium isotopes at ISOLDE has been shown in Figure 1. Spectroscopy with the weak beam intensities between 10^8 and 10^4 atoms/s can only be achieved in strong resonance lines. Two of these lie in the visible spectral range: $7s^2\ {}^1S_o \rightarrow 7s7p\ {}^1P_1$ for the neutral atom (RaI) at 4827 Å and the "D_1" line $7s\ {}^2S_{1/2} \rightarrow 7p\ {}^2P_{1/2}$ for the alkali-like ion (RaII) at 4683 Å. We have performed measurements in both these lines using a slightly modified version of our apparatus in which the tuning voltage can be applied alternatively to the charge-exchange cell or the detection chamber. Figure 7 gives an example of the respective hyperfine structure patterns for ^{223}Ra.

The measurements cover the isotopic range $208 \leqslant A \leqslant 232$[25,26]. The isotope ^{232}Ra has been observed here for the first time by

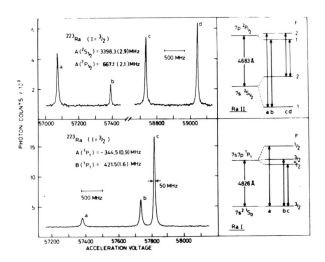

FIGURE 7   Hyperfine structures in the Ra II and Ra I resonance lines for the example of ^{223}Ra (I = 3/2).

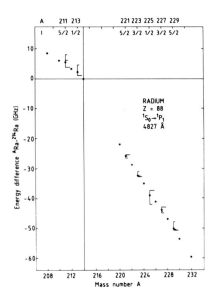

FIGURE 8 Relative positions of the single resonance (even isotopes) or hyperfine components with their centres of gravity (odd isotopes) in the Ra I line at 4827 Å. The spin values determined from these structures and the corresponding ones in the Ra II line are given at the top.

laser spectroscopy. The half-lives range from 1600 y (^{226}Ra) to 20 ms (^{220}Ra). The gap between A = 214 and 220 is due to the rapid α-decay of nuclides just above the N=126 neutron-shell closure. These nuclides have half-lives of the order of µs and decay during diffusion through the target matrix. Figure 8 gives a survey of the observed single resonances of the even isotopes and the hyperfine structure patterns of the odd isotopes in the RaI line. The distinct change of the isotope shifts at N=126 is related to the minimum of nuclear deformation at the neutron-shell closure (cf. N=82 in the rare-earth region).

The hyperfine structures yield the nuclear spins and the magnetic dipole and electric quadrupole coupling constants A and B which in turn are proportional to the respective nuclear moments.

As shown in the introduction, the evaluation of absolute moments requires the knowledge of the magnetic field and the electric field gradient of the electrons at the site of the nucleus. For lack of direct measurements or accurate calculations, we use an established semi-empirical procedure dating back to the early days of hyperfine-structure studies[14]. The effective sharges and quantum numbers describing the term values of an alkali-like atom in analogy to hydrogen give the electron density $|\Psi(0)|^2$ at the nucleus in the $^2S_{1/2}$ ground-state of RaII (Goudsmit-Fermi-Segrè formula), from which the contact hyperfine field - including relativistic and nuclear-volume corrections - can be calculated within a few percent. The evaluation of quadrupole moments from the B-factors of the excited $^1P_1$ state in RaI has to account for the singlet-triplet mixing. This requires the additional study of hyperfine structures in the ^3P states. We are also aiming at measurements in the 7p $^2P_{3/2}$ state of RaII for which the analysis is less complex.

The semi-empirical value of $|\Psi(0)|^2$ can also serve as a basis for evaluating the changes of the mean-square nuclear charge radii $\delta \langle r^2 \rangle$ from the isotope shifts. In this case, the ground and excited states are involved and taken care of by the so-called shielding factors[16]. The consistency of this precedure can be checked by a comparison between the s-p transition in RaII and the s^2-sp transition in RaI. Alternatively, a theoretical ab initio approach using Dirac-Fock atomic wave functions is being performed for both the electronic factor of the isotope shift and the magnetic hyperfine field[27,28].

Without anticipating a final analysis of the electronic factors by atomic-structure calculations we can already give conclusive results[25,26]. In the strongly-deformed region (A > 220), the spectroscopic quadrupole moments are close to the strong-coupling limit as compared to the intrinsic quadrupole moments known from

40 LASERS IN ATOMIC, MOLECULAR AND NUCLEAR PHYSICS

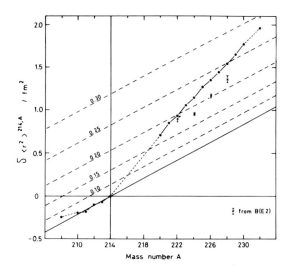

FIGURE 9   Development of the mean square charge radii in the range $^{208-232}$Ra, as compared to the droplet model prediction with quadrupole deformations β_2 from experimental B(E2) values.

B(E2) values. The spin sequence and the corresponding magnetic moments support the hypothesis of a static intrinsic octupole deformation gradually decreasing from ^{221}Ra to ^{229}Ra^{29,30}. This is also reflected in an inversion of the odd-even staggering of the isotope shift for ^{221,223,225}Ra, and in the absolute values of the mean square charge radii with respect to the nearly-spherical ^{214}Ra (N=126). In Figure 9 the $\delta \langle r^2 \rangle$ values are compared with a droplet model prediction including the quadrupole deformation term, with β_2 taken from the B(E2) values in the heavy deformed isotopes. The obvious inconsistency is removed essentially by the assumption of the additional octupole deformation, β_3 = 0.1, postulated by theory, and a small contribution from β_4.

Below the magic neutron number N=126, the moments of the odd-A isotopes ^{211}Ra and ^{213}Ra are well described by $(f_{5/2})^{-1}$ and $(p_{1/2})^{-1}$ neutron configurations of the shell model, and the isotope shifts confirm that they are nearly spherical.

4. REMARK ON THE SENSITIVITY

The detection sensitivity is the crucial point in spectroscopy on unstable isotopes far from stability. This has been illustrated by Figure 1. The weakest beam of an even radium isotope, used in this experiment, was about 10^4 atoms/s. Odd isotopes require 10 to 100 times stronger beams, depending on their hyperfine structure. For ^{208}Ra, the recording of a resonance with a signal-to-noise ratio of 10 in 50 channels took 10^3 s, which means that about 10^7 atoms, or 4 fg of radium, have passed through the apparatus during the measurement. The signal was 200 counts/s, arising from about 10 excitations per atom and a total photon-detection efficiency of 0.2%. This has to be compared to a background of 10^4 counts/s of which the dominant part is due to stray light.

A careful optimization beyond these typical "on-line-running" conditions may improve the sensitivity by a factor of 10. More dramatic improvements, however, can only be expected from new detection schemes, of which a rather promising and simple one will be described in the next section.

5. SENSITIVE ION-DETECTION AFTER STATE-SELECTIVE STRIPPING

The general applicability of the collinear laser-fast atomic beam technique is essentially based on the neutralization of ion beams by charge exchange with alkali vapours. The beams are prepared in (ground or metastable) atomic states about 4 - 6 eV below the ionization threshold from where resonance excitation is possible with visible light (1.5-3 eV), corresponding to the spectral range of cw dye lasers[8]. This is particularly important for rare gases whose first excited states lie about 10 eV above the np^6 ^1S_o ground states. The metastable J = 2 states of the configuration $np^5(n+1)s$ —designated $(n+1)s\,[3/2]_2$ — about - 4 eV from the ionization threshold are efficiently populated in charge exchange with caesium.

42 LASERS IN ATOMIC, MOLECULAR AND NUCLEAR PHYSICS

They are connected to the states of the configuration $np^5(n+1)p$ by strong transitions on the red.

As pointed out in Section 4, the counting of fluorescence photons involves a relativeley low detection efficiency and background problems mainly due to stray light, molecular excitation in the alkali vapour and radioactivity. The practical sensitivity limits (typically between 10^4 and 10^6 atoms/s) depend on the strenghts of the transitions and the complexity of the hyperfine structures.

A considerable gain in sensitivity requires non-optical detection schemes, for example the ion counting after stepwise laser ionization. This technique is well developed for pulsed lasers with high peak power but limited duty cycle[31]. For cw lasers, the problems of power limitation remain to be solved.

We have designed an alternative and much simpler scheme which - at least for the rare gases - should reach the same goal. It is based on the selective electron stripping from the metastable atoms as they pass through a molecular gas. An analogous scheme of state-selective charge-transfer in ion-beam spectroscopy is well-established since the early work of Wing et al.[32] and Gaillard et al.[33], although its full advantages have been shown only recently[34].

The beam, after neutralization in the caesium-vapour cell and removal of the remaining ions, interacts with the laser light and then passes a differentially pumped gas cell after which the re-ionized part is deflected into a Faraday cup or an ion counter. If the optical excitation to one of the $np^5(n+1)p$ states has a strong decay branch to the ground-state, the net effect at resonance is a quantitative optical pumping to the ground state. Since the cross-section for stripping from the ground state (-14 eV) is much smaller than from the metastable state (-4 eV), this pumping leads to a

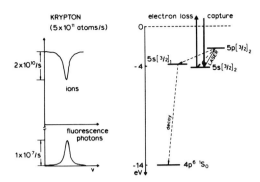

FIGURE 10   Simplified energy-level diagram of Kr I and the resonance 5s $[3/2]_2$ - 5p $[3/2]_2$ in ^{84}Kr detected by fluorescence as well as selective collisional ionization of Kr atoms in the metastable state.

reduction of the ion signal which can thus be used to detect the optical resonance.

In test measurements, mainly on ^{84}Kr, we have shown that this scheme is rather promising. We used the laser excitation from the 5s $[3/2]_2$ metastable to the 5p $[3/2]_2$ state at 7602 Å. The predominant decay channel leads to the ground state via 5s $[3/2]_1$. Among various stripping gases, SO_2 gave the highest re-ionization efficiencies of 10 - 20%, mainly limited by secondary processes, such as charge-transfer neutralization. At resonance, the flop-out signal in the ion current was about 40%. In Figure 10 a direct comparison with the simultaneously recorded fluorescence signal shows a gain in detection efficiency of more than a factor of 10^3. In principle, this ion signal is free of background.

However, for the planned experiments on weak radioactive isotopes it will be essential to have low contamination of the beams with molecular ions or strong neighbouring masses from the isotope separator. Tests have shown that these conditions can be met at ISOLDE at least for the heavier rare gases. Background from long-lived radioactivity can be removed by a tape transport intergrated in the ion counter. In favourable cases, the radioactive decay may even serve for detection, thus removing the possible background from stable beams. The proposed scheme is a high-efficiency and low-background alternative to the conventional fluorescence detection capable of improving the sencitivity by several orders of magnitude. Further tests have to explore the possible extensions to atoms with two-valence-electron spectra (e.g. cadmium and mercury) in which the energy differences between ground and metastable states are considerably smaller.

ACKNOWLEDGEMENT

These lectures are based on the experimental results obtained at the ISOLDE facility (CERN) during the past four years. I am indebted to all my colleagues who shared in these experiments (see references) and to those who supplied us with the intense high-quality beams of so many isotopes.

REFERENCES

1. H. L. RAVN, Phys. Reports, 54, 201 (1979)
2. E. W. OTTEN, in Laser Applications in Atomic, Molecular and Nuclear Physcs, 2nd Int. School, Vilnius 1981 (Nauka, Moscow, 1983), p. 228.
3. H.-J. KLUGE, in Lasers in Nuclear Physics, edited by C. E. Bemis, Jr. and H. K. Carter, Nuclear Science Research Conference Series, Vol. 3 (Harwood Scientific Publishers, 1982), p. 137.
4. C. THIBAULT, F. TOUCHARD, ibid., p. 113.
5. H. REBEL, G. SCHATZ, ibid., p. 197.
6. V. N. FEDOSEEV, V. S. LETOKHOV, V. I. MISHIN, G. D. ALKHASOV, A. E. BARZAKH, V. P. DENISOV, A. G. DERNYATIN, V. S. IVANOV, Optics Comm., 52, 24 (1984)
7. A. C. MUELLER, F. BUCHINGER, W. KELMPT, E. W. OTTEN, R. NEUGART, C. EKSTROM, J. HEINEMEIER, Nucl. Phys., A403, 234 (1983)
8. F. BUCHINGER, A. C. MUELLER, B. SCHINZLER, K. WENDT, C. EKSTROM, W. KLEMPT, R. NEUGART, Nucl. Instr. Meth., 202, 159 (1982)
9. S. A. AHMAD, G. HUBER, W. KLEMPT, H. LOCHMANN, R. NEUGART, G. ULM, K. WENDT, to be published

10. P. BRIX, H. KOPFERMANN, Z. Phys., 126, 344 (1949)
11. R. NEUGART, in Lasers in Nuclear Physics, edited by
    C. E. Bemis, Jr. and H. K. Carter, Nuclear Science
    Research Conference Series, Vol. 3 (Harwood Scientific
    Publishers, 1982), p. 231
12. R. HEUGART, K. WENDT, S. A. AHMAD, W. KLEMPT, C. EKSTROM,
    Hyperfine Interactions, 15/16, 181 (1983)
13. S. A. AHMAD, W. KLEMPT, C. EKSTROM, R. HEUGART, K. WENDT,
    submitted to Z. Phys. A
14. H. KOPFERMANN, Nuclear Moments (Academic Press, 1958)
15. Y. TANAKA, R. M. STEFFEN, E. B. SHERA, W. REUTER,
    M. V. HOEHN, J. D. ZUMBRO, Phys. Rev. Lett., 51, 1633 (1983)
16. K. HEILIG, A. STEUDEL, At. Data Nucl. Data Tables,
    14, 613 (1974)
17. W. D. MYERS, K. H. SCHMIDT, Nucl. Phys., A410, 61 (1983)
18. S. E. LARSSON, G. LEANDER, I. RAGNARSSON,
    Nucl. Phys., A307, 189 (1978)
19. R. F. CASTEN, D. D. WARNER, D. S. BRENNER, R. L. GILL,
    Phys. Rev. Lett., 47, 1433 (1981)
20. G. D. ALKHASOV, A. E. BARZAKH, E. Ye. BERLOVICH,
    V. P. DENISOV, A. G. DERNYATIN, V. S. IVANOV,
    A. N. ZHERIKHIN, O. N. KOMPANETS, V. S. LETOKHOV,
    V. I. MISHIN, V. N. FEDOSEEV, JETP Lett., 37, 274 (1983)
    and Zh. Eksp. Teor. Fiz., 86, 1249 (1984)
21. K. DORSCHEL, W. HEDDRICH, H. HUHNERMANN, E. W. PEAU,
    H. WAGNER, G. D. ALKHASOV, E. Ye. BERLOVICH, V. P. DENISOV,
    V. N. PANTELEEV, A. G. POLYAKOV, Z. Phys., A312, 269 (1983)
    and A317, 233 (1984)
22. F. G. VAN DEN BERG, W. VAN RIJSWIJK, W. J. HUISKAMP,
    Phys. Lett., 120B, 67 (1983)
23. T. I. KRACIKOVA, S. DAVAA, M. FINGER, V. A. DERYUGA,
    Hyperfine Interactions, 15/16, 73 (1983)
24. E. RASMUSSEN, Z. Phys., 86, 24 (1933) and 87, 607 (1934)
25. S. A. AHMAD, W. KLEMPT, R. NEUGART, E. W. OTTEN, K. WENDT,
    C. EKSTROM, Phys. Lett., 133B, 47 (1983)
26. S. A. AHMAD, W. KLEMPT, R. NEUGART, E. W. OTTEN, K. WENDT,
    to be published in Nucl. Phys.
27. T. P. DAS, private communication
28. B. FRICKE, A. ROSEN, private communication
29. I. RAGNARSSON, Phys. Lett., 130B, 353 (1983)
30. G. A. LEANDER, R. K. SHELINE, Nucl. Phys., A413, 375 (1984)
31. V. I. BALYKIN, G. I. BEKOV, V. S. LETOKHOV, V. I. MISHIN,
    Sov. Phys. Usp., 23, 651 (1980)
32. W. H. WING, G. A. RUFF, W. E. LAMB, Jr., J. J. SPEZESKI,
    Phys. Rev. Lett., 36, 1488 (1976)

33. F. BEGUIN, M. L. GAILLARD, H. WINTER, G. MEUNIER,
J. de Physique, 38, 1185 (1977)
34. R. E. SILVERANS, private communication

1.3 Photoionization Laser Spectroscopy of Short-Lived Isotopes

V. I. MISHIN

Institute of Spectroscopy, USSR Academy of Sciences, 142092 Troitzk, Moscow Region, USSR

1. INTRODUCTION

The measurement of the isotope shifts and the hyperfine structure of atomic lines provides us with data which allow defining some nuclear parameters important for nuclear physics. These are the variation of charge distribution mean-square radii $\delta \langle r^2 \rangle$, the nuclear spin I, the magnetic dipole μ, and electric quadrupole Q moments. The technique of production and mass separation of radioactive nuclides and the use of high-resolution and high-sensitivity laser optical methods have made it possible at present to carry out such measurements for long chains of radioactive isotopes extending beyond the band of nuclear beta-stability up to the limits of nuclear stability.

The hyperfine structure (HFS) and the isotope shift (IS) of atomic lines are small effects in atomic spectra and are usually screened by the spectral line Doppler broadening. Therefore they should be investigated using Doppler-free spectroscopy methods. In the case of radioactive nuclides the problem becomes more complicated due to the fact that the flows of mass-separated isotopes produced by the proton and ion accelerators in nuclear reactors are law. The maximum isotope flow of rare-earth elements produced, for example, by the ISOLDE complex at CERN is 10^8 to 10^{11} isotope/s and reduced to 10^3 - 10^4 isotope/s for isotopes with their half-life of about a minute[1]. Besides, nuclides with a short lifetime can be investigated using only high-sensitivity methods which permit "on-line" operation with a source of radioactive nuclides.

In[2] the method of optical pumping of atoms in a beam by laser rediation and magnetic detection was used to investigate long chains of isotopes of alkali elements from Na to Fr. The isotopes of Hg[3] and Cd[4] were studied using the method of excitation of atomic vapour fluorescence by laser radiation in the cell. The isotopes of Ba, Yb, Er, Dy, and Eu were studied using the collinear laser technique[15]. The same method was used to investigate the isotopes of Sm and Eu[6]. More detailed information on HFS and IS of nuclei and on the new methods can be found in review[7] and in the proceedings of the conference[8].

To study the HFS and IS of radioactive muclides we applied for the first time the method of step photoionization of atoms in a beam by laser radiation proposed at the Institute of Spectroscopy, USSR Acad. Sci.[9]. The method consists in the following. The multy-frequency radiation excites resonantly the atom successively from level to level, when excited the atom becomes ionized. Since the intermediate transitions are resonant, they can be easily saturated. A considerable part of atoms, nearly 100%, turns out to be excited here. From the excited state the atom may be ionized with high efficiency either through autoionization states[10] or through Rydberg states[11]. The resulting particles, electrons and ions, can be detected with an efficiency similar to unity. The HFS and IS of absorption lines of any transition are studied with the frequency of the laser radiation acting on this transition being varied.

The laser photoionization method for studying radioactive

nuclei forms the basis for the laser-nuclear facility designed with the Institute of Spectroscopy, USSR Acad. Sci. This facility is based on a proton accelerator and a mass separator[12]. It should be noted that our method differs from other optical methods because it allows not only studying the HFS and IS of atomic lines but also separating isotopes, isomers and isobars selectively converting them from the neutral state to the ionized one.

As the first subjects of investigation we used rare-earth elements with their nuclei in the vicinity of the filled neutron shell with N = 82. It is of interest to investigate the nuclei in this region because, first, systematic measurements of the charge radii of isotope chains including magic nuclei (with N = 82) will enable understanding shell effects for the neutron-deficient nuclei removed from the β-stability band. Secondly, as the number of neutrons increases from 82 to 90 the shape of nucleus must change from spherical to highly deformed. And, finally, with N < 82 there may be a new region of deformation[13]. The results obtained from "off-line" measurement of the variation of $\delta \langle r^2 \rangle$ of Eu isotopes are presented in[14]. At the present time "on-line" measurements have been also taken for a chain of Eu isotopes up to ^{138}Eu.

2. SCHEME OF LASER MULTISTEP PHOTOIONIZATION OF EUROPIUM ATOMS

For lantanoids including europium the three-step ionization scheme is the most optimal since in this case it is possible to reach the ionization limit (for Eu E_i = 4573.9 cm^{-1})[15] using radiation at the basic generation frequency. Figure 1 shows a partial level scheme of EuI. The transitions from the ground $6s^2\ ^8S^o_{7/2}$ state to $6s6p\ ^{10}P$ studied for stable isotopes, as far as the HFS and IS are concerned, can be chosen as a transition of the first excitation step. The wavelengths of transitions to the levels of the ^8P and ^{10}P terms lie between 600 and 710 nm. If these transitions are used at the first excitation step, it is quite possible to realize

50 LASERS IN ATOMIC, MOLECULAR AND NUCLEAR PHYSICS

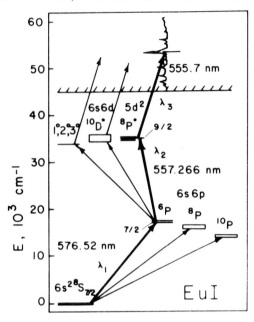

FIGURE 1 Schematic energy level diagram for stimulated transitions in the three-step photoionization of Eu atoms

three-step excitation and ionization of Eu atoms through Rydberg states. If, however, at the first excitation step the transition to $^6P_{7/2}$ (λ_1 = 576.520 nm) or to $^6P_{5/2}$ (λ_2 = 564.580 nm) is used, the atom needs radiation with λ_3 > 700 nm to be excited to a Rydberg state. When we performed the experiment, no effictive dyes generating in this spectral region, with the pumping done by Cu laser radiation, were known. So the only reasonable way of ionizing in this case was to excite the Eu atom to an autoionization state.

When choosing the first transition at which the HFS and IS were studied we took into account the hyperfine structure width and the value of isotope shift. For some of the transitions it is known[16]. The hyperfine structure of these transitions is mainly defined by the upper level splitting. For the $6s6p\,^6P_{5/2,7/2}$ states

ATOMS AND NUCLEII 51

its width is abnormally small (about 300 MHz) compared to the isotope shift between the stable isotopes ^{151}Eu and ^{153}Eu, $\Delta\nu$(151-153) = 3619 MHz ^{16} at λ_1 = 576.520 nm. Therefore, the excitation lines λ_1 = 564.580 nm and λ_1 = 576.520 nm are very convenient for measuring isotope shifts because it is not necessary in this case to identify the hyperfine structure components and to determine the centre of gravity of line. On the basis of these considerations the transition with an abnormally small hyperfine structure was used at the first excitation step. In this case the atom can be effectively ionized only through autoionization states.

The optimal photoionization scheme was chosed after 1) studying the spectra of excitation to autoionization states, 2) measuring the saturation energies and lifetimes of the intermediate states from which the strongest autoionization resonances can be obtained, 3) measuring the frequencies of the transitions between excited states.

The investigations were carried out with a spectrometer consisting of three dye lasers pumped by N_2-laser radiation[17]

Taking into consideration the data obtained (the hyperfine structure width, the saturation energy of the second transition, the lifetime of the second transition, the relative intensity of autoionization transitions) we chose the most effective sequence of atomic transitions in europium suitable for measuring the isotope shifts of its radioactive isotopes (Figure 1).

3. MEASUREMENT OF ISOTOPE SHIFTS

The isotope shifts of radioactive isotopes were measured with a spectrometer consisting of three dye lasers pumped by two Cu-lasers. The average radiation power of each Cu-laser was 5 W at two lines - 510.6 and 578.2 nm. The Cu-laser generation pulse rate was 10^4 pulse/s. The radiation wavelength of the first-step dye laser was varied by changing the pressure in the chamber where the grating and the Fabry-Perot interferometer of dispersion cavity of this laser were placed. The radiation line width of this laser was 0.02 cm^{-1}. From the $^6P_{7/2}$ state the europium atoms were excited by two dye lasers the wavelengths of which were resonant with the chosen transitions and the line width was 0.8 cm^{-1}. The laser radiation power at the first, second, and third excitation steps was respectively 2.120 and 250 mW.

The laser beams cut at the right angle two atomic beams produced in two high-temperature ovens. According to the mode of operation ("off-line" or "on-line") one oven contained tantalum foil with radioactive isotopes or implanted ions of radioactive isotopes. The other oven contained metal europium. The atomic beam with stable isotopes ^{151,153}Eu was used as a reference beam as well as to tune the laser wavelengths to the atomic transitions. Part of narrow-band laser radiation was directed to the confocal interferometer with a free spectral region of 0.125 cm^{-1}. The ions produced selectively were drawn out of the region of interation of laser and atomic beams and detected by an electron channel multiplier.

As the frequency of the first-step laser was scanned, three spectra appeared: a photoionic spectrum of excitation of radioactive isotopes, a photoionic spectrum of excitation of stable isotopes and an iterferometer transmission spectrum. The free dispersion region of interferometer was determined from two lines of stable isotopes, and the number of fractional and whole interference bands made it possible to define the position of the lines of radioactive isotopes about the line of ^{151}Eu. The position of isotope lines was measured accurately to ±70 MHz and detemined by laser line width instability and nonlinearity of frequency tuning.

Long-lived radioactive isotopes with their half-decay $T_{1/2}$ larger than 100 hours were investigated during "off-line" opera-

ATOMS AND NUCLEII

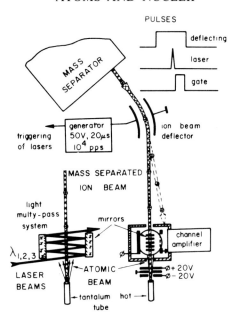

FIGURE 2 Experimental setup for on-line studies of isotopes exit from mass-separator

tion. In this case isotopes accumulated in the tantalum foil at the mass-separator outlet. After accumulation a piece of foil with the studed isotope was cut off and placed into the high-temperature oven to be studied. Since the separated isotopes were implanted into the tantalum foil for a depth of several hundreds of Angströms, they were completely preserved during all the operations with the foil.

Short-lived radioactive isotopes with $T_{1/2} <$ 10 hrs were investigated during "on-line" operation with an accelerator and a mass-separator. They produced in a tantalum foil with its mass of 20 g irradiated by a proton beam with an intensity of 10^{12} s^{-1}. The ion beam of separated isotopes with its energy of 30 keV was directed into a tantalum tube closed at one end (50 mm long, 5 mm in diameter). This tube could be heated to 1500°C. At such temperature the isotopes implanted into the bottom of the tube flew out

54 LASERS IN ATOMIC, MOLECULAR AND NUCLEAR PHYSICS

in the opposite direction as a collimated beam (Figure 2). A system of metal diaphragms at different potentials arrested ions and electrons of thermal origin. The atoms were excited to an autoionization state by three laser beams brought together at a small angle and crossing the atomic beam at a near-right angle. A system of two plane mirrors which allowed the laser beams to pass through the region of interaction several times was used to increase the ion yield. The number of passes was 10.

"On-line" measurements were carried out in two ways. Isotopes with rather a long lifetime ($T_{1/2} \geqslant 1$ min) accumulated in a cold tantalum tube throughout $3 \cdot T_{1/2}$. Then the ion beam of the mass-separator was turned off and the tube was heated quickly. The evaporation of isotopes from the tube lasted about 20 s and was equal to the time of one laser frequency scanning across the investigated spectrum region. In this mode of measurement the background pulse level was about a pulse per ten seconds.

Isotopes with $T_{1/2} \leqslant 1$ min were studed in a next way. The tantalum tube was heated all the time. 10 µs before laser pulses appeared (duration 17 ns, rate 10^4 pulse/s), the ion beam from the mass-separator was interrupted for 20 µs. After it was turned off, the ion cloud formed by the ionization of the residual gas in the vacuume chamber by a beam of high-energy ions from the mass-separator was dispersed (the residual pressure in the chamber was $5 \cdot 10^{-6}$ Torr). The background in this case was a pulse per second. This mode of measurement was used to investigate the isotopes Eu(144, 141, 139, and 138).

In the experiments with a multipass optical system the efficiency of detection of Eu atoms defined as the ratio of the flow of ions from the mass-separator at a fixed frequency of first-step laser radiation tuned to the centre of the atomic line was 0.03%. This value was basically determined by the losses caused by the

ATOMS AND NUCLEII 55

formation of a collimated atomic beam (the fraction of atoms in the region of irradiation was 3%), the losses due to laser radiation duty cycle (the fraction of atoms irradiated by laser pulses was 20%) and the atomic photoionization efficiency (5%). To illustrate a high detection sensitivity of Eu isotopes we should note that with a turned-off separated ion beam, when the background was 10^{-1} pulses/s, we could detect 3000 Eu atoms with a signal-to-noise ration of about 10 if these atoms were placed into the tantalum tube.

4. RESULTS AND DISCUSSIONS

This table presents isotope shifts and $\lambda^{151,A}$ measured with relation to ^{151}Eu.

TABLE

A	N	$T_{1/2}$	$\Delta\nu^{151,A}$, GHz	$\lambda^{151,A}$, fm^2	$\langle\beta^2\rangle^{1/2}$
154	91	16 y	4.34(0.027)	0.744	0.36
153	90	-	3.619(0.02)	0.620	0.35
152	89	12.7 y	3.342(0.024)	0.571	0.35
151	88	-	0	0	0.25
150	87	12.6 day	-1.48(0.08)	-0.254	0.2
149	86	93 day	-1.85(0.08)	-0.320	0.19
148	85	54 day	-2.98(0.08)	-0.518	0.14
147	84	22 day	-3.39(0.08)	-0.586	0.14
146	83	4.6 day	-4.19(0.1)	-0.725	0.09
145	82	5.9 day	-4.9(0.1)	-0.820	
144	81	10.2 s	-5.35(0.3)	-0.895(50)	0.06(6)
143	80	2.6 min	-5.20(0.20)	-0.874(33)	0.12(2)
142m	79	1.2 min	-4.94(0.15)	-0.833(24)	0.16(1)
141	78	37 s	-4.6(0.15)	-0.78(24)	0.19(1)
140	77	1.3 s	-4.4(0.3)	-0.75(50)	0.22(2)
139	76	22 s	-4.15(0.2)	-0.71(33)	0.24(1)
138	75	12 s	-4.4(0.35)	-0.75(57)	0.24(2)

The isotope shift $\Delta\nu^{A-A'}$ in the spectra of heavy atoms is described by a well-known formula

$$\Delta \nu^{AA'} = b \lambda^{AA'} + \Delta \nu_M$$

where b is a coefficient associated with the atomic wave function variation, $\lambda^{AA'}$ is determined by the charge distribution in the nucleus, $\Delta \nu_M$ is a specific isotope mass shift.

The transitions between the $6s^2$ and 6s6p configurations in EuI can be regarded as transitions between pure configurations[18]. For such transitions $\Delta \nu_M$ is evaluated in a standard way[19]

$$\Delta \nu_M = \frac{\nu_0}{1836} (1 \pm 0.5) \frac{A-A'}{AA'}$$

where ν_0 is the transition frequency, A and A' are the atomic numbers of the isotopes. This evaluation agrees well with the theoretical calculation of the isotope mass shift and with the results of calibration of optical shifts against the X-ray and μ-mesoatomic shifts for the next element.

To calculate the variation of electron density on the nucleus $|\Psi(0)|^2$ and the "b" coefficient one can use the Goudsmith-Fermi-Segre formula relating $|\Psi(0)|^2$ to the level energy of the $n\,S_{1/2}$ series in the atom or the ion in this atom and, also, the formula relating $|\Psi(0)|^2$ to the a_{ns} constant of magnetic hyperfine splitting for the ns-electron. We used data from literature for a_{ns} of the configurations $4f^7 6s6p$[20], $4f^7 5d6s$[21] in EuI and $4f^7 6s$ in EuII[22] as well as data for the $4f^7 ns$ series in EuII and $4f^7 5dns$ in EuI[23]. Averaging all the values obtained for b we have $\bar{b}$ = 6.12(30) GHz/fm^2, where the error is calculated as a root-mean-square deviation.

The value $\lambda^{AA'} = \Delta \langle r^2 \rangle (1 + C_1 \frac{\Delta \langle r^4 \rangle}{\Delta \langle r^2 \rangle} + C_2 \frac{\Delta \langle r^6 \rangle}{\Delta \langle r^2 \rangle} + ...)$. The C_1 and C_2 coefficient are small and given in[24]. In the simplest model of a liquid drop with a uniform charge distribution and a sharp edge the relation

$$\Delta \langle r^2 \rangle^{AA'} = \lambda^{AA'}/0.958$$

is valid.

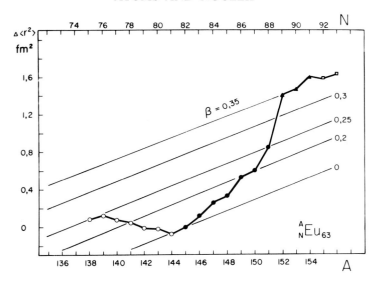

FIGURE 3   Variation of the mean-square charge radius of Eu
nuclides (measured with reference at the MSCR of
^{145}Eu) as a function of the mass number

Figure 3 shows an experimental dependence of $\Delta\langle r^2\rangle$ on the mass number as well as similar dependences calculated with the drop model with different deformation parameters β. A smooth growth of $\langle\beta^2\rangle$ for the nuclei of Eu(145-151) shows that the deformation increases gradually near the nucleus of ^{151}Eu with N = 88. This fact supports the conclusions drawn in[25]. One can clearly observe an even-odd effect leading to a relative decrease of $\langle\beta^2\rangle$ for odd-neutron nuclei. The behaviour of the MSCR curve changes in the case of the nucleus of ^{152}Eu which corresponds to a sharp increase of deformation for this nucleus. As a whole, the behaviour of MSCR varying with N for Eu isotopes corresponds to that of MSCR for the adjacent elements Gd and Sm and differs considerably form the behaviour of MSCR for Ba and Yb when N = 90 [5] the curves of which have not characteristic changes. For N ≤ 82 the shell effect is well pronounced - the slope of the MSCR curve decreases drastically. For instable nuclei this effect was observed for the first

58 LASERS IN ATOMIC, MOLECULAR AND NUCLEAR PHYSICS

time. The increase of nuclear size points to the fact that near N = 82 the nucleus is deformed. According to^{26}, the character of deformation in this region is dynamic. With a further decrease of N, near the limit of proton stability of Eu nuclei, the deformation grows greatly which may point to a new region of static deformation.

REFERENCES

1. H. J. KLUGE, R. NEUGART, E.-W. OTTEN, in Laser Spectroscopy IV, edited by H. Walter and K. W. Rothe, Proceedings of Internation Conf., Rottach-Egern, FRG, June 11-15, 1979 (Springer Verlag, Berlin, Heidelberg, New York, 1979), p. 517.
2. G. HUBER, C. THIBAULT, R. KLAPISCH, et al., Phys. Rev. Lett., 34, 1209 (1975); C. THIBAULT, F. TOUCHARD, S. BUTTGENBACH, et al., Phys. Rev., C23, 2720 (1981)
3. T. KUHE, R. DABKIEWICZ, C. DUKE, et al., Phys. Rev. Lett., 39, 180 (1977)
4. F. BUCHINGER, R. DABKIEWICZ, H. J. KLUGE, et al., Hyperfine Interactions, 9, 165 (1981)
5. R. NEUGART, Proc. Conf. on "Lasers in Nuclear Physics", edited by C. E. Bemis and H. C. Carter (Harwood Acadeimic Publ., Chur, London, N. Y., 1982)
6. G. D. ALKHASOV et al. and K. DORSKHEL et al., Atomic Masses and Faundamental Constants, Proc. of the 7th Inter. Conf., edited by Ol Klepper (Darmstadt, 1984), p. 327.
7. E.-W. OTTEN, Nucl. Phys., A354, 471 (1981)
8. Lasers in Nuclear Physics, Pros. Conf., edited by C. E. Bemis and H. C. Carter (Harwood Academic Publ., Chur, London, N. Y., 1982)
9. V. S. LETOKHOV, V. I. MISHIN, A. A. PURETZKY, Progress in Quantum Electr., edited by J. H. Sanders and S. Stenholm, Vol. 5, part 3, 1977, p. 139; N. V. KARLOV, B. B. CRYNEZKIY, A. A. MISHIN, A. M. PROCHOROV, Usp. Fiz. Nauk, 127, 593 (1979); G. S. HURST, M. G. PAYNE, S. D. KRAMER, T. P. YOUNG, Rev. Mod. Phys., 132, 293 (1979); V. I, BALYKIN, G. I, BEKOV, V. S. LETOKHOV, V. I. MISHIN, Usp. Fiz. Nauk, 132, 293 (1980)
10. G. I. BEKOV, V. S. LETOKHOV, O. I. MATVEEV, V. I. MISHIN, Pis'ma Zh. Eksp. Teor. Fiz., 37, 231 (1983)
11. G. I. BEKOV, V. S. LETOKHOV, V. I. MISHIN, Zh. Eksp. Teor. Fiz, 73, 157 (1977)
12. E. Ye. BERLOVICH, E. I. IGNATENKO, Yu. N. NOVIKOV, Proc. VIII Intern. EMIS Conf., Skövele, Sweden, 1973, p. 349

13. G. A. LINDER, P. MOLER, Phys. Lett., 110B, 17 (1982)
14. G. D. ALKHAZOV, et al., Pis'ma Zh. Eksp. Teor. Fiz., 13, 305 (1971)
15. W. L. MARTIN, R. ZALUBAS, L. HAGAN, Atomic Energy Levels, The Rare-Earth Elements, NSRDS-NBS60, Washington, 1978.
16. G. J. ZAAL, W. HAGERVARST, E. R. ELIEL, et al., Zs. Phys., A290, 339 (1979)
17. A. N. ZHERIKHIN, et al., Zh. Eksp. Teor. Fiz., 86, 1249 (1984)
18. F. SMITH, B. J. WYBOURNE, JOSA, 55, 121 (1965)
19. K. HEILIG, A. STEUDEL, At. Data Nucl. Data Tables, 14, 613 (1974)
20. E. R. EHELL, K. A. H. VAN ZEENWEN, W. HAGERVARST, Phys. Rev., A22, 1491 (1980)
21. H.-W. BRANDT, et al., Zs. Phys., A302, 291 (1981)
22. G. GUTHORLEIN, Zs. Phys., 214, 332 (1968)
23. J. SUGAR, J. READER, JOSA, 55, 1286 (1965)
24. E. C. SELZER, Phys. Rev., 188, 1916 (1969)
25. E. Ye. BERLOVICH, Izv. AN SSSR, Ser. Fiz., 29, 2177 (1965)
26. D. HABS, et al. Zs. Phys., 267, 149 (1974)
27. K. HEILIG, A. STENDEL, ADNDT, 14, 613 (1974)

1.4 Laser Cooling and Localization of Atoms

V. G. MINOGEN

Institute of Spectroscopy, USSR Acad. Sci., 142092, Troitzk, Moscow Region, USSR

1. INTRODUCTION

Nowadays there is a growing interest to the methods of control over the atomic motion based on the resonant laser radiation pressure. The basic property of the resonant laser radiation pressure which attracts the research interest is its velocity selectivity. Because of the velocity selectivity the laser light pressure effectively accelerates (or decelerates) only the atoms being in resonance with the radiation. Velocity-selective acceleration or deceleration of atoms in their turn results in effective deformation of the velocity distribution of atomic ensembles.

Since the value of the optical photon momentum is small, the variation in the atomic velocity under the action of the resonant light pressure is usually smaller than or comparable to the average thermal velocity of atoms at room temperature. Therefore the most effective application of resonant light pressure is the deceleration of atoms and the cooling of atomic ensembles. The atoms in the resonant radiation field are decelerated automatically when the radiation frequency is smaller than the atomic transition frequency. In this case only the atoms whose velocities are directed towards the radiation wave vector are in resonance with the radiation. The latter means that the direction of the light pressure

62 LASERS IN ATOMIC, MOLECULAR AND NUCLEAR PHYSICS

force acting on an atom is opposite to that of the atomic velocity.

The deceleration of atoms by laser radiation pressure makes it possible in some cases to produce ensembles of cold atoms whose temperature may reach ultralow values of about $10^{-3} - 10^{-4}$ K. Such cold atomic ensembles are of interest in many fields of physical research and particularly for superhigh-resolution spectroscopy since in the latter case the use of cold atoms allows eliminating such fundamental sources of spectral line broadening and shift as, for example, the second-order Doppler effect and flight broadening. Along with this direct application, cold atoms can be kept long in atomic traps. In this case it is quite possible to carry out spectroscopic studies with ultimate sensitivity, up to a single atom.

The present paper summarizes the methods of laser cooling of atoms, considers the ways of localization of cold atoms and discusses some applications of cold localized atoms in spectroscopy and atomic physics.

2. LONGITUDINAL COOLING OF ATOMIC BEAMS

The basic method of production of cold atoms at present is velocity-selective deceleration of atomic beams by laser radiation. Qualitatively, the longitudinal cooling of an atomic beam consists in the following (Figure 1). The light pressure force acting on the atoms in the beam from the resonant laser radiation decelerates the resonant atoms which gives rise to a dip in the velocity distribution centred at the velocity $v_{res} = (\omega_o - \omega)/k$, where ω_o is the atomic transition frequency, ω is the laser radiation frequency. The formation of the dip is followed by the formation of the peak of decelerated atoms centred at a velocity smaller than v_{res}. When the interaction time is long enought, almost all the atoms are decelerated and, since the velocity dependence of the light pressure force F (Figure 1) has a nonlinear belled form,

ATOMS AND NUCLEII 63

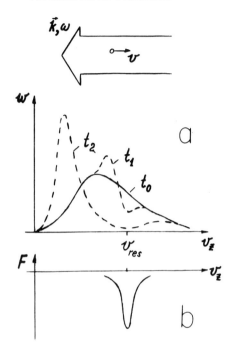

FIGURE 1   The effect of velocity monochromatization in atomic
            beam decelerated by laser radiation. The solid curve
            is the initial velocity distribution of atoms
            $w = w(v_z)$, the duched curves show the atomic distribu-
            tion for times t_1 and t_2 ($t_1 < t_2$) (a).
            The velocity dependence of the light pressure force (b).

the initial wide distribtution always turns to a narrow monove-
locity distribution.

The longitudinal cooling of atomic beams was experimentally
investigated with a Na atomic beam decelerated by CW dye laser
radiation[1-8]. Below we shall dwell on the results of work[3], where
the velocity monochromatization was first observed and the experimental velocity distribution was compared with the distribution
calculated from the kinetic equation.

Figure 2 schematically shows the scheme of an experiment in
which the velocity destribution monochromatization of a Na atomic
beam decelerated by laser radiation was observed. A narrow atomic

64 LASERS IN ATOMIC, MOLECULAR AND NUCLEAR PHYSICS

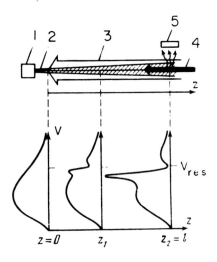

FIGURE 2 The scheme of an experiment on longitudinal cooling of atomic beam. Botton: the deformation of velocity distribution at different distances from the point where the atoms enter the light beam.

beam (2) coming from an oven (1) was irradiated by a laser beam (3) being in resonance with the $3S_{1/2} - 3P_{3/2}$ transition of Na atom. Since the ground state of Na atom is split into two hyperfine structure sublevels, the laser radiation had two frequencies to make the atom-radiation interaction cyclic. One frequency was in resonance with the $3S_{1/2}(F=1) - 3P_{3/2}$ transition, the second one with the $3S_{1/2}(F=2) - 3P_{3/2}$ transition. The difference between these two frequencies was taken to be equal to the frequency of the hyperfine strucrture interval between the sublevels $3S_{1/2}(F=1)$ and $3S_{1/2}(F=2)$ that comes to 1772 MHz. The atomic velocity distribution along the beam axis was determined by detecting the fluorescence excited by a probe laser beam (4). The probe beam was monofrequency and propagated to meet the atomic beam, too. Its frequency could be tuned within the absorption line of the atomic beam. In order that the powerful two-frequency laser beam could not disturb the fluorescence signal excited by the probe beam, the

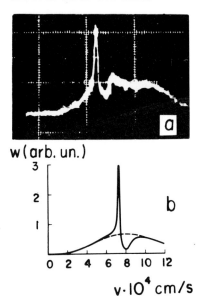

FIGURE 3 The experimental (a) and theoretical (b) profiles of velocity distribution of a Na atomic beam decelerated by laser radiation.

two-frequency laser beam was periodically interrupted by a mechanical chopper. The fluorescence signal was recorded by a detector (5) only at the time intervals when the powerful beam was switched off by the chopper.

One of the experimental dependences of fluorescence intensity on probe beam frequency is shown in Figure 3. The curve determines the longitudinal atomic velocity distribution in the beam resulting from the nonlinear atomic deceleration by the laser radiation. This curve $w = w(v_z)$ was obtained with the laser beam tuned to resonance with the atoms being at the maximum of the initial thermal velocity distribution.

Since in the experiment[3] it was the light pressure force that made the main contribution to the velocity distribution deformation, the results of the experiment are close to the results obtained from the Liouville equation. The curve in Figure 3a corresponds to the calculated dependence shown in Figure 3b[3].

In the first experiments[3] the ratio between the initial velocity distribution width and the narrow peak width of monochromatized atoms was $\Delta v_{in} / \Delta v_{fin} \simeq 20$. This degree of monochromatization was consistent with the decrease in temperature of relative atomic motion from its initial value $T_{in} = 573$ K to $T_{fin} = 1.5$ K. In the experiments[4,5] atomic beams of Na with the absolute temperature $T = 0.07$ K were produced.

Consider the basic parameters which characterize the longitudinal cooling of atomic beams. For definiteness, we assume that the atoms are the two-level ones. The probability of spontaneous relaxation of atom from the upper state to the lower ground state is denoted 2γ, the recoil velocity of atom is $v_r = \hbar k / M$, were k is the laser radiation wave vector. The saturation parameter is $G = I/I_s$, where I_s is the atomic transition saturation intensity. The detuning of the laser radiation frequency relative to the atomic transition frequency is $\Omega = \omega - \omega_o$.

According to work[9] the atomic beam is decelerated to the zero average velocity during the time

$$t_d \simeq (|\Omega|/\gamma)^3 / 3 G k v_r \qquad (1)$$

covering the distance

$$l_d \simeq \lambda (\gamma / k v_r)(\Omega/\gamma)^4 / 12 G. \qquad (2)$$

The temperature of atoms at the end of the deceleration path is

$$T_o^d \simeq \hbar |\Omega| / k_B \qquad (3)$$

where k_B is Boltzmann's constant. When the saturation parameter $G \simeq 100$, the deceleration time is 10^{-3} s, the deceleration path is about 100 cm and the temperature ranges from 10^{-1} to 10^{-2} K.

3. TRANSVERSE COOLING OF ATOMIC BEAMS

Along with the one-dimensional longitudinal cooling described

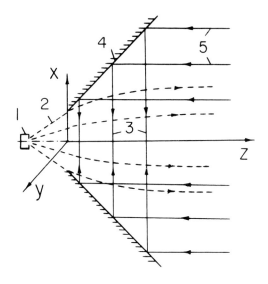

FIGURE 4

above, two-dimensional transverse cooling of atomic beams is possible, too. One of the schemes for such cooling is shown in Figure 4. The atomic beam (2) running out of the source (1) is irradiated by an axisymmetric light field (3) the frequency of which is red-shifted about the atomic transition frequency ω_o[10]. The axisymmetric field is assumed to be formed by the reflection of the laser beam (5) from the internal surface of a hollow meatal cone (4)

Physically, the effect of transverse cooling of atomic beams can be explained in the following way. In the axisymmetric field (3) with its frequency $\omega < \omega_o$ the light pressure force is directed towards the radial atomic velocity (Figure 5a). Because of this, the light pressure force makes the velocity distribution across the beam axis narrower, that is, it cools the atomic beam (Figure 5b). The transverse cooling of the atomic beam is followed by its collimation, that is, by a decrease of its divergence.

The basic parameter which determine the transverse cooling of

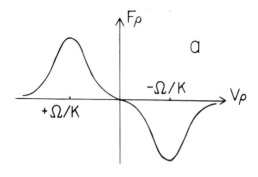

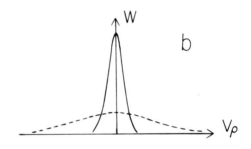

FIGURE 5 a) The light pressure force F in an axisymmetric field as a function of the radial atomic velocity v. b) The narrowing of the transverse velocity distribution of an atomic beam under the action of the force F.

an atomic beam is the effective temperature which characterizes the width of atomic velocity distribution across the beam axis. With the detuning $\Omega = \omega - \omega_o$ this temperature is [10]

$$T_{tr}^o \simeq \frac{\hbar \gamma}{4k_B}(|\Omega|/\gamma + \gamma/|\Omega|) \quad (4)$$

when $\Omega = -\gamma$, the temperature reaches its minimum

$$T_{min} \simeq \hbar \gamma /k_B \quad (5)$$

that ranges from 10^{-4} to 10^{-3} K.

The scheme shown in Figure 4 was used in work[11] to realize transverse cooling of sodium atomic beam. As a result, in the

ATOMS AND NUCLEII 69

experiments[11] a thermal beam of Na atoms was cooled to a record temperature, 3.5 mK.

4. THREE-DIMENSIONAL COOLING OF ATOMIC ENSEMBLES

The above-considered narrowing of the velocity distribution of atomic beams (along or across the beam axis) is not, in a strict sense, cooling of atoms, since it is necessary to decrease the modulus of the velocity v and not the modulus of one or two velocity projections for true cooling of an atomic ensemble. In may be seen, however, that radiative cooling can be applied to three-dimensional space. Indeed, one can take four beams and direct them from the corners to the centre of a right tetrahedron. In this configuration it is possible to decrease the absolute values of all the three projections of atomic velocity, that is, to realize true cooling of an atomic ensemble. More beams can set up more complex light fields providing rediative cooling of atoms. Six beams, for example, may be directed from the centres of the faces to the centre of a cube. In the case of eight beams they must be directed from the corners of the cube to its centre.

The main question arising in analysis of the effect of radiative cooling of atoms is the width of stationary velocity distribution, that is, the cold atomic ensemble temperature. To answer this question one should take into account the fact that a stationary velocity distribution narrowing due to the light pressure force by the broadening on account of velocity diffusion. The inclusion of these two factors in any scheme of three-dimensional cooling results in a temperature close to (4). The minimum value of temperature for any cooling scheme is accordingly close to (5).

5. LOCALIZATION OF COLD ATOMS IN LASER FIELDS

In some works it was suggested that, in order to produce cold atoms, they should be localized in laser fields. Some of these

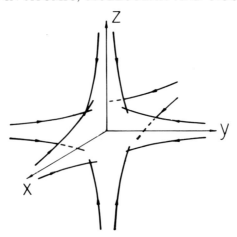

FIGURE 6

suggestions are reviewed in[12]. The recent investigations show that it is impossible to realize stable localization of atoms in light fields[13,14]. In order to understand what forbides the stable localization of atoms in light fields, we may consider the configuration of a light field consisting of six divergent laser beams directed along the ±x, ±y, and ±z axes (Figure 6). The frequencies of the beams are assumed to satisfy the condition of laser cooling of atoms: $\omega < \omega_o$.

In this centrosymmetric field the atoms can be cooled only at low intensities of the beams when the saturation parameter G satisfies the condition (for $|\Omega| > \gamma$)[14]

$$G < |\Omega|/\gamma \qquad (6)$$

At large saturation parameters, when condition (6) is not satisfied, the light pressure force heats the atoms instead of cooling them.

The total force acting on an atom in the field of three pairs of counter-running beams consists of a light pressure force and a gradient force. Because of the condition (6) the light pressure force is proportional to the Poynting vector. By this reason this

ATOMS AND NUCLEII 71

force does not set up an absolute potential minimum^{13}. The gradient force sets up an absolute minimum. But by virtue of condition (6) the gradient force potential is limited by the value^{14}

$$U_{GR} \leq \hbar \gamma \qquad (7)$$

that is close to the minimum energy of cold atoms (compare with (5)). Therefore, it is also impossible to realize stable localization of cold atoms on account of the gradient force.

As a result of the above factors it is impossible to set up a light field where stable localization of cold atoms would be realized.

6. <u>LOCALIZATION OF COLD ATOMS IN A MAGNETIC TRAP</u>

A real way of long localization of cold atoms is to confine them in magnetic traps.

One of the simplest magnetic traps is toroidal-shaped (Figure 7). In this scheme six derect currents set up a nonuniform magnetic field the magnitude of which increases from the centre to the periphery of the torus cross-section proportionaly to the square of the distance from the axial line

$$|H(\vec{r})| = H_m r^2/a^2 \qquad (8)$$

where a is the small radius of the tore, H_m is the field on the torus surface.

The motion of a cold atom injected into the magnetic field of a toroidal trap depends on the orientation of the dipole moment $\vec{\mu}$ of atom with respect to the vector $\vec{H}$. At antiparallel direction of $\vec{\mu}$ and $\vec{H}$ the atom is pulled into the centre of the tore, at parallel directions of $\vec{\mu}$ and $\vec{H}$ it is pushed out of the magnetic field. In the case of Na^{23} atoms in a toroidal trap only the atoms being in the ground state $3S_{1/2}$ at the hyperfine structure sublevels $F = 1$, $m_F = -1$; $F = 2$. $m_F = 2,1$ can be confined. The lowest confining force atcs on the atoms in the state $F = 1$, $m_F = -1$.

72 LASERS IN ATOMIC, MOLECULAR AND NUCLEAR PHYSICS

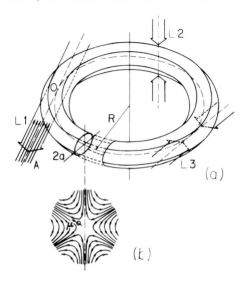

FIGURE 7 a) The scheme of injecting cold atoms into a torroidal magnetic trap. The laser beam L_1 is used to decelerate the injected atomic beam A. The counter-running beams L_2, L_3 can be used to cool the atoms in the trap.
b) The magnetic field lines transverse to the trap.

For stable localization of the atoms near the axial line of the torus it is necessary that two basic conditions should be fulfilled[15]. One of the conditions is that the potential well depth must exceed the average kinetic energy of the cold atoms moving across the axial line of the torus. The other necessary condition is that the centrifugal shift of atoms should be smaller than the small radius of the torus. Let the average kinetic energy of the cold atoms moving transverse to the axial line of the torus be consistent with the effective temperature $T_{tr} = 10^{-3}$ K. The average energy of the atoms moving along the axial line is an order higher and corresponds to the effective temperature $T_1 = 10^{-2}$ K. The temperature T_{tr} here is close to the ultimate theoretical value that can be attained at radiative cooling of atoms. T_1 is taken to be an order higher because the atoms being injected into the trap may have a nonzero average velocity.

With these temperatures, given the ratio $\alpha = \mu H_m/k_B T_{tr}$, the first condition allows determining the value of magnetic field on the torus surface. For example, when $\alpha = 20$,

$$H_m = \alpha(2k_B T_{tr}/\mu_o) \simeq 600 \text{ G} \qquad (9)$$

where μ_o is the Bohr magneton. Given the ratio $\beta = \mu H_m/k_B T_1$, the second condition restricts the ratio a/R. Assuming that the value of centrofugal displacement is bound to

$$\rho \simeq a^2/\beta R \ll a \qquad (10)$$

we may choose a/R = 0.1. When a = 3 cm, the long radius of the torus must be R = 30 cm.

It should be noted that the function of the magnetic trap is not restricted only to confining atoms at the temperatures of the beam being injected. If the atoms circulating in the trap are irradiated by counter-propagating light beams, their temperature can be reduced to a minimum (5). For Na^{23} atoms the ultimate temperature may be $3.4 \cdot 10^{-4}$ K.

7. CONCLUSION

Finally, we note some applications of cold atoms in atomic and molecular physics, superhigh-resolution spectroscopy and quantum metrology. The use of cold atoms in atomic and molecular physics is of particular interest for studying collisional processes, effects of formation of chemical bonds and condensation. In this field the laser methods of atomic cooling are able to complement essentially the methods of gas cooling in supersonic nozzles now in use. The use of cold atoms in spectroscopy and quantum metrology makes it possible to remove the fundamental factors of broadening and shift of narrow spectral resonances due to the final velocity: the first- and second-order Dopler effect, time-of-flight broadening. Because of this the application of cold atoms can drastically increase both the resolution of spectroscopic investigations and the accuracy of quantum frequency standards.

Other applications follow from the fact that cold atoms can be accumulated and confined for a long time in magnetic traps. The accumulation and storage of localized atoms enable us to solve the problem of increasing the sensitivity of some methods applied in atomic physics, spectroscopy and in nuclear fields, for example, allows spectroscopic studies of radioactive atoms produced in countable quantities by nuclear reactions.

Finally, due to frequency selectivity of resonant light pressure, a combination of radiative deceleration of atoms with their subsequent localization is of much interest for isotope-selective detection of single atoms[16].

REFERENCES

1. V. I. BALYKIN, V. S. LETOKHOV, V. I. MISHIN, Pis'ma ZhETF, 29, 614 (1979); ZhETF, 78, 1376 (1980)
2. V. I. BALYKIN, V. S. LETOKHOV, V. G. MINOGIN, ZhETF, 80, 1779 (1981)
3. S. V. ANDREEV, V. I. BALYKIN, V. S. LETOKHOV, V. G. MINOGIN, Pis'ma ZhETF, 34, 463 (1981); ZhETF, 82, 1429 (1982)
4. W. D. PHILLIPS, H. METCALF, Phys. Rev. Lett., 48, 596 (1982)
5. J. V. PRODAN, W. D. PHILLIPS, H. METCALF, Phys. Rev. Lett., 49, 1149 (1982)
6. W. D. PHILLIPS, J. V. PRODAN, H. J. METCALF, in Laser-Cooled and Trapped Atoms, edited by W. D. Phillips (NBS Special Publication, Washington, 1983), p. 1.
7. R. BLATT, W. ERTWER, J. L. HALL, Ibid., p. 142.
8. V. I. BALYKIN, V. S. LETOKHOV, A. I. SIDOROV, Opt. Comm., 49, 248 (1984); ZhETF, 86, 2019 (1984)
9. T. V. ZUEVA, V. S. LETOKHOV, V. G. MINOGIN, ZhETF, 81, 84 (1981)
10. V. I. BALYKIN, V. S. LETOKHOV, V. G. MINOGIN, T. V. ZUEVA, Appl. Phys., B35, 149 (1984)
11. V. I. BALYKIN, V. S. LETOKHOV, A. I. SIDOROV, Pis'ma ZhETF, 40, 251 (1984)
12. V. S. LETOKHOV, V. G. MINOGIN, Phys. Rep., 73, 1 (1981)

13. A. ASHKIN, J. P. GORDON, Opt. Lett., 8, 511 (1983)
14. V. G. MINOGIN, Opt. Lett., (1985) to be published
15. V. S. LETOKHOV, V. G. MINOGIN, Opt. Comm., 35, 199 (1980)
16. V. I. BALYKIN, V. S. LETOKHOV, V. G. MINOGIN, Appl. Phys., B33, 247 (1984)

1.5 New Trends in the Structure and Classification of the States of Atoms and Ions

Z. B. RUDZIKAS and J. M. KANIAUSKAS

Institute of Physics, Academy of Sciences of the Lithuanian SSR, K. Pozelos 54, Vilnius, 232600, USSR

1. INTRODUCTION

The famous postulates, formulated in 1913 by N. Bohr, served as the basis for the quantum mechanics of the hydrogen atom, which has been pretty soon generalized for the case of many-electron atoms. The character of the interatomic interactions is known fairly exactly. Therefore, one may think that the theory of many-electron atoms, in fact, has reached its completion and that here it is impossible to get any new results. In this paper we shall try to show that it is not true, that recently in atomic spectroscopy - both theoretical and experimental - a number of remarkable effects and phenomena have been revealed; and atom now, when we are going to celebrate N. Bohr's, the founder's of its theory, a hundred year jubilee, still preserves a series of important secrets.

We should mention the qualitative changes in atomic spectroscopy, achieved by using laser and beam-laser methods[1]. By these methods it became possible to investigate in laboratories atoms and ions in the states, which earlier had been accessible (and even not always) only in astrophysical studies. The usage of the laser equipment allows one to get the unprecedented ionization degrees and the high selectivity of excitation (including autoionizing and Rhydberg states) of atoms and ions. Thanks to this it was accumulated a fairly large number of new experimental data concerning the spectra of such systems. However, there are some specific difficulties in classifying and theoretically interpreting these spectra as the peculiarities of the structure of the energy levels are caused by the ratio of the fine relativistic and correlation effects. The latters' relative role essentially depends both - on the ionization degree of an atom and on the character of its excitation: with the increase of the ionization degree the role of the relativistic effects grows, whereas with the excitation of the atoms and ions the correlation corrections often begin to prevail. The role of the relativistic effects was discussed in detail in[2,3], therefore, here we shall pay the main attention to the problems of symmetry, accounting for correlation effects, peculiarities in the structure and properties of many-electron atoms and ions, including very highly ionized ones.

2. PECULIARITIES OF THE STRUCTURE AND THE SYSTEMATIZATION OF THE STATES OF MANY-ELECTRON ATOMS AND IONS

Requirements of other fields of science (e.g., plasma physics, fusion problems, astrophysics, etc.) and technology (e.g., laser technique, spectral equipment, etc.) stimulate rapid experimental spectroscopic investigations - experiments in laboratories as well as measurements in space ships. So there rise new problems to be solved by theorists.

At the initial stage of the development of atomic spectroscopy it turned out from the simplest calculations that even a non- spherical part of the electrostatic interaction energy of the atomic electrons is larger than that of the spin-orbit interaction. This explained the experimental fact that the distances between terms were much larger than those between the levels of the multiplet under consideration, i.e. the fine structure of terms was really a small quantity and described small splitting of terms caused by magnetic interactions. Thus here rose the idea of the LS coupling widely spread among theorists and, especially, among experimentalists, what is proved by its ixtensive usage at present even in those cases when it is obviously out of place.

However, penetration to the higher ionized atoms as well as to the higher and higher excited states showed that there occur vast regions of ions and their states for which the LS coupling and the description of their spectra in the notion of terms and their fine splitting are absolutely impossible.

Complete sets of quantum numbers, with the help of which it is possible to classify the energy spectrum of the atom or ion under consideration, may be found only from theoretical studies. Therefore, the task was set for the theory of many-electron atom to develop a generalized method for classifying atomic states, covering as wide as possible stages of ionization and excitation. This led to various new coupling schemes (LK, JK, and JJ). These schemes illustrate the fact that the order of coupling of angular momenta of separate shells or even electrons reflects a relative value of interactions between corresponding momenta: at first the most strongly interacting momenta are coupled, then - the following momenta according to the strength of interaction, and so on. E.g., for the JJ coupling the orbital angular momentum of each electron interacts more strongly with its own spin momentum than with the orbital angular momenta of the other electrons, i.e. in this case the spin-orbit interactions prevail over the non-spherical part of Coulomb interactions.

Thanks to the extensive studies of the energy spectra of atoms and ions of various ionization and excitation degrees it has turned out that every abovementioned coupling scheme has its own regions of ionization stages and excitation types. Moreover, sometimes even for the classification of the levels of one and the same configuration one has to use different coupling schemes. However, the extensive calculations reveal that every pure coupling sceheme, which has to be the closest to the real one. Its quantum numbers, though being more or less approximate, are used for identifying and classifying the energy spectra of atoms and ions. Therefore, the urgent problem is to develop the efficient methods to optimize coupling schemes to transform the wavefunctions from one coupling scheme to another one. The latest achievements in this field are discussed in detail in[3].

About the change of a coupling scheme it is possible to judge by studying the qualitative picture of the energy spectrum (e.g., the appearance of the doublet structure in the case of the JJ coupling instead of singlets and triplets for the LS coupling), as well as by investigating the regularities of the changes of spectral characteristics along isoelectronic sequences.

When the traditional LS coupling or similar to it coupling schemes are valid, then the non-relativistic approach and non-relativistic wavefunctions may be efficiently used. And magnetic interactions (spin- orbit and spin-spin), as well as relativistic corrections to Coulomb interaction may be accounted for in the frameworks of the first order of the perturbation theory making use of non-relativistic wavefunctions[3]. Calculations show that such an approach suits fairly well for theoretical evaluation of the energy spectra and electronic transitions in the case of a wide region of atoms and ions, including even highly ionized atoms.

However, for very highly ionized atoms, as well as for studying small effects (e.g., hyperfine splitting, spectroscopic evidencies of parity violation effects, a number of processes connected with the electrons from deep inner shells, etc.) the relativistic effects can not often be treated as corrections. Then at the very beginning one has to start with relativistic Hamiltonian of the system and relativistic wavefunctions. Moreover, as we are going to see later, the structure of atomic electronic shells and the classification of its states considerably change. The corresponding mathematical apparatus of the theory in both cases is described in great detail in^3.

Relativistic approach is valid when the interactions of the orbital momentum of an electron with its spin one exceeds the non-spherical part of interelectronic Coulumb interactions in a shell of equivalent electrons (the jj coupling within the shell). Then the usual shell $l^N \alpha LSJ$ splits into a set of the subshells $nlj_1^{N_1} j_2^{N_2} \alpha_1 J_1 \alpha_2 J_2 J$, the number of electrons in every of which changes according to the values of the quantum numbers J_1, J_2, and J. Here $j_1=l-1/2$, $j=l+1/2$, $N=N_1+N_2$, and l is preserved to indicate the parity of the configuration.

Thus, while taking the relativistic approach, we loose even the notion of electronic (shell-like) configuration in the common sense of this word, because the total set of the levels of the non-relativistic shell l^N in a relativistic case groups into a number of the j-configurations (subshells). Their maximally possible number equals N+1. Really, while considering the energy spectra of extremely highly ionized atoms, one can see^3, that their levels form separate groups corresponding to certain subshells or their sets.

Thus, in a relativistic case we come across a more complicated structure of electronic configurations (the increase of a number of open shells (subshells), the non-diagonality of the main terms of the energy operator with respect to the quantum numbers of separate subshells; therefore, it is necessary already in the usual single configuration approach to take into account the non-diagonal, with respect to configurations, matrix elements). However, the compensation of these difficulties is the comparative simplicity of the relativistic Hamiltonian and, especially, the smallness of the subshells, as well as a good capability to classify their levels. Really, as the shell is filled when $N=2j+1$, then the subshells $[1/2]^N$, $[3/2]^N$, $[5/2]^N$, and $[7/2]^N$, which are the analogues of the filled shells s^N, p^N, d^N, and f^N, will consist of 2, 4, 6, and 8 electrons respectively. For a simple classification of the levels of all the abovementioned subshells it is enough to adopt the relativistic variant of the seniority quantum number v and the total momentum J. Therefore, in a relativistic approach even all levels of the shell f^N will be simply classified, what is not true in a non-relativistic case.

In conclusion the development of the methods, in which the relativistic effects are accounted as corrections and in the relativistic approach, has considerably enriched the spectroscopy of many-electron atoms. It has also enabled one to consider, in a general approach, the spectra of atoms and ions, including very highly ionized atoms, as well as to study a number of very precise spectroscopic effects, greatly to help to inentify and classify the measured experimentally of observed astrophysical spectra of atoms and ions of various ionization degrees. Moreover, the studies of the long isoelectronic sequences, accounting for the increasing along them relativistic effects, illustrate the appearance of the qualitatively new structure of the ions and their spectra. In this way they breakdown the traditional worldwide accepted spectroscopic notions and regularities (the break-down of the LS coupling, decay of the usual electronic shells and even, as we shall see below, electronic configurations, etc.).

3. NEW SYMMETRY PROPERTIES IN MANY-ELECTRON ATOM

The validity of some coupling scheme in atom is caused by the realization of certain symmetry properties, usually related with the transformations in some space, most often - with rotations. Taking into consideration the symmetry properties of a many-particle system one is able to simplify the problem of its theoretical description, as well as to obtain a large ammount of valuable information about its structure and properties. This is obtained without solving the corresponding equations, only studying their behaviour grounded by corresponding transformations. Thus, many properties of the angular orbital momentum L depend on the symmetry of the rotation of the three dimensional space. The seniority quantum number v is connected with the rotation in the so called quasispin space, whereas the quantum numbers $(u_1 u_2)$ and $(w_1 w_2 w_3)$, used for the classification of the states of the shell f^N, follow from the theory of representations of the groups G_2 and R_7 respectively.

However, sometimes one succeeds in finding new symmetry types in many- particle system. It is highly efficient to consider them as they come handy to reveal new peculiarities in the structure and properties of many-electron atom, to give a possibility to present new modifications of the model of an atom. Let us present several examples to illustrate this.

One of these models was proposed in[4] for configurations, consisting of two shells of equivalent electrons $n_1 l_1^{N_1} n_2 l_2^{N_2}$ when $l_1 = l_2$. The analysis of the matrix of the Coulomb interaction shows that there are cases, when for these configurations the non-spherical part of the Coulomb interaction of the electrons from different shells is similar to the order of magnitude of the same sort of the interaction of the electrons in a shell. As the relative values of the non-spherical parts of the Coulomb interactions define the order of vectorial coupling of the orbital angular momenta of separate electrons, therefore, we have to look for such theoretical model, in which the momenta of the electrons from different shells would interact as the momenta of the electrons from one shell.

For this purpose there is introduced the isospin formalism, in which the principal quantum numbers n_1 and n_2 are considered as an additional degree of freedom in isospin space. Using the isospin quantum number T, which transformation properties are analogical to those of the spin momentum, we may refuse the wavefunctions, obtained with the help of vectorial coupling of the momenta of separate shells and to adopt wavefunctions, characterized by the eigenvalues of the operators N, T^2, T_z

$$|n_1 n_2 (ll)^N \alpha \, TLSM_T M_L M_S\rangle \equiv |n_1 n_2 l^{N_1} l^{N_2} \alpha \, TLSM_L M_S\rangle \quad (3.1)$$

where $M_T = (N_2 - N_1)/2$, $T = \frac{N}{2}, \frac{N}{2} - 1, \ldots, \frac{1}{2}(0)$, $0 \leqslant N = N_1 + N_2 \leqslant 8l+4$, and α represents the rest quantum numbers necessary for a simple classification of the states.

Thus, in the isospin basis two shells with the same l are united into one, similar to the nuclear shell. The shortcoming of this basis is the appearance of the not simply classified levels, because here we loose the former intermediate quantum numbers and gain only one T. To solve this problem one has to make use of the groups of the higher ranks.

The operators, representing the physical quantities, may be expressed in terms of the irreducible tensors in the isospin space. Obviously, the isospin T will be the exact quantum number in the case, when in such an expression only scalar terms will be present. In a non-relativictic case for this there ought to be valid the follwing equality between radial integrals of the Coulomb interaction:

$$F_k(n_1 l, n_1 l) = F_k(n_2 l, n_2 l) = F_k(n_1 l, n_2 l) + g_k(n_1 l, n_2 l),$$

$$(k > 0) \quad (3.2)$$

In general this equality, of course, does not hold. However, for excited states of this configuration less strong restrictions are

ATOMS AND NUCLEII 85

enough. So, for the configuration $n_1 l n_2 l^{N_2}$ it is sufficient that the equality $F_k(n_1 l, n_2 l) + g_k(n_1 l, n_2 l) = F_k(n_2 l, n_2 l)$, $k > 0$ holds.

Calculations show that for a number of configurations with $n_2 = n_1 + 1$, especially of the type $n_1 l n_2 l^{N_2}$, this equality holds fairly well, i.e. there are cases when a new basis is preferable for classifying atomic states. The analogical approach for the relativistic case is developed in[5].

The calculations of the energy matrices in an intermediate coupling scheme, when one starts with the usual and isospin bases, also prove in a number of cases the preference of the isospin basis and the real confluence of two shells of the type $n_1 l^{N_1} n_2 l^{N_2}$ in a "supershell" $n_1 n_2 (l l)^N$. E.g., such calculations with the help of the relativistic wavefunctions for Xe^{47+} in the configuration $1s^2 2s^2 2p[3/2]^1 3p[3/2]^2$ ($J = 3/2$), show that in the intermediate coupling the wavefunction consists of two functions of the pure coupling scheme. Moreover, their weights in the usual basis are equal to 0.916 and -0.400, whereas in the isospin basis they are equal to 1.000 and -0.009. So, paying attention to the appearance of the new selection rules, one may interpret anew some peculiarities of the autoionizing states, including those obtained with the help of laser excitation.

Up till now it has been assumed that atomic states may be described in a single-configuration approximation. However, it has been noted while studying energy spectra, that there are cases when the distribution of the electrons between shells cannot be described exactly with the help of one certain configuration, i.e. it has turned out that the electronic configuration is not an exact quantum number. To obtain the real picture one may use the superposition-of-configurations method in which the so called correlation effects are considered. Usually this may be achieved by adding to the energy matrix the non-diagonal (with respect to configurations) matrix elements and by the further diagonalization of the matrix obtained.

The superposition-of-configurations method is asymptotically exact as when the number of configurations tends to the infinity we come to the exact solution, which does not depend on the zero order approximation. The shortcoming of this method is its slow convergancy as well as the absence of the exact criteria to select the most important configurations from the fairly large number of the possible ones.

In the cases when the non-diagonal (with respect to configurations) matrix elements are comparable by the order of magnitude with the difference between the corresponding non-diagonal elements then the definite electronic configuration can not be useful even as an approximate quantum number. Such a situation is not an exception[6]. It may be illustrated by the iron group elements as well as the doubly excited states of helium.

The complexity of elaboration of new theoretical models, allowing to introduce new quantum numbers, is due to the fact that to refuse the single-configuration approximation means to refuse, as well, the model, in which the electrons move independently in the central field of a nucleus - at the very beginning one has to take into account their correlation.

In some other regions of the many-body quantum theory, i.e. in the theory of superconductivity, solid state, Fermi gas and nucleus, while describing correlation effects it has turned out to be a very fruitful idea of quasiparticles - various collective excitations of the system, the physical manifestations of which have a strongly pronounced character. However, in complex atoms the attempts to distinguish the regular collective excitation failed. Therefore, the models, based on this conception, are rarely used. However, it has turned out that in some particular cases such models describe the correlation effects in atoms and ions fairly well.

One of them is the model of vectorial coupling of quasispin angular momenta of different shells.

The quasispin formalism is usually used for describing tensorial properties of the space of occupation numbers of a shell. In the particle-hole formalism the shell of equivalent electrons may be regarded as a completely filled one for any configuaration 1^N, because there are in it N electrons and $_hN = 4l + 2 - N$ holes. The Z-projection of the operator of the quasispin angular momentum may be written as follows:

$$Q_z = (\hat{N} - _h\hat{N})/4 \tag{3.3}$$

It shows that for every occupied one-particle state in the configuration 1^N we can ascribe the value of the Z-projection of the quasispin angular momentum equal to 1/4, and for not occupied (hole) state − −1/4. If the number of electrons in a shell changes by one, then the projection of the quasispin momentum of the wavefunction on the Z axis changes by 1/2. Requiring the wavefunctions of the configuration 1^N to be the eigenfunctions of the squared quasispin mementum of the shell

$$\vec{Q}^2 |1^N \alpha QLSM_L M_S\rangle = Q(Q+1) |1^N \alpha QLSM_L M_S\rangle \tag{3.4}$$

we came to the quantum number Q, simply connected with the seniority quantum numer v, introduced already by Racah for additional classification of the states of a shell of equivalent electrons, namely, $Q = (2l + 1 - v)/2$. The vectorial coupling of the quasispin momenta of two separate shells $\vec{Q} = \vec{Q}_1 + \vec{Q}_2$ leads to the wavefunction:

$$|(1_1 + 1_2)^N \alpha_1 Q_1 L_1 S_1 \alpha_2 Q_2 L_2 S_2 QLSM_L M_S\rangle = \sum_{M_{Q_1} M_{Q_2}} \begin{bmatrix} Q_1 & Q_2 & Q \\ M_{Q_1} & M_{Q_2} & M_Q \end{bmatrix} \times$$

$$\times |1_1^{N_1} 1_2^{N_2} \alpha_1 Q_1 L_1 S_1 \alpha_2 Q_2 L_2 S_2 LSM_L M_S\rangle , \tag{3.5}$$

$$M_{Q_i} = (N_i - 2l_i - 1)/2$$

Thus, the new quantum number of the total quasispin Q defines the weights of separate configurations of the multiconfigurational wavefunction, the physical suitability of which is defined by their degree of relations to the weights, obtained by the direct diagonalization of the energy matrix of the atomic system.

According to the definition the seniority quantum number v is equal to the number of not-paired particles in the state considered. (Two electrons are called paired (or the Cooper pairs) if their total orbital and spin momenta are equal to zero). Therefore, at the presence of the shortrange pairing interaction (this is the case, e.g., in the theories of nucleus and superconductivity), the quasispin model describes fairly well the pair correlations of the superconducting type in the ground state of many-particle system.

In atoms there is no shortrange pairing interaction because there are the Coulomb repulsing forces. Therefore, in atomic systems the Cooper pairs are possible only in strongly excted states. Let us consider the case of doubly excited states of helium. If the wavefunctions of such states are build up from the hydrogen-like orbitals, then the energies of the states $(ns^2, np^2,...)LS$ in the zero approximation will be degenerated. In non-relativistic theory one can avoid this degeneration while accountiong for the Coulomb interaction between electrons. Moreover, these configurations are strongly mixed. The wavefunctions, obtained after diagonalizing the energy matrix of the Coulomb interaction of the electrons, are expressed by the wavefunctions $|2l^2\ ^1S\rangle$ or $|2(s+p)^2 v\ ^1S\rangle$ as follows:

$$\psi_1 = 0.4756 |2s^2\ ^1S\rangle + 0.8796 |2p^2\ ^1S\rangle =$$
$$= 0.9988 |2(s+p)^2 v=0\ ^1S\rangle - 0.0483 |2(s+p)^2 v=2\ ^1S\rangle ;$$
$$\psi_2 = 0.8796 |2s^2\ ^1S\rangle - 0.4756 |2p^2\ ^1S\rangle =$$
$$= 0.0483 |2(s+p)^2 v=0\ ^1S\rangle + 0.9988 |2(s+p)^2 v=2\ ^1S\rangle \quad (3.6)$$

It is obvious from (3.6) that in the basis (3.5) the energy matrix of the Coulomb interaction is almost diagonal. Hence the quantum number of the total quasispin in this case is fairly accurate.

A peculiar case of symmetry is the symmetry of the operators of electric multipole radiations with respect to the gauge condition K of the electromagnetic field potential^{3,7}. Then the physical quantity - the electric multipole transition probability - depends on the non-physical parameter K. This dependence in a general case has the form of a parabola. It can be adopted to estimate the quality of the wavefunctions used because in the case of the exact wavefunctions the transition probabilities will not depend on K - the parabola will turn into the straight line, paralel to the axis K.

4. CONCLUSION

It is obvious from what has been said above that the theory of an atom, the main principles of which were formulated by N. Bohr, was being developed and improved all the time. Qualitatively it is pretty good and in many cases even quantitatively it describes the structure and properties of any atom and ion of periodic table. With the help of this theory one can interpret the results of experimental measurements, plan new experiments more succesfully.

In spite of the rapid development and the successes achieved, atomic spectroscopy still remains one of the most important regions of physics, strongly influensing the other branches of science and technology^{8}. One who works in atomic spectroscopy can use very efficient means for experiments, such as: lasers, the sources of synchrotron radiation, ion traps, the beams of rapid ions, cosmic technique, etc. The free single atoms and ions are being investigated, as well as the effects of their collisions, the influence of the electric and magnetic fields, photoionization and the other types of the interaction between atoms and their surrounding. In practice they are very important to explain the vacuum ultraviolet spectra, to create the lasers, generating in the shortwave

90 LASERS IN ATOMIC, MOLECULAR AND NUCLEAR PHYSICS

length region of the spectrum, to model the astrophysics - including laser and thermonuclear ones, to solve the problem of the controlled thermonuclear reaction.

However, there is still a lot of unsolved problems in atomic spectroscopy. One of them is a further progress in systematizing atomic spectra and the other spectroscopic constants. Discovered and explained by N. Bohr the first regularities in the spectrum of the most simple element - hydrogen atom, later on stimulated the search for the regularities of this sort in the, as if, randomly arranged lines of the spectra of complex atoms.

To develop further atomic spectroscopy successfully it is necessary for theorists and experimentalists to cooperate. Modern experimental technique allows one to collect a large amount of experimental data whereas the theoretical modelling of these objects allows to get a deeper insight in their structure, to reveal new regularities and tendencies in spectrodcopic data.

The urgent question is to unificate and standardize the methods of measurements and calculations, to represent the data obtained, to organize them after critical analysis in the form of data base by modern computers in order to satisfy operatively the needs of those who require any atomic constants.

Studies of the regularities in atomic spectra and, what sometimes is even more important, deviations from them help to reveal new peculiarities in the structure of many-particle systems, to discover new fine effects. Thus the studies of the collapse (rapid change of the location) of the orbit of the outer excited electron (in particular, the mean separation of the 4f electron on the configuration $4d^9 4f(^2D)LS$ of BaI for terms ^3D and ^1P equals 0.8 and 0.007 a.u., correspondingly) enable one, e.g., with the help of the tunable laser to control the features of the object studied. In this sense it is impossible to overestimate the role of the progress of the quantum electrodynamical theory of an atom, the practical verification of its fundamental principles. The achieved accuracy of the measurements and calculations allows to use atoms to investigate the structure of atomic nucleus (by studying the hyperfine interactions), to discover the parity violation effects, and, in this way, to unify the weal and electromagnetic interaction, to study various exotic atoms, highly-excted (Rydberg) atoms, to separate isotops, to detect single atoms.

The investigations of the regularities along the isoelectronic, isonuclear and homologic sequences are highly fruitful. There is a permanent advancement towards the achievement of the higher and higher ionization degrees up to the hydrogenlike ones. So in [10] it is reported on the obraining the ions $H^{88+} - H^{92+}$.

REFERENCES

1. V. S. LETOKHOV, in Laser Applications in Atomic, Molecular and Nuclear Physics (Nauka, Moscow, 1979), p. 24
2. Z. B. RUDZIKAS, in Laser Applications in Atomic, Molecular and Nuclear Physics (Nauka, Moscow, 1979), p. 105
3. A. A. NIKITIN, Z. B. RUDZIKAS, Foundations of the Theory of the Atomic and Ionic Spectra (Nauka, Moscow, 1983), p. 320 (in Russian)
4. Z. B. RUDZIKAS, J. M. KANIAUSKAS, Quasispin and Isospin in the Theory of an Atom (Nauka, Moscow), p. 190 (in English, in press)
5. V. C. SIMONIS, J. M. KANIAUSKAS, Z. B. RUSZIKAS, Int. Journ. Quant. Chem., 25, 57 (1984)
6. M. I. BOGDANIVICIENE, Lietuvos Fizikov Rinkinys (English translation: Soviet Physics - Collection), 20, No.5, 59 (1080)
7. J. M. KANIAUSKAS, G. V. MERKELIS, Z. B. RUDZIKAS, Lietuvos Fizikos Rinkinys, 19, 475 (1979)
8. J. J. WYNNE (Ed.) Current Trends in Atomic Spectroscopy (National Academy Press, Washyngton, 1984), p. 82
9. R. J. KARAZIJA, UFN, 135, 79 (1981)
10. H. GOULD, et al., Abstracts of Int. Conf. on the Physics of Highly Ionized Atoms (University of Oxford, England, 2-5 July, 1984), p. 43

1.6 Spectroscopy of Highly-Excited States of Rare Earth Elements

E. VIDOLOVA-ANGELOVA

Institute of Solid State Physics, Academy of Sciences, Bulgaria

1. INTRODUCTION

In this report, a review of our works devoted to the investigation of high-lying states of complex atoms, particularly rare earths, is presented. We have investigated the high-lying bound states with one excited Rydberg electron - Rydberg states (RS), as well as the autoionization states (AS) lying above the ionization limit of the atom.

The intensive investigation of high-lying atomic RS and AS is connected not only with atomic spectroscopy problems, by also with their different applications, for example in the multistep laser ionization method[1], the use of RS as detectors of far IR[2] or microwave radiation, etc. The atoms in RS and AS are widely used in the investigation of metastable nuclei by the photoionization spectroscopy method[3]. In the last years, methods for trapping and long-term keeping of one-charge ions in electro-magnetic traps were developed[3], that allow one to investigate the spectral properties of the atoms in very small quantities. All of the above mentiond problems and tasks explain the interest ot the highly-excited RS and AS of complex atoms.

The problems of effective excitation and detection of highly-excited atoms have been successfully solved by the laser spectroscopy methods. Nowadays, there are different methods for the excitation and registration of excited atoms. Using these laser methods it is possible to study even the high-lying states of such complex atoms as lanthanides and actinides[4-6]. As a result, a large amount of information for the spectra of rare earths have been obtained. Up to now, however, uniquely determined identification of the observed levels of these atoms has not been made. This is due to the fact that the rare earths represent very complex manyelectron systems. To provide reliable analysis of the experimental data a careful theoretical study is neened. With regard to this, the low-lying states of rare earths have been mainly examined[7]. For the hyghlying states (RS and AS) there are only some fragmentary or at least incomplete data. Only for some elements, particularly Yb, a detailed analysis of high-lying states has been performed, new data on RS and AS have been obtained, and some known results have been made more precise[8,9].

In order to perform not only an experimental research but also a profound theoretical consideration, and to obtain complete spectroscopic information, we have investigated relatively simple rare earths-atoms Tm, Yb, Lu, as well as ions Yb^+, Lu^+, Lu^{2+}. The atoms and ions under investigation have ground electron configurations and nuclear charges Z, as follows:

Tm: $4f^{13}6s^2$ Z = 69
Yb: $4f^{14}6s^2$ Z = 70
Yb^+: $4f^{14}6s$
Lu: $4f^{14}6s^2 5d$ Z = 71
Lu^+: $4f^{14}6s^2$
Lu^{2+}: $4f^{14}6s$

It is easy to notice that in some approximation all these systems can be considered as containing a core of closed shells and a max-

ATOMS AND NUCLEII 95

imum of three quasiparticles above the core. (A quasiparicle is an electron above the core or a vacancy inside the core.) The structure of these atoms and ions allows us to construct a perturbation theory (PT), the zero-order approcimation of which describes the movement of the outside quasiparticles in the field of the core.

The main part of this report will be devoted namely to this PT description. As a illustration some main qualitative and quantitative results on the investigated spectra will be presented. These results have been confirmed by the experiments, using the multistep laser ionization method.

2. THEORETICAL ANALYSIS

In the theoretical consideration of manyelectron systems, the relativistic and correlation effects have to be correctly and simultaneously taken into account. This is the basic difficulty when a manyelectron system is examined. In spite of the small bond energy of the outside electrons in heavy atoms, they should be regarded as relativistic systems. For example, the fine structure intervals for the rare earths are of the same order of magnitude as the energy difference between the nonperturbated terms. As a rule, the value of the relativistic effects increases as Z^4. The correlation effects' contribution is very large, too.

At present, there are three popular methods for calculation of manyelectron systems' spectra: i) the Hartree-Fock method with different modifications; ii) the PT of the total interelectron interaction (I/Z-expansion); iii) the PT with Hartree-Fock zero-order approximation. All these methods allow a relativistic generalization. However, these methods possess certain disadvantages.

We have used another method, based on the relativistic PT with model zero-order approximation. The method is formally exact, i.e. its accuracy is determined by the manyelectron equation used. The main advantage of our method consists in the possibility of

96 LASERS IN ATOMIC, MOLECULAR AND NUCLEAR PHYSICS

obtaining a very great accuracy, unattainable by the traditional semiempirical methods, with a relatively simple calculation procedure. Another attractive peculiarity of the method is determined by the introduction of a definite empirical information in the frames of formally exact procedure. For example, in the case of systems with several quasiparticles above the core, information on the states with minimum number of quasiparticles is used. The model potential is characteristic only of the core, so that a given analytical form of the model potential can be used for the systems with different number of quasiparticles but with the same core. These are the main advantages of the method, which make it very suitable for the investigation of manyelectron systems.

In this approach the electron state of the manyelectron system examined is described by the manyelectron relativistic equation with Hamiltonian:

$$H = \sum_i h(r_i) + \sum_{i>j} (r_{ij}^{-1} + V_{ij}) \tag{1}$$

In (1), the sum is extended over all electrons of the system. $h(r_i)$ is the Dirac Hamiltonian for one electron in the nucleus Qoulombic field. V_{ij} is the relativistic interelectron interaction, and $1/r_{ij}$ is the Qoulombic interaction between the electrons.

In the manyelectron equation with Hamiltonian (1), all one-electron relativistic effects (except of the radiative corrections) are completely taken into account. The considerably weaker two-electron relativistic effects are taken into account with an accuracy of α^2 (α is the fine structure constant).

The Hamiltonian (1) can be presented as:

$$H = H_o + H_{int} \tag{2}$$

$$H_o = \sum_i [h(r_i) + V_{el}(r_i)] \tag{3}$$

$$H_{int} = \sum_{i>j} [r_{ij}^{-1} + V_{ij}] - \sum_i V_{el}(r_i) \tag{4}$$

ATOMS AND NUCLEII 97

where H_o is the zero-order approximation Hamiltonian, H_{int} is the interaction Hamiltonian (perturbation in PT), V_{el} is the one-electron central potential, describing the interelectron interaction (model potential).

The first stage in the model potential method consist in the construction of the zero-order approximation. The model potential has to be constructed so as to describe correctly the spectrum of one valent electron above the core. We have constructed one-parameter model potential[10], describing well the high-lying RS as well as the states with excited valent shell of rare earths. The analytical form of the model potential is as follows (Qoulombic units are used):

$$V_{el}(r) = \int dr' \rho(\vec{r}')/|\vec{r} - \vec{r}'| = V_1 + V_2 + V_3 \qquad (5)$$

$$V_1(r) = (2/rZ)[1 - (1+r)\exp(-2r)] \qquad (6)$$

$$V_2(r) = (8/rZ)[1 - (1+0.6r+0.16r^2+0.032r^3)\exp(-0.8r)] \qquad (7)$$

$$V_3(r/\beta) = [(N-10)/rZ][1 - (1+\beta r+(\beta r)^2+\gamma r^3)^{-1}] \qquad (8)$$

$$\gamma = 0.01(N-10)/Z^3$$

$\rho(\vec{r})$ is the electron density of the core in the one-particle approximation, N is the number of the core electrons, and β is an adjusting parameter.

$V_1(r)$ and $V_2(r)$ approximate the K-shell and L-shell potential respectivelly. $V_3(r)$ ensures Thomas-Fermi form of V_{el} in the interatomic region. The parameter β is determined by the experimental energy of the electronic affinity to the core. It is easy to obtain the asymptotes of the model potential:

$$V_{el} \longrightarrow \begin{cases} \text{const, at } r \longrightarrow 0 \\ N/rZ, \text{ at } r \longrightarrow \infty \end{cases} \qquad (9)$$

$$\frac{dV_{el}}{dr} \longrightarrow 0, \text{ at } r \longrightarrow 0 \qquad (10)$$

98 LASERS IN ATOMIC, MOLECULAR AND NUCLEAR PHYSICS

The atomic matrix elements, depending on the multiple oscillations of the one-electron wave functions, are ctrongly sensitive to the behaviour of $V_{el}(r)$, at $r \to 0$. This is mainly related to the investigation of the decay processes of RS and the states of the valent shell of manyelectron atoms.

On the basis of the model potential so constructed, a PT is developed. Its zero-order approximation gives a true qualitative picture of the spectrum. This important property of PT is connected with the fact that the large contribution of one-particle relativistic effects, as well as the greatest part of the correlation effects have been accounted for in the zero-order approximation.

In the approach used, a group of near-lying states is selected, including the given investigated state and the energy matrix is constructed in jj-scheme of representation. The transition to the intermediate scheme of representation is realized by diagonalization of the energy matrix.

It should be noted that the discussed method gives a possibility to calculate simultaneously the energies and widths (radiative and autoionization) of the levels investigated. For this purpose, the complex secular matrix M has to be constructed:

$$M = E - i\Gamma/2 \qquad (11)$$

$$\text{Re } M = E, \quad \text{Im } M = -\Gamma/2 \qquad (12)$$

Its real part coincides with the positions (energies E) of the levels and the imaginary part, with their widths Γ.

The system with three quasiparticles above the core is the most complex system we examined. For such a system, the matrix elements M_{ik} can be written in the following way:

$$M_{ik} = M_{ik}^{(0)} \delta_{ik} + M_{ik}^{(1)} \delta_{ik} + M_{ik}^{(2)} + M_{ik}^{(3)} \qquad (13)$$

$M^{(0)}$ is a real matrix, proportional to the unity matrix. Its elements are equal to the core energy, but do not influence the spectrum structure. By setting $M^{(0)} = 0$, we have calculated the entire

ATOMS AND NUCLEII 99

spectrum with regard to zero core energy. $M^{(1)}$ is a diagonal matrix, which elements are equal to the energy of all quasiparticles of the system, i,e. $\text{Re } M^{(1)}$ can be presented as a sum of the ionization energies of the outside quasipartivles in the core considered:

$$\text{Re } M^{(1)} = \varepsilon_1^{(0)} + \varepsilon_2^{(0)} + \varepsilon_3^{(0)} \tag{14}$$

Therefore, we take $\text{Re } M^{(1)}$ from the experiment[7].

$\text{Im } M^{(1)}$ takes into account the contribution of the one- quasiparticle transitions $n'l'j'$ — $n"l"j"$. The total contribution of these transitions has been obtained by summation over all possible final states $n"l"j"$ and all possible multiplicities of the transitions:

$$\text{Im } M^{(1)} = \sum_\lambda \sum_{n"l"j"} P(n'l'j' \longrightarrow n"l"j") \tag{15}$$

(P is the transition probability). We have restriced our consideration to the transitions of lowest λ for every quasiparticle.

$M^{(2)}$ is a nondiagonal matrix, which takes into account the two-quasiparticle interactions - direct and through the polarizable core. For the valent electron, only the Qoulombic part of the interaction has been accounted for. The relativistic effects of this interaction have been taken into account in the wave functions of the valent electrons and in experimental energies $\varepsilon_i^{(0)}$, composing $M^{(1)}$. In the case of two-quasiparticle system, the first order of PT has been completely accounted for in $M^{(2)}$ $\langle 2 \rangle$. These first order corrections have been expressed by the relativistic Slater integrals F_k and G_k and angular parts. The high-order contributions to the two-quasiparticle interactions have been accounted for effectively. One can divide the high-order corrections into:
i) corrections connected with the core polarization and
ii) high-order corrections to the direct interaction of the valent electrons. The first kind of corrections express the influence of

the electron configurations with excited core, and the second kind of corrections, the contribution of the electron configurations with an excited valent shell. The high-order effects can be accounted for by additional experimental information for simpler systems, by modification of the inter-quasiparticle interaction, or by introducing additional screening potential. For example, the high-order corrections to the direct interaction in the case of Yb atom have been presented by means of the equation:

$$\Delta M_{ik}^{(2)} = \alpha [M_{ik}^{(2)}]^2 / |\mathcal{E}_{>}| \qquad (16)$$

($\mathcal{E}_{>}$ is the energy of the elctron with more diffuse orbital). The parameter α is determined by the experimental energy of the ground atomic state and is common for the whole spectrum. Eq. (16) imitates the second-order correction connected with the screening of the core field by one of the valent electrons.

In the case of calculating the high-lying RS spectra, the same effect has been taken into account by modifying the interelectron interaction:

$$V_{el}^{scr} = \int dr' \, \rho(\vec{r}'/6s_{1/2}) / |\vec{r}-\vec{r}'| \qquad (17)$$

$\rho(\vec{r}/6s_{1/2})$ is the electron density for the orbital 6s. This correction accounts for the radial quasiparticle correlations to a certain degree. A large part of the angular correlations can be considered by taking into account the interaction between near-lying states of suitable symmetry.

The core polarization is also expressed by modifying the interelectron interaction potential. This leads only to redetermination of the radial integrals in the equations for the first-order corrections.

In the case of three-quasiparticle system, the matrix $M^{(2)}$ has been expressed in terms of $M^{(2)}\langle 2 \rangle$ for two-quasiparticle system:

$$M^{(2)}\langle 3 \rangle = 3C'(3/2)M^{(2)}\langle 2 \rangle C(3/2) \qquad (18)$$

where C is the genealogical coefficients matrix, C' is the transposed matrix.

$\text{Im}M^{(2)}$ describes the 2^λ-pole radiations transition $\alpha_1 \alpha_2 \to \alpha_1 \alpha_2'$. α_2 is the active particle, accomplishing the transition. According to the Pauli principle, the presence of the quasiparticle α_1 influences kinematically the possibility of this transition. The dynamic influence (the deformation of the active quasiparticle orbital) is not taken into account in the first order of PT.

$M^{(3)}$ is also a nondiagonal matrix, which expresses the three-quasiparticle interactions (direct and through the polarizable core). The three-quasiparticle interactions emerge in second order of PT. We have effectively taken into account $\text{Re}M^{(3)}$ as a correction to the two-quasiparticle interactions. This correction is connected with the screening of the core potential by a third electron. If the matrix $M^{(2)} \langle 3 \rangle$ for the two-quasiparticle interactions has been calculated in such a screening potential, the contribution of the three-quasiparticle interactions has been effectively taken into account. It is possible to prove that

$$M^{(3)} \langle 3 \rangle = 3C'(3/2) \Delta M^{(2)} \langle 2 \rangle C(3/2) \qquad (19)$$

$\text{Im}M^{(3)}$ has been completely neglected in our calculations.

Among the high-order corrections to RS, we have acconted for only the effects of mutual screening (antiscreening) of the quasiparticles. The effects of the core polarization have been completely neglected in this case. The mutual screening of the quasiparticles is manifested in the form of their radial orbitals only. The analytical formulae for the secular matrix elements conserv the form of the low-order corrections of PT. Only the radial integrals have been modified.

3. REVIEW OF THE MAIN RESULTS AND DISCUSSION

3.1. Results on the Rydberg States.

We have investigated the RS with excited valent electron which converge to the first ionization limit of the atom. Sush states can be excted easily by visible laser radiation in two or three steps.

The excited Rydberg electron is located in an unit effective charge. This fact is used in the quantum defect method, in which the excited Rydberg electron is considered in Qoulombic field of a unit charge. From this follows the hydrogen-likeness of the Rydberg atom. Recently, another semiempirical method was developed for the analysis of RS. This is the multichannel quantum defect theory[11], based on the quantum defect method. Up to now, this method has been used for the consideration of RS with one or two quasiparticles above the core. Obviously, the use of the method for three-quasiparticle systems leads to essential complication of the calculation procedure.

It ought to be noted that our calculation method is asymptotically exact when the energies of RS are calculated, i.e. at $n \to \infty$ the method gives asymptotically exact energies. The calculation error decreases faster than $1/n^3$. We have calculated the energies of some Rydberg series[12,13]: $4f^{14}nl$, $l = 0,1,2$ for Yb^+ and Lu^{2+}; $4f^{14}6s^2nl$, $l = 0,1,2$ for Lu; $4f^{14}6snl$, $l = 0,1$ for Yb and Lu^+. The convergence limit of all these series is the first ionization limit of the corresponding system. In the case of the Tm atom, the series $4f^{13}6snl$, $l = 0,1$ have been considered, which converge to the four states of Tm^+: $4f^{-1}_{7/2}6s_{1/2}[4;3]$ and $4f^{-1}_{5/2}6s_{1/2}[2;3]$. Most of these energies are obtained for the first time.

We have also investigated the decay processes in the Tm atom. The RS of Tm, which converge to an excited state of Tm^+ and are located above the first ionization limit of the atom, are autoionization states. Two kinds of decay exsist for these ARS - radiation decay and autoionization decay:

$4f_j^{-1}6s_{1/2}(J')nlj[J] \xrightarrow{rad} 4f_{j''}^{-1}6s_{1/2}(J'')n_1l_1j_1[J_1]$, $J_1=J-1,J,J+1$.

$4f_j^{-1}6s_{1/2}(J')nlj[J] \xrightarrow{autoion} 4f_{j'''}^{-1}6s_{1/2}(J''')kl_kj_k[J]$.

The set of quantum numbers (j",J", n_1j_1,J_1) defines the radiation decay channels, whereas the quantum numbers (j'",J'",l_kj_k) determine the autoionization decay channels. The contribution of different channels to the total width of the level change radically with the variation of n. This fact is very interesting from the point of view of the use of these states in the multiphoton laser spectroscopy methods.

We have discovered some new essentially non-Qoulombic properties of RS, non-vanishing at n $\rightarrow \infty$, for example:

i) the fine-structure splitting of the levels is higher by several orders of magnitude compared to the H-like ion with the some charge of the core Z*;

ii) the dependence of the fine-structure splitting on Z* is essentially weaker than Z^{*4};

iii) the quantum defects of the states of a given electron configuration differ by more than unity, even at n $\rightarrow \infty$. This leads to overlapping of the localization regions of levels with different n.

The availability of quantum defect expresses certain non-Qoulombic properties of RS, which can be accounted for in the frame of the H-like approximation. We have in mind, however, finer effects, which cannot be described by means of the H-like approximation. These peculiarities of RS of complex atoms express essentially manyelectron behaviour of the system. In contrast ot the Qoulombic potential, the core potemtial of a complex atom is mainly localized in the core region. This resuts in the overlapping of all orbitals.

The peculiarities found of the high-lying RS have been completely confirmed by our experiments[14,15]. As a illustration, the relative location (obtained theoretically and experimentally) of

104 LASERS IN ATOMIC, MOLECULAR AND NUCLEAR PHYSICS

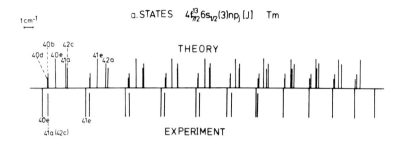

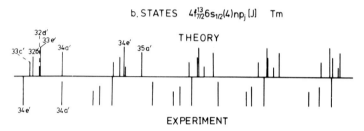

FIGURE 1   Relative positions of the Tm Rydberg states.
a) $4f_{7/2}^{13}6s_{1/2}(3)npj$ states, b) $4f_{7/2}^{13}6s_{1/2}(4)npj$ states.

the Tm levels investigated is presented in Figure 1 [14]. The experiment showed and unusual grouping of the levels of different Rydberg series. (a,b,s,... in Figure 1 mean different Rydberg series.) Moreover, the ralative location of RS changes drastically with the change of n. A large change of the quantum defect values is observed too, which confirms the non-Qoulombic properties theoretically predicted.

The comparison between the experimental and theoretical data shows that the calculation accuracy is of the order of 2% for the bond energy of the valent electron. In Table 1, a comparison of the experimental and calculation results for the most complex investigated system - Tm atom - is given. Some experimental results by other authors[7] are also presented. The excellent agreement of the results is seen.

TABLE 1 Energies of the Tm $4f_{7/2}^{13}6s_{1/2}(3)np_{1/2}[5/2]$ Rydberg states (in cm^{-1})

n	E_{exp}	E_{th}	E_{exp}^{7}
43	50040.6	50042.1	50040.71
44	50044.3	50045.7	50044.77
45	50048.0	50049.1	50048.38
46	50051.4	50052.3	50052.09
47	50054.2	50055.3	50054.66

3.2 Results on the Autoionization States

Autoionization states represent quasistationary states lying above the first ionization limit of the atom. These states interact very strongly with the states of the continuum, which results in their autoionization decay. The autoionization decay is a radiationless transition with detachment of one electron. Because of their strong bond with the continuum, the AS widths are very large (10^3-10^5 cm^{-1}) in comparison with the discrete bond states' widths ($\sim 10^{-2}$ cm^{-1}). In certain cases, however, anomalously narrow AS emerge. This is possible if the decay of AS is forbidden in nonrelativistic appoximation by some selection rules. In the last years, as a result of the investigation carried out by means of the laser spectroscopy methods, groups of narrow AS near the first ionization limit of complex atoms have been observed. Their widths are less than $10\,cm^{-1}$. The availability of a large number of closely located narrow AS proposes that the stability of these AS is not connected with the decay prohibition. Our studies proved that there is an other reason for tha existence of narrow AS in the complex atoms' spectra. These narrow AS arise through excitation of the valent shell of the atom.

In the case of two valent electrons, these are with certainty doubly excited states of the valent shell. For these states $\Gamma \sim 1/z^2$, where Γ is the autoionization width of the state.

The stability of this kind of AS is not connected with any prohibitions, but is determined by the manner of their formation, and is connected with the behaviour of the electrons taking part in their decay process. These are electrons in states with large values of the principal quantum number n. The wave functions of these electrons are characterized with multiple oscillations in the main region of core-electron interaction. This results in decreasing of the matrix elements giving the width of the AS. If the narrow AS can decay only through a three-electron transition, their widths may decrease by an order of magnitude: $\Gamma < 1$ cm^{-1}. A further narrowing of these AS is also caused by the additional action of some decay prohibitions. For example, if the decay is only due to the relativistic interaction with the continuum, these AS are anomalously narrow ($\Gamma < 0.01$ cm^{-1}). Anomalously narrow AS have been predicted in the Yb ($5d^2$ $^3P_{0,1}$) and Lu ($5d^3$ $^2P_{1/2}$, $^4P_{3/2}$, $^2H_{11/2}$) spectra.

Among the narrow AS with excited valent shell, there are low-lying states, situated near the first ionization limit of the complex atom. All this makes these AS suitable for application in the multistep spectroscopy methods.

By means of the PT with zero-order model approximation, we have calculated the energies and widths of the doubly excited states of Yb^{10,16}(5d6p,6p^2,5d^2,7s6p) and Lu^{17}(6s6p^2,6p^{2}5d, 6p5d^2, 5d^3). The interaction of nearly degenerated AS leads to considerable redistribution of the decay probabilities. This explains the large scatter in the widths' obtained.

Experimentally research of narrow AS has been carried out for the Yb^{16} and Lu^{15} atoms. Table 2 illustrates the well coincidence of the treoretical and experimental results for the 7s6p-states of Yb.

TABLE 2   Energies and widths of the 7s6p - AS Yb states (in cm^{-1})

Term	Theory		Experiment	
	E	Γ	E	Γ
$^3P^o_0$	59800	0.70	59130.5	1.2
$^3P^o_1$	60000	3.00	60428.7	0.95
$^3P^o_2$	62600	0.70	62529.1	1.6
$^1P^o_1$	63600	1.80	63655.8	2.6

4. CONCLUSION

New spectroscopic information has been obtained as a result of the theoretical and experimental research of the highly-excited spectra of rare earth atoms. The discovery of essentially non-Qoulombic properties of the highly-excited Rydberg atoms and the clarifying the nature of the narrow low-lying AS should be especially underlined.

The performed inverstigation gave possibility to make some conclusions for the very structure of the considered systems. For the excited valent shell' states of Yb the contribution of the configurations with excited 4f-shell is decisive. This leads to decreasing the interaction of the states $4f^{14}n_1l_1n_2l_2$.

In the case of Lu^+ with the same electron configuration and the next atomic number, it has been observed an absolutely other situation. The contribution of the core polarization to the valent states' energies is negligibly small. The same effect is valid for the neutral Lu atom. The role of the angular correlations is considerable for Lu^+. This may be taken into account by mixing the near-lying configurations with equal parity. Therefore, the increase of Z with one unit leads to a considerable rise of the core hardness in this case.

The results obtained are important not only for the atomic spectroscopy but also for all the fields of physics, in which the highly-excited states of complex atoms are applied.

In conslusion, the author express gratitude to Dr. L. N. Ivanov, Dr. G. I. Bekov, and Dr. V. I. Mishin from the Institute of Spectroscopy Acad. Sci. of USSR with whose participation the considered results have been obtained, as well as to Prof. V. S. Letokhov for the supporting of the work and useful discussions.

REFERENCES

1. G. I. BEKOV, E. VIDOLOVA-ANGELOVA, V. S. LETOKHOV, V. I. MISHIN, in Laser Spectroscopy IV, edited by H. Walter, Proc. 4th Int. Conf. on Laser Spectroscopy (Springer, Berlin, 1979), p. 283.
2. P. GOY, C. FABRE, M. CROSS, S. HAROSHE, J. Phys. B: At. Molec. Phys., 13, 183 (1980)
3. V. I. VALIKIN, G. I. BEKOV, V. S. LETOKHOV, V. I. MISHIN, Uzp. Fiz. Nauk, 132, 293 (1980)
4. E. F. WORDEN, R. W. SOLARZ, J. A. PAISNER, J. G. CONWAY, J. Opt. Soc. Am., 68, 52 (1978)
5. J. A. PAISNER, L. R. CARLSON, E. F. WORDEN, S. A. JOHNSON, C. A. MAY, R. W. SOLARZ, Preprint Lawrence Livermore Lab.: UCRL-78034 (1976)
6. J. A. PAISNER, L. R. CARLSON, R. W. SOLARZ, C. A. MAY, S. A. JOHNSON, Preprint Lawrence Livermore Lab.: UCRL-77537 (1975)
7. W. C. MARTIN, R. ZALUBAS, L. HAGAN, Atomic Energy Levels. The Rare Earth Elements. National Bureau of Standards, April 1978.
8. M. Aymar, A. DEBARRE, O. ROBAUX, J. Phys. B: At. Molec. Phys., 13, 1089 (1980)
9. E. VIDOLOVA-ANGELOVA, L. N. IVANOV, V. S. LETOKHOV, J. Opt. Soc. Am., 71, N 4 (1981)
10. E. VIDOLOVA-ANGELOVA, E. P. IVANOVA, L. N. IVANOV, Opt. Spect., 50, 243 (1981)
11. J. J. WYNNE, J. A. ARMSTRONG, IBM J. Res. Dev., 23, 409 (1979)

12. E. VIDOLOVA-ANGELOVA, L. N. IVANOV, V. S. LETOKHOV,
    J. Phys. B: At. Molec. Phys., 15, 981 (1982)
13. L. N. IVANOV, E. P. IVANOVA, E. VIDOLOVA-ANGELOVA, D. ANGELOV,
    Izd. Com. Spectr. (1984)
14. E. VIDOLOVA-ANGELOVA, G. I. BEKOV, L. N. IVANOV, V. FEDOSEEV,
    A. A. ATAKHADJAEV, J. Phys. B: At. Molec. Phys., 17, 953 (1984)
15. G. I. BEKOV, E. VIDOLOVA-ANGELOVA, Quant. Electr., 8, 227
    (1981), in Russian
16. G. I. BEKOV, E. VIDOLOVA-ANGELOVA, L. N. IVANOV, V. S. LETO-
    KHOV, V. I. MISHIN, JETP, 80, 866 (1981), in Russian
17. E. VIDOLOVA-ANGELOVA, L. N. IVANOV, E. P. IVANOVA, V. S. LE-
    TOKHOV, Izv. AN SSSR, Ser. Fiz., 45, 2300 (1981), in Russian

1.7 The Atomic Collisions in Laser Fields

A. M. BONCH-BRUEVICH, S. G. PRJIBELSKY and V. V. KCHROMOV

From its origin and up to now the physics of atomic molecular collisions naturally includes both fundamental and applied investigations. Evidently this and the multitude of collision forms are responsible for everlasting interest to this field of physics.

Optical spectroscopy of atomic collisions also probably is still developing. It is especially evident now when thanks to the perfection of spectroscopy technique and especially thanks to lasers there appeared new possibilities. It is enough to noted kinetics spectroscopy, in particular, photon echo or Doppler-free spectroscopy[1] in order to see that these are qualitatively new methods. Although there may be difference in details the relation of spectrum forms of different optical processes with interacting atom potentials are the basis of all methods of collision spectroscopy. Traditionally spectra in neighbourhood of an atomic line are conditionally devided into two regions: an impact one and a static one. The impact region is mainly connected with distant collisions which (as one might think) induce the phase jump of atomis oscillator and damping of its movement coherence. These collisions lead to a shift and broadening of an atomic line which are usually hidden because of Doppler effect. These lattent collision characteristics can be discovered by laser spectroscopy method, the laser spectroscopy method being until recently used mainly for the impact region investigation. The far line wings (statical region) are formed by near collisions as a result of which there appear the so called quasimolecules or exiplexes. The use of lasers for far wing investigations and for obtaining from it some information about close atom collisions began latter than the analogeous investigations for the impact region. In given lecture there are going to be discussed only some aspects of the far wing spectroscopy and are given only some results of collision investigations connected in general with the interests of the authors of this paper. It should be emphasised that the questions discussed here recieved a more detailed description in the publications$^{2-4}$.

In order to make more clear the questions discussed later on let us examine Figure 1 which is a scheme of optical processes in colliding atom systems. For making it more simple let us assume that there are only two terms: the quasimolecule A basic state and quasimolecule B excited state. The dependence of A, B atom interaction energy $U_{A,B}$ on the distance between them R is at the same time a dependence of electronic state energy on R. It means that to each R corresponds $[U_B(R) - U_A(R)]h^{-1} = \Delta(R)$ - the frequency of the atomic line shifted by the interaction. Colliding atoms move and therefore Δ is changed in time at collisions. The quasi-

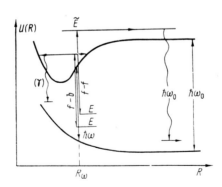

FIGURE 1   The scheme of excitation by the radiation with the frequency ω of atomic (f - f) and molecular (f - b) emissions.

molecule excitation by a monochromatic field with ω frequency in far wing takes place efficiently in the neighbourhood of $R \simeq R_\omega$ where R_ω is determined by the resonance condition $\Delta(R_\omega) = \omega$. As a matter of fact there are the Frank-Condon rules.

The following processes are possible in this scheme. For atoms with the relative movement energy lower than E^x the photon absorption transforms the system from A into a bound state of B term. The quantum with energy $\hbar\omega'$ is emmited from state B during average time γ^{-1}. This is a process of the excitation of molecular emission, the initial energy of the atom relative movement changing into $\hbar\omega - \hbar\omega'$.

For the atoms with $E > E^x$ the absorption of the photon with the energy $\hbar\omega$ transforms the system in E state which is higer than the dissociation energy. Then the atoms quickly, during 10^{-12} s leave the interaction region and emit one quant during average time $\gamma^{-1} \simeq 10^{-8}$ s at the frequency ω_o of the nondisturbed atomic transition. That is an excitation of atomic emission. The energy $\hbar\omega - \hbar\omega_o$ is transformed into the atom movement.

In gases collisions take place with different E, therefore atomic and molecular emission are excited simultaneously.

We are giving a simple quantitative description of quasimolecule excitation by the emission field $Fe^{i\omega t} + c.c.$. In the frame of the adopted representation the dynamics of the transition into a field is determined by an equation system for the amplitude of the lower a(t) and upper b(t) states:

$$i\hbar\dot{a} = U_A a + DFe^{i\omega t}b; \quad i\hbar\dot{b} = U_B b + DFe^{-i\omega t}a \qquad (1)$$

Here $U_{A,B}$ depends on the time t in R(t). The dipole matrix element of allowed transition is assumed to be constant D = const.

The solution (1) in accordance with the perturbation theory allows to determine in the first approximation the probability of excitation $p = |b(\infty)|^2$. Taking into consideration the fact that the excitation takes place nearly $R = R_\omega$, determined by the resonance condition $U_B(R_\omega) - U_A(R_\omega) = \hbar \omega$ and that in the neighbourhood of $R = R_\omega$ the potentials $U_{A,B}(t)$ may be approximated by the linear function t there is obtained a well known formula Landau-Zinner for one passing of resonance:

$$p = (DF/\hbar)^2 \cdot 2\pi / |\Delta'_\omega| v_\omega = (DF/\hbar)^2 \tau^2 \qquad (2)$$

where $\Delta'_\omega = \partial \Delta / \partial R_\omega$ and $v_\omega = dR_\omega/dt$ - a relative velocity of atoms in R_ω. It is convenient ot interpret formula (2) as the product of the squares of transition frequencies $DF/\hbar$ and τ -the duration of the colliding atom staying in the resonance with exitation. The typical meanings $\tau \simeq 10^{-13}$ s and the linear sizes of the resonance region $v_\omega \tau \simeq 10^{-9}$ cm are small in comparison with the radius of the action of interatomic forces.

The relation p with $\Delta'(R)$ allows to determine the potentials U_A and U_B, in principal, according to the data of absorption (emission) spectra. It is possible to it thanks to the following. The absorption coefficient K_ω at frequency ω is proportional to the frequency of transition from A to B. This frequency in its turn is proportional to an average (over a collision ansemble) of the product of the possibility of transition during one collision - p - and collision frequency $\nu_\omega = 4\pi R_\omega^2 n_\omega v_\omega$ - the flow across the R_ω radius sphere. Namely,

$$K_\omega \sim \langle \nu_\omega p \rangle \sim \Delta'(R_\omega) \langle n_\omega \rangle \sim \Delta'_\omega e^{-U_A(R_\omega)/T} \qquad (3)$$

In the relations (3) is ommited everything that is not connected with potentials and it is assumed that at the temperature T in equilibrium an average concentration of non excited atoms at the distance R_ω from the given particle $\langle n_\omega \rangle \sim e^{-U_A(R_\omega)/T}$ - Boltzmann factor. Thus thanks to the temperature connection with U_A the interaction potentials of several alkali metal atoms with inert gases have been determined[5]. The laser specificity in these investigations was expressed only in the technical aspect because there was no necessity either in short pulses, or in spectral narrowing of excitation, or in its extrimely high intensity.

Laser allow to introduce into methods of far wing investigations new ones connected with nonlinear optical phenomena. At first sight the situation we are interested in makes us expect that nonlinear phenomena may become noticible when the saturation condition $p \gtrsim 1$ is satisfied, i.e. for typical parameters at exitation intensity $I \gtrsim I_c = 10^{11}$ W/cm^2. This magnitude is attainable but, hardly suitable for our purposes.

Yet, it appeared[6], that at $I \ll I_c$, the nonlinear phenomena in the far wings can be observed. In particular, nonlinear dependences of intensity S_M of molecular emission on I have been found. Figure 2 and 3 show experimental results and scheme of processes in the investigated systems. Special checks have shown that molecular emission was due to primarely colliding atom excitation. It is evident that the dependencies become noticibly nonlinearity begining with $I \gtrsim I_s \simeq 10^8$ and 10^6 W/cm^2 for Cs and Rb correspondingly. It should be mentioned for comparison[7], that the Cs$_2$ and Rb$_2$

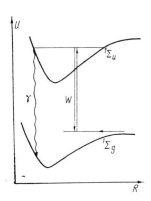

FIGURE 2 The scheme of molecular emission excitation
Rb + Rb (Cs + Cs).

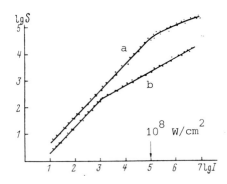

FIGURE 3 The dependences anti-Stocks emission intensity in the molecule band Cs_2 (a) and Rb_2 (b). molecule band saturation occurs at $I \simeq 10^2$ W/cm^2.

For explanation of the discovered phenomenon let us pay attention to the fact that the intensity of the spontaneous emission is proportional to the quantity of the excited molecules, i.e. $S_M \sim \gamma \tilde{n}$. The stationary number of $\tilde{n}$ may be determined from the balance equation

$$0 = p \mathcal{V}_\omega - \tilde{n}(\gamma + 2p/\Theta) \qquad (3)$$

where Θ - a period of classical oscillations of atoms in an excited molecule. Here the first term describes molecule formation and the second one - molecule distraction by spontaneous and stimulated processes. The equation solution (3) shows, that the dependence of

$$S_M \sim \gamma p(\gamma + 2p/\Theta)^{-1} \qquad (4)$$

becomes substantially nonlinear at $p \gtrsim p_s = \gamma\Theta \simeq 10^{-3} \div 10^{-4}$. This p_s corresponds to experimental one. However the predicted dependence (4) and the experimental one $S_M \sim \sqrt{I}$ are different in the extremely nonlinear region $I \gg I_s$.

The root dependecies are well known in nonlinear optics and testify to nonhomogeneity of collision ensemble. However the case of free-bound transition analysed here differs substantially from

ATOMS AND NUCLEII 117

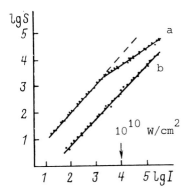

FIGURE 4 The dependence of atomic emission intensity on the excitation power for the system Cs (a) and Cs + Ne (b).

the known ensembles^{1,7}, that are unhomogeneous ones. The probable cause of the square root dependence in this case will discuss below.

The nonlinear dependence of atomic emission $S_a(I)$ has also been registered in the Cs vapours^8. Here (Figure 4) the deviation from the linear dependence becomes noticeable at $I \gtrsim I_s = 4 \cdot 10^9$ W/cm^2 and in an extremely nonlinear region can be well approximated by the law $S_a \sim I^{0.7 \pm 0.05}$. At first sight it may seem that at such low intensity as $I \ll I_c$ there is no reason for a noticeable nonlinearity in the free-free transition excitation process.

However the probable cause of this phenomenon may be^9 the delay of the scattering of atoms in quasimolecule excited state. This delay is due to centrofugal barrier which is formed on an attrackted potentials of interatomic interaction. In the frame of this model the decompositon of the excited quasimolecule is possible (Figure 5) either by radiation decay or by tunnel one, the later without quenching the atomic emission. Before disintegration by the tunnel way the molecule with $E^* < E_B$ passes many time the neighbourhood R_ω where the laser radiation deexcites it. It is evident that this process is a like to the process of molecule

118 LASERS IN ATOMIC, MOLECULAR AND NUCLEAR PHYSICS

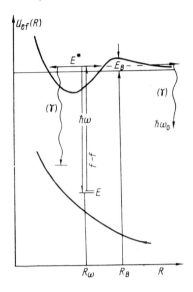

FIGURE 5   The scheme f - f transitions in the system with the centrofugal bareer.

emission excitation discussed above. The elementary description of the atomic excitation is similar too:

$$0 = p\mathcal{V}_\omega - n^*(\gamma + \tilde{\gamma} + 2p/\Theta) \tag{5}$$

where $\tilde{\gamma}$ - the frequency of the quasimolecule tunnel disintegration. The solution of this equation shows that the criterion of nonlinearity $p_s = \bar{\gamma}\Theta \ll 1$, where $\bar{\gamma}$ - the most possible frequency of quasimolecule disintegration. From the registered value of I_s the value of $\bar{\gamma} \simeq 10^{10}$ s $\gg \gamma \simeq 10^8$ s.

In passing it should be noted that in the linear region when $p \ll p_s$ the intensity of atomic emission S_a does not depend on the presence of barrier factor.

The system of Rb + Xe vapours has been investigated to check our assumptions about the role of centrofugal barrier in nonlinear emission dependence phenomenon in f - f transition excitation scheme. This system has been studied in detail and its peculiarity is a monotone dependence R_ω on ω : a smaller ω corresponds to a

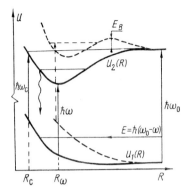

FIGURE 6 The scheme of the atomic emission excitation in Rb + Xe system. The section line - an effective potential with $l \gg 1$.

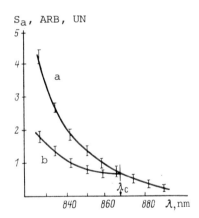

FIGURE 7 The longwave spectra of atomic excitation in Rb + Xe system at $I = 10^6$ W/cm^2 and $I = 10^9$ W/cm^2 (b).

smaller R_ω.

The scheme of atomic emission excitation and an action spectrum at two meanings of I are given in Figures 6 and 7. The emission with $\omega < \omega_c$ excites only quickly colliding states with $E > E_B$ because at any pulse moment $\hbar l$ the minimum energy of scattering $E(l) = U_2(R_\omega) + \hbar^2 l(l+1)/2\mu R_\omega^2$ is always greater than the height of the barrier $E_B = U_2(R_B) + \hbar^2 l(l+1)/2\mu R_B^2$. This follows from fact that $R_\omega < R_B$ and $U_2(R_\omega) > U_2(\infty) > U_2(R_B)$ and leads to conclusion that the excitation with $\omega < \omega_c$ is not blocked up by the barrier. Figure 7 shows that this is revealed in the fact that in the region $\lambda > \lambda_c$ atomic emission is not changed therefore in the region $\lambda < \lambda_c$ the saturation is observed.

The centrofugal barrier in its fundamental state also may provoke nonlinear phenomenon in emission at the excitation with $I \ll I_c$ [10]. These phenomena in contrast with ones discussed above are not connected with a stimulated excitation quenching. Their mechanism consists in the competition of the processes deplenishing the basic state region containing R_ω. There are two such processes: the tunnel penetration trough the barrier and photoexcitation into a state from which emission occurs. It is clear that when photoexcitation is prevailing then the emission intensity stops being dependent on I, because the quickness of excited quasi-molecule formation is limited by the quickness of their formation in the basic electronic state i.e. by the quickness of the tunnel penetration of the colliding atoms to the R point. Quasidiscret structure of energy spectrum answers the quasibound movements in the basic and excited states. The transitions between these states form transition frequency ensemble which is like the molecular one which can in principal be the cause of the power laws mentioned above.

Blocking up barriers must not obligatorily centrofugal. It is known[11] that in the excited state bareers can be of polarized nature. Generally speaking quasibounded movements are known in spectroscopy as predissociation. In the light of mentioned above it seems to be advantages the application of saturation spectroscopy for the investigation of quasibounded complexes.

Let us analyze some more nonlinear optical phenomena in the

ATOMS AND NUCLEII 121

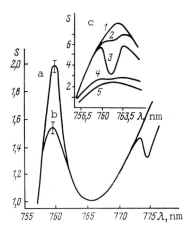

FIGURE 8 The spectra of atomic emission excitation in the system Rb + Xe at I = 10^5 (a) and $3 \cdot 10^8$ W/cm^2 (b). Separately the change of the form of the top at I decrease subsequently begining with (b) three times is shown.

far wing region. At first let us pay attention to the significant difference between the values of I established for Rb and Cs in experiments with molecule emission (Figure 3). This difference is easely explained by a great disparity $\Delta'(R_\omega)$ at the excitation with the wavelength $2\pi c/\omega$ = 1060 nm, $\Delta'(R_\omega)$ of Rb being near 0. In linear spectroscopy the proximity $\Delta'(R)$ to zero is revealed as a spectral peculiarity (the satelite of atomic line). In this case it is possible in the same way as before to determine in the frames of the classical relative atom movement the duration of excitation near to the resonance.

Taking into consideration the fact that in the neighbourhood of R_ω the term divergence depends quadratically on R: $\Delta(R) \simeq \Delta''(R_\omega)(R - R_\omega)^2/2$ it is easy to obtain the evaluation of this period $\tau \simeq (\Delta''_\omega v_\omega^2)^{-1/3}$ which exceeds the evaluation defined above (2) usually up to order times.

The satelite may be revealed in different optical processes, but not obligatorely in absorption. The Figure 8 shows the results

of the investigation of the satelite form dependence on the pump
intensity in the process of the atomic emission excitation in the
Rb + Xe system. It is clear that the deformed satelite has a still
more saturated core. This points to the presence of especial group
in nonhomogeneous ensemble of colliding atoms. The possibility of
excitation of this group by the irradiation with λ = 760 nm is
noticeable greater than of the other ones. The excitation possibility is maximum for those collisions, the trajectories of which
yield the greatest time of the system being in the resonance with
the emission. It is known that these trajectories touch to the
spheres with R = R$_\omega$. The observed dip in the satelite has been
connected[12] with the collision group with touching trajectories.
The parameters of this dip are connected with the characteristics
of the observed system terms.

Let us dwell at last on the nonlinear phenomenon at $I \gtrsim I_c$ in
case when both terms of the basic and the excited state are repel-
led. It should be emphasize that at every collision the point Rω
is passed twice: the first time at convergence and then at divergence of collisions. The possibility of one time excitation at
$I > I_c$ tends to 1 according to the law $1 - e^{-p}$. This results in
the fact that an effective excitation at convergence is accompanied by an equally effective de-excitation at divergence and it follows that the excitation of colliding atoms should decrease with
the increase of the intensity at $I > I_c$. This effect differs qual-
itatevely from the known saturation one because it comes to the
drop emission with the increase I. This conclusion has been experimentally checked. In the system Tl + Ar the drop of emission intensity of Tl atoms at $I > 10^9$ W/cm^2 has been registered. It should
be stressed that for the observation of the mentioned effect of
the medium bleaching without saturation it is necessary to have
special system characteristics, because the processes masking

ATOMS AND NUCLEII 123

those which interest us at great intensities becomes substantial.

In our opinion the given results show the possibility of obtaining information additional to the traditional one given by the far wing spectroscopy and the efficiency of laser usage for close atomic collision investigation. The analysed processes and phenomena are far from exhausting the possibility of laser usage in collision investigation. It appears possible to apply a picosecond technique because the duration of collision $\simeq 10^{-12}$ s and during this time the transition frequency changes determinately. The other aspect of application may be seen in the polarized far wing spectroscopy in intensive fields.

Summing up, we may state that nowadays the most suitable and probably the most interesting for investigating by the collision laser methods appears to be the long living molecule complexes which are formed in gas media.

REFERENCES

1. V. S. LETOKHOV, V. P. CHEBOTAYEV, Prinzipy nelinejnoj lasernoy spektroskopii (Nauka, Moscow, 1975).
2. S. I. JAKOVLENKO, Radiazionno-stolknovitelnije javlenija (Energoatomizdat, Moscow, 1984).
3. A. M. BONCH-BRUJEVICH, S. G. PRJIBELSKY, V. V. KHROMOV, Jr. priklad. spektroskopii, $\underline{33}$, 980 (1980)
4. A. M. BONCH-BRUJEVICH, S. G. PRJIBELSKY, V. V. KHROMOV, S. I. JAKOVLENKO, Izv. Akad. Nauk SSSR, ser. fiz., $\underline{48}$, 587 (1984)
5. A. GALLAGHER, Proc. of the 4th Internat. Conf. on Atomic Phys. (Plenum Press, New York, London, 1975)
6. A. M. BONCH-BRUJEVICH, S. G. PRJIBELSKY, V. V. KHROMOV, GETF, $\underline{72}$, 1738 (1977)
7. N. N. KOSTIN, M. P. SOKOLOVA, V. A. KHODOVOJ, V. V. KHROMOV, GETF, $\underline{62}$, 476 (1972)
8. A. M. BONCH-BRUJEVICH, S. G. PRJIBELSKY, A. A. FEDOROV, V. V. KHROMOV, GETF, $\underline{71}$, 1733 (1976)
9. T. A. VARTANJAN, S. G. PRJIBELSKY, GETF, $\underline{74}$, 1733 (1978)
10. M. Ja. AGRE, L. P. RAPOPORT, Opt. i spektr., $\underline{48}$, 1023 (1980)
11. G. W. KING, J. H. VAN VLECK, Phys. Rev., $\underline{55}$, 1165 (1939)

12. T. A. VARTANJAN, Ju. N. MAKSIMOV, S. G. PRJIBELSKY,
    V. V. KHROMOV, Pis'ma v GETF, 29, 281 (1979)
13. A. M. BONCH-BRUJEVICH, T. A. VARTANJAN, V. V. KHROMOV,
    GETF, 78, 538 (1980)

2. MOLECULES AND LASER INDUCED PROCESSES

2.1 State-Selected Photodissociation Dynamics of Formaldehyde

C. BRADLEY MOORE

Department of Chemistry, University of California, Berkeley, California 94720, USA

and

DOUGLAS J. BAMFORD

Chemical Physics Laboratory, SRI International, 333 Ravenswood Avenue, Menlo Park, California 94025, USA

1. INTRODUCTION

The study of photofragmentation dynamics allows unimolecular chemical reactions to be examined in a state-resolved manner. Such state resolution can reveal a great deal about the potential energy surfaces which control these elementary events, and can serve as a crucial test for theories of reaction dynamics on those surfaces. The internal energy (v,J) distributions of photofragments have often been determined by laser-induced fluorescence, in cases where the fragments absorb in the visible or near-ultraviolet regions accessible with commercial tunable dye lasers[1,2]. If the fragments do not have such convenient electronic spectra, the techniques of nonlinear optics must be used to provide the necessary sensitivity and spectral resolution. In the research presented here, such techniques have, for the first time, completely determined the distribution of energy among all the possible degrees of freedom in the photodissociation of a tetratomic molecule. In addition, rotational state resolution has been obtained in both entry and exit channels for the photofragmentation.

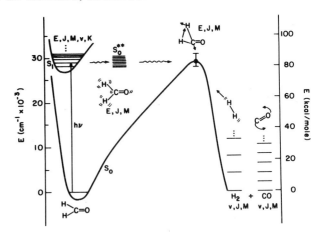

FIGURE 1 Schematic diagram of potential surfaces and photophysical processes for formaldehyde[4].

Formaldehyde is an ideal system in which to study photofragment (v,J) distributions for several reasons. The relevant energy levels and dissociation limits are shown in Figure 1. Its well-resolved and well-understood UV spectrum[3] allows for state-specific excitation with a narrowband UV dye laser. The mechanism of the dissociation, after decades of work, is now well established[4]. The molecule internally converts to high vibrational levels of the ground electronic state and then dissociates in the absence of collisions. The fragments, H_2[5-7] and CO[8], are both detectable in low concentrations using available laser spectroscopic techniques. Early work on this photodissociation measured a vibrational distribution of the CO[9]. Later the translational energy distribution was measured by time-of-flight mass spectroscopy[10], the most recent experiments are reviewed here and complete the picture by giving (v,J) distributions for both fragments[11-14].

2. CO (v,J) DISTRIBUTIONS BY VACUUM ULTRAVIOLET LASER-INDUCED FLUORESCENCE

A. Experimental

To carry out state-selected photodissociation of formaldehyde followed by state-specific product detection of CO, two separate pulsed tunable laser systems were needed[11]. The experimental concept was simple. The photolysis laser was tuned onto a formaldehyde absorption peak. A short time after the photolysis pulse, the probe laser was fired. The probe laser wavelength was then scanned to obtain a fluorescence excitation spectrum of the photochemical CO. Figure 2 shows the overall schematic.

Tunable ultraviolet radiation for state-selective formaldehyde photolysis was provided by a commercial system consisting of an excimer laser (Lumonics TE-861, 10 Hz, 15 ns) operating on XeCl at 308 nm which pumped a dye laser (Lambda-Physik 2002E). The dye laser was etalon-narrowed at 0.09 cm^{-1} bandwidth and pressure-tuned to produce tunable radiation in the 339 nm region. Formaldehyde absorption lines were identified by fluorescence excitation spectra[15], using line lists provided by Ramsay[16].

The vacuum ultraviolet probe laser system was patterned after

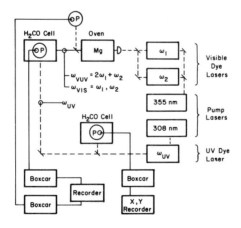

FIGURE 2   Schematic of photofragmentation experiment for CO(v,J) measurements.

the original Toronto system[17]. The third harmonic of a Nd:YAG laser (Quanta-Ray DCR-1A) pumped two dye lasers. One of these lasers (Lambda-Physik 2002) was fixed at a wavelength of ~ 430.9 nm, corresponding to a two-photon resonance in magnesium (ω_1). The other (ω_2) dye laser (Quanta-Ray PDL-1) was tuned between 470 -510 nm or between 565 - 600 nm, depending on the band of CO being studied. Energy from each of the dye lasers was typically 1.0 - 1.5 mJ per 5 ns pulse. The dye beams were combined in a Glan Prism and focused into the center of magnesium heat-pipe oven. Tunable VUV radiation ($\omega_8 = 2\omega_1 + \omega_2$) was generated in the focal region by resonantly enhanced four-wave mixing[18]. Carbon monoxide was excited via the $A\ ^1\pi \leftarrow X\ ^1\Sigma^+$ transition[19]. Fluorescence at right angles to the plane of intersection of photolysis and probe lasers was detected by a solar-blind photomultiplier tube (EMR 542G-09-18). From spectra of room temperature CO, a detection limit of ~ 10^8 molecules per cm^3 per quantum state was inferred. Absolute pulse energies were not measured, but the observed CO detection efficiency corresponded to between 10^{11} and 10^{12} photons per pulse. To obtain a spectrum of the photochemical CO, the photolysis laser was tuned onto the desired formaldehyde absorption. Then, the lasers were synchronized to give a delay of 150 - 500 ns between photolysis and probe pulses. At the formaldehyde pressures of 0.05 - 0.10 Torr used, this was a sufficiently short delay to allow unrelaxed rotational destributions of CO to be obtained. The probe laser was then tuned to obtain a laser-induced fluorescence spectrum of the photochemical CO. A typical spectrum is shown in Fig.3.

B. Results

Carbon monoxide rotational distribution following H_2CO photolysis are shown in Figure 4. The CO has a remarkable amount of rotational excitation, with no detectable population in J states below 20. Room temperature CO, by contrast, has a rotational distribution

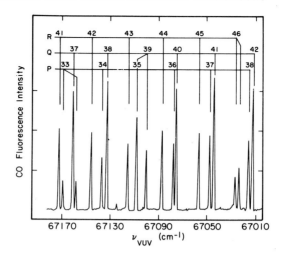

FIGURE 3 Portion of laser-induced fluorescence spectrum of carbon monoxide in the (2,0) band produced in the photolysis of 0.05 Torr of H_2CO, with the photolysis laser fixed on the $rQ_1(3)E + rQ_1(4)0$ transitions at 29515.2 cm^{-1}, and 150 ns delay (from Ref. 11, with permission).

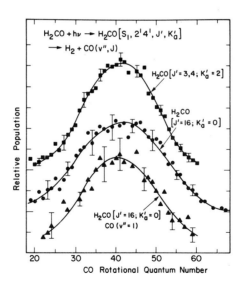

FIGURE 4 CO photofragment rotational distributions for H_2CO. Top trace is for CO(v=0) produced by photolysis on the $rQ_1(3)E + rQ_1(4)0$ transitions of H_2CO. Pressure was

0.05 Torr and delay 150 ns. Middle trace is for CO(v=0) produced by photolysis on the $pR_1(15)0$ transition of H_2CO at 29490.5 cm^{-1}. Conditions identical to those in the top trace. Bottom trace is for CO(v=1) following $pR_1(15)0$ photolysis of 0.10 Torr of H_2CO, with 150 ns delay. The three traces have been displaced by 2 vertical units. The curves have all been normalized to the same area, with one vertical unit corresponding to a probability of 0.0178 for formation of a given J state of CO (from Ref. 11, with permission).

peaking at J = 7. CO (v=1) has nearly the same rotational distribution as CO(v=0). The CO has very little vibrational excitation, in agreement with earlier infrared experiments[9]; no signals were seen for CO(v>1). Figure 5 shows that increased rotational excitation of H_2CO leads to a slightly broader CO rotational distribution, without changing the most probable J value. As Figure 6 demonstrates, the CO rotational distribution cannot be characterized by a temperature.

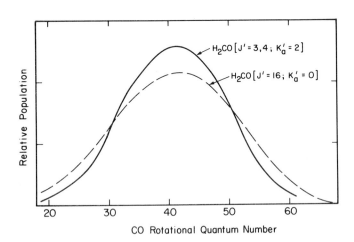

FIGURE 5 The effect of initial H_2CO angular momentum on the CO(v=0,J) distribution. The hand-drawn lines through the data points in the top two traces of Figure 4 show a broader distribution for the higher initial angular momentum (from Ref. 11, with permission).

MOLECULES AND LASER INDUCED PROCESSES 131

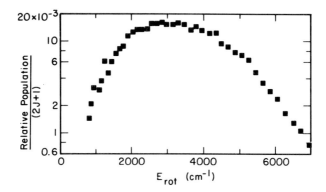

FIGURE 6 Boltzman plot for the data in the top trace of Fig. 4. The distribution cannot be characterized by a temperature (from Ref. 11, with permission).

3. HYDROGEN (v,J) DISTRIBUTION BY CARS SPECTROSCOPY

The vibrational and rotational distribution of the H_2 was determined using Coherent Anti-Stokes Raman Scattering(CARS), in a series of experiments carried out in France[12,13]. The experimental concept was the same as that of the CO experiments described above, with the tunable VUV system replaced by the CARS spectrometer[20]. The spectrometer has a detection limit for H_2 of approximately 10^{11} molecules per cm^3 per quantum state, making photolysis at the low pressures of formaldehyde used in the CO experiments impossible. Fortunately, it was observed that the (v,J) states of H_2 formed in formaldehyde photolysis have slow relaxation rates, so that distributions obtained from about 3 Torr of formaldehyde with 7 Torr of helium, with a 150 ns delay between photolysis and probe lasers were almost unrelaxed. The photolysis laser was a Quantel YAG-pumped, frequency-doubled dye laser.

The complete (v,J) distribution following H_2CO photolysis at 29496 cm^{-1} is shown in Figure 7. The hydrogen has substantial vibrational and modest rotational excitation. Because the photolysis laser only excited formaldehyde molecules in odd K states (ortho-

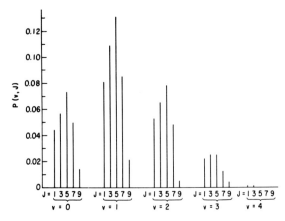

FIGURE 7   Quantum-state distribution for the $H_2(v,J)$ fragment from formaldehyde dissociation as determined by CARS spectroscopy (Ref. 13).

formaldehyde), only odd J states of H_2 (ortho-hydrogen) were produced, as earlier Raman experiments[12,21] had demonstrated. Nuclear spin is conserved during the dissociation.

4. DISCUSSION

The available data can be combined to give a complete account of the energy disposal in this photodissociation. For each degree of freedom the experimental energy distribution is summarized in Table I.

Because the angular momentum of the CO is much larger than both the initial angular momentum of the H_2CO and the angular

TABLE I   Energy distribution in $H_2CO \rightarrow H_2(v,J) + CO(v,J)$ as percent of 29500 cm^{-1} total energy.

	Trans-lation	Rotation		Vibration	
		CO	H_2	CO	H_2
Minimum (10%)	38	4(23)	1/2(1)	0(0)	0(0)
Peak (Σ = 97)	65	12(42)	6(5)	0(0)	14(1)
Maximum	91	22(58)	10(7)	7(1)	40(3)

Quantum numbers in parentheses.

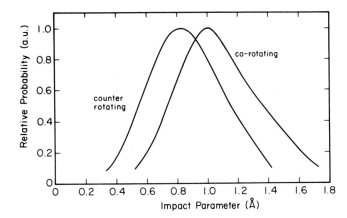

FIGURE 8   Calculated impact parameter distributions assuming uncorrelated (v,J) states of the two products. Because the distance from the carbon atom to the center of mass is ~ 0.6 Å in both the transition state and the free CO molecule, most of the impact parameters correspond to a point of impact outside the carbon atom.

momentum of the H_2 product, the fragments must fly apart with a large amount of orbital angular momentum. The data presented above can be used, along with simple angular momentum conservation, to calculate a distribution function for the product impact parameter, b. The result is shown in Figure 8. Inspite of the uncertainty about the relative directions of CO and H_2 rotation, the most probable impact parameter is clearly quite large, indicative of a bent transition state[22] with the hydrogen having a point of impact a fraction of Angstrom outside the nucleus of the carbon atom.

5. CONCLUSION

The experimental characterization of the dynamics of formaldehyde photodissociation is now nearly complite. The use of well-established techniques of nonlinear optics has made this the most thoroughly studied and well-understood polyatomic photofragmentation. Application of these techniques to other problems in reaction dynamics will provide a sensitive test for both theoretical

134 LASERS IN ATOMIC, MOLECULAR AND NUCLEAR PHYSICS

potential energy surfaces and dymanical calculations using those surfaces.

REFERENCES

1. S. R. LEONE, Adv. Chem. Phys., 50, 255 (1982)
2. J. P. SIMONS, J. Phys. Chem., 88, 1287 (1984)
3. D. J. CLOUTHIER, D. A. RAMSAY, Ann. Rev. Phys. Chem., 34, 31 (1983)
4. C. B. MOORE, J. C. WEISSHAAR, Ann. Rev. Phys. Chem., 34, 525 (1983)
5. E. E. MARINERO, C. T. RETTNER, R. N. ZARE, A. H. KUNG, Chem. Phys. Lett., 95, 486 (1983)
6. E. E. MARINERO, R. VASUDEV, R. N. ZARE, J. Chem. Phys., 78, 692 (1983)
7. M. PEALAT, J.-P. E. TARAN, J. TAILLET, M. BACAL, A. M. BRUNETEAU, J. Appl. Phys., 52, 2687 (1981)
8. J. W. HEPBURN, F. J. NORTHRUP, G. L. OSRAM, J. C. POLANYI, J. M. WILLIAMSON, Chem. Phys. Lett., 85, 127 (1982)
9. P. L. HOUSTON, C. B. MOORE, J. Chem. Phys., 65, 757 (1976)
10. P. HO, D. J. BAMFORD, R. J. BUSS, Y. T. LEE, C. B. MOORE, J. Chem. Phys., 76, 3630 (1982)
11. D. J. BAMFORD, S. V. FILSETH, M. F. FOLTZ, J. W. HEPBURN, C. B. MOORE, J. Chem. Phys., 82, ... (1985)
12. M. PEALAT, D. DEBARRE, J.-M. MARIE, J.-P. E. TARAN, A. TRAMER, C. B. MOORE, Chem. Phys. Lett., 98, 299 (1983)
13. D. DEBARRE, M. LEFEBVRE, M. PEALAT, J.-P. E. TARAN, D. J. BAMFORD, C. B. MOORE (in preparation)
14. D. J. BAMFORD, Ph. D. Thesis, University of California, Berkeley, 1984.
15. J. C. WEISSHAAR, C. B. MOORE, J. Chem. Phys., 70, 5135 (1979)
16. D. A. RAMSAY (private communication).
17. J. W. HEPBURN, D. KLIMEK, K. LIU, R. G. MCDONALD, F. J. NORTHRUP, J. C. POLANYI, J. Chem. Phys., 74 6226 (1981)
18. S. C. WALLACE, G. ZDASIUK, Appl. Phys. Lett., 28, 449 (1976)
19. J. C. SIMONS, A. M. BASS, S. G. TILFORD, Astrophys. J., 155, 345 (1969)
20. S. DRUET, J.-P. E. TARAN, Progr. Quantum. Electron., 7, 1 (1981)
21. B. SCHRAMM, D. J. BAMFORD, C. B. MOORE, Chem. Phys. Lett., 98, 305 (1983)
22. J. D. GODDARD, Y. YAMAGUCHI, J. F. SCHAEFER, III, J. Chem. Phys. 75, 3459 (1981)

2.2 Inverse Electronic Relaxation at IR Multiple-Photon Excitation of Molecules

A. A. PURETZKY

Institute of Spectroscopy, Academy of Sciences USSR, 142092, Troitzk, Moscow Region, USSR

The radiationless transitions of electronic energy to vibrational one in isolated molecules are well known at present and under intense study[1,2]. These transitions result in a decrease in visible luminescence quantum yield and are often used as a method of preparing vibrationally excited molecules[3-5]. The process in which, on the contrary, the strong vibrational excitation of a molecule transfers to its electronic excitation that, in its turn, may induce light in the visible or UV has become the subject of investigation rather recently[6-10]. Such a process was called inverse electronic relaxation (IER)[6,7]. The discovery and development of IR multiple-photon excitation (MPE), a universal method for preparation of vibrationally excited molecules, stimulated a search for this process and its investigation. MPE allows preparing molecules almost with any desired average vibrational energy, which, depending on the molecular size, is limited only by the channel of its dissociative decay[11]. At sufficiently high levels of vibrational excitation MPE results in IR multiple-photon dissociation (MPD) of molecule via the ground electronic state. It should be noted that the process of MPD itself was experimentally disclosed in 1971 by observing the visible luminescence induced by IR laser radiation[12]. Since then a great number of molecules has been found whose MPE is

136 LASERS IN ATOMIC, MOLECULAR AND NUCLEAR PHYSICS

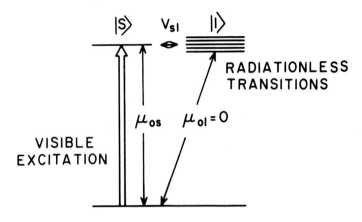

FIGURE 1a To the explanation of direct electronic relaxation.

followed by visible or UV emission[13]. But in most cases the nature of visible luminescence remains to be cleared up. This is mainly caused by the fact that it is difficult to identify the luminescent species. It is this problem that explains the basic difficulties in search of IER and its investigation.

1. DIRECT ELECTRONIC RELAXATION

Before discussing the process of inverse electronic relaxation we must consider briefly the basic conditions under which direct relaxation of electronic energy to vibrational one occurs in isolated molecule. In more detail this point is discussed in[1,2].

Let a vibrational level of the excited electronic state $|s\rangle$ be coupled, for example by the nonadiabatic interaction v_{sl}, with a certain group of isoenergetic high-lying vibrational levels $|l\rangle$ of the ground electronic state (Figure 1a) so that the full wave function of the molecular state $|j\rangle$ is

$$|j\rangle = a_{sj}|s\rangle + \sum_{l} a_{lj}|l\rangle \qquad (1)$$

Let the molecule at the time $t = 0$ be in the $|s\rangle$ state. It is essential that the $|s\rangle$ state is nonstationary and hence the probability of the molecule being found in this state will be varied in

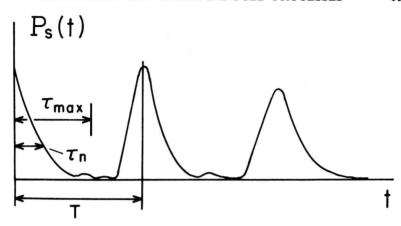

FIGURE 1b The qualitative time behaviour of the P_s probability to find the molecule in the initial state $|s\rangle$.

time. The probability of the molecule being found in the $|s\rangle$ state P_s at the time $t > 0$ is

$$P_s(t) = |\sum_j |a_{sj}|^2 \exp(-iE_j t/\hbar)|^2 \qquad (2)$$

The qualitative character of the time behaviour of $P_s(t)$ is given in Figure 1b. It can be seen that the drastic fall of probability with a characteristic time τ_n is followed again by successive recurrences of P_s with a certain recurrency time T. This means that the molecule will be found again in the initial state $|s\rangle$ in the time T. The recurrency time T is apparently defined by the level density ρ_1 of the $|1\rangle$ state

$$T = 2\pi\hbar\rho_1 \qquad (3)$$

The value τ_n can be expressed through the parameters of the molecule system within a specific model. For example, one of the simplest model of Bixon and Jortner gives

$$\tau_n = \hbar/2\pi v^2 \rho_1 \qquad (4)$$

where the matrix elements v_{sl} are assumed to be the same and equal to v.

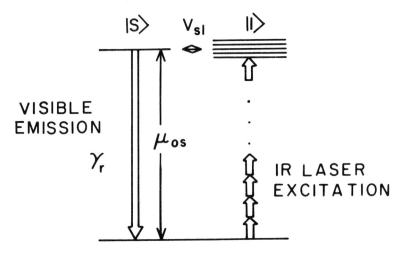

FIGURE 1c To the explanation of inverse electronic relaxation.

We can speak about the relaxation of the initial state $|s\rangle$ only if the process is irreversible, that is there is no recurrence in $|s\rangle$. Such irreversibility will take place when, for example, the $|s\rangle$ level is coupled with the continuum of the $|l\rangle$ states (T = ∞). But, if the time τ_{max} during which we observe the molecule is actually limited for some reason (finite experimental time) and the relation

$$\tau_{max} \ll T \tag{5}$$

takes place, the excitation transfer from the $|s\rangle$ level will be considered as an irreversible too. In this case it is ussually to speak of so-called practical irreversibility. If the condition (5) is not fulfilled, the relaxational transition of electronic energy to vibrational one is impossible. In this case only quantum beats between the $|s\rangle$ and $|l\rangle$ states take place.

Formally the finite time of experiment means that the $|l\rangle$ levels may have the width

$$\gamma_{max} = 2\pi\hbar/\tau_{max} \tag{6}$$

MOLECULES AND LASER INDUCED PROCESSES

so as their energy will be

$$\tilde{E}_1 = E_1 - i \gamma_1 \qquad (7)$$

It can be seen from (2) and (7) that in this case an exponentially damping factor appears in $P_s(t)$.

Thus, before speaking about a relaxational process in isolated molecule (direct or inverse) two conditions will be fulfilled:

1) The initially prepared state must be nonstationary;
2) the irreversibility has to take place.

2. INVERSE ELECTRONIC RELAXATION

Assume that the group of high-lying vibrational levels of the ground electronic state $|1\rangle$ is excited in the process of MPE by IR laser radiation (Figure 1c). As may be seen, the basic conditions of relaxation, of the vibrational excitation to excited electronic term are fulfilled here: 1) nonstationary states are usually prepared at MPE; 2) the irreversibility is provided by a loss of vibrational excitation due to electronic relaxation, that is, because of visible emission from the $|s\rangle$ state.

There are two approaches in the theoretical description of IER in terms of Born-Oppenheimer (BO) states or in terms of the molecular states. The both approaches are fully equivalent and only give different names of the process. Figure 2 schematically shows the levels corresponding to these descriptions[7]. The $\{|j\rangle\}$ states here are eigenstates of the molecular Hamiltonian, $H_M = H_{BO} + V$ where V is the coupling operator; $\{|G\alpha\rangle\}$, $\{|S\beta\rangle\}$ are the eigenstates of the BO Hamiltonian H_{BO} for ground $|G\rangle$ and excited $|S\rangle$ electronic states, respectively. The lower molecular states $|j\rangle$ ($E < E_o$) are identical to the BO states $\{|G\alpha\rangle\}$. These levels can be excited by IR laser field but the spontaneous transition to the ground state $|G0\rangle$ with one photon emission is forbidden, that is, the rate of this transition $\gamma_j = 0$. Starting from the E_o energy

140 LASERS IN ATOMIC, MOLECULAR AND NUCLEAR PHYSICS

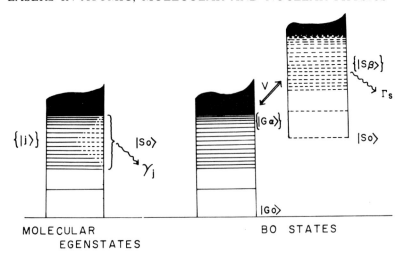

FIGURE 2 The explanation of IER in terms of molecular and Born-Oppenheimer states.

corresponding to the onset of the upper electronic term, the molecular states represent a linear combination of the state $\{|G\alpha\rangle\}$ and $\{|S\beta\rangle\}$. The character of the admixture of $\{|G\beta\rangle\}$ to $\{|G\alpha\rangle\}$ here is determined by the overlap region of these levels. In the simplest case when just one state, for example $|S0\rangle$, is mixed with the ground state vibrational quasi-continuum $\{|G\alpha\rangle\}$, a Loretzian distribution of the admixture of $|S0\rangle$ state with its centre at E_o and half-width Δ takes place[7]. Because of the molecular states with $E \geqslant E_o$ are of mixed type a spontaneous single-photon transition from these states to the ground state $|G0\rangle$ with a certain rate γ_j is possible. In the BO basis set the vibronic levels of the excited state $\{|S\beta\rangle\}$ may radiatively decay to the ground state $|G0\rangle$ with a certain rate Γ_s. But the coupling term V decreases the decay rate.

Let us consider how Γ_s changes in the presence of coupling term V. In the simplest case when the coupling of each particular vibrational level of the excited electronic state, for example

MOLECULES AND LASER INDUCED PROCESSES 141

|S0>, with the vibrational quasi-continuum of the ground electronic state $\{|G\alpha\rangle\}$ can be considered separately the introduction of the coupling V will give rise, instead of |S0>, to a Lorentzian contour with its width $\Delta = 2\pi \langle S\beta|V|G\alpha\rangle^2 \rho_G$, where ρ_G is the vibrational level density of the ground electronic state [6,7] (see Figure 2). As a result of such broadening the oscillator strength of the transition which corresponds to the radiative decay rate Γ_s spreads over all the states in the Lorentzian contour and this causes Γ_s to decrease. The number of such states N in the Lorentzian contour can be estimated if it is replaced by a rectangle with its width Δ being equal to that of the Lorentzian contour. In this case

$$N = 2\pi \langle S\beta|V|G\alpha\rangle^2 \rho_G^2 \qquad (8)$$

where N is called the dilution factor. Thus, the observed radiative decay rate in our case is

$$\gamma_j \simeq \Gamma_s/N \qquad (9)$$

Such an estimation is valid in the case when the vibrational level density of the excited electronic state in the region of coupling with the ground state is sufficiently low so that $\rho_s \Delta < 1$. Otherwise, when $\rho_s \Delta \geqslant 1$, the coupling of two vibrational quasicontinua of the ground and excited electronic states should be considered. In this case the dilution factor will be determined by a statistical expression as a ratio of vibrational level densities of the corresponding electronic states

$$N = \rho_G/\rho_s \qquad (10)$$

So, when there is an upper electronic state below the dissociation limit of the ground electronic state of molecule, which is coupled by a radiative transition with the ground one, this coupling will lead to a decrease in rate of this transition according to (8) - (10). Therefore, the first indication of collisionless UV or visible luminescence in IER is its radiative lifetime, which has to be longer than, that for the dipole-allowed electronic transition in a molecule.

Besides the luminescence lifetime, different dependences of the peak luminescence intensity, that is dependences of the luminescence pulse peak, are usually studied in experiments. The peak luminescence intensity I, apart from the radiative transition rate, will be defined by the population of the corresponding vibrational sublevels as well. For example, when $\rho_s \Delta \geq 1$,

$$I = n \Gamma_s \int_0^\infty [\rho_s(E-E_o)/\rho_G(E)] \cdot f_G(E) dE \qquad (11)$$

where n is the total number of vibrationally excited molecules, E_o is the energy gap between the ground and excited electronic states, $f_G(E)$ is the vibrational distribution function of molecules after the IR laser pulse excitation. In the case of Boltzman distribution of vibrational energy with a corresponding vibrational temperature T the peak luminescence intensity is expressed as

$$I = n \Gamma_s (Q_s/Q_G) \exp(-E_o/kT) \qquad (12)$$

where Q_G, Q_s are the vibrational partition functions of the ground and excited electronic states respectively. From (11) it follows that in order to observe the visible luminescence arising in IER one must choose such a molecule that the ρ_s/ρ_G factor is not too small. Among them there are, first of all, small molecules consisting of three to five atoms.

Let's consider now the energy position of the |s⟩ state. Generally speaking, this state must lie below the dissociation limit D of the ground electronic term. Otherwise, a competitive process, unimolecular dissociation, with its rate[14]

$$W(E) \simeq A(\Delta E/E)^{s-1} \qquad (13)$$

may occur, where A is the constant, E is the total vibrational

MOLECULES AND LASER INDUCED PROCESSES   143

energy of molecule, ΔE is the excess of vibrational energy over the dissociation limit, s is the number of vibrational degrees of freedom. For intermediate size molecules (s $\simeq$ 10) the unimolecular decay does not allow the molecule to be overexcited considerably and hence it cannot reach the high-lying electronic states $|s\rangle$. Besides, in the accessible energy range ΔE over the dissociation limit such decay effectively competes with IER. Molecules with a large number of vibrational degrees of freedom, however, can be greatly excited over the dissociation limit. For such molecules a range ΔE with a small value of $W(E)$ can be chosen. In this case the $|s\rangle$ state may lie over the dissociation limit within ΔE.

3. SEARCH FOR AND EXPERIMENTAL INVESTIGATION OF IER

Collisionless visible luminescence with its radiative lifetime required by the process of IER (see (8) - (10)) is revealed at present for a large number of molecules excited by IR laser radiation, for example: SO_2[15-17]; F_2CO[10]; OsO_4[8,18-22]; CrO_2Cl_2[9,23-26]; $VOCl_3$[27]; C_2H_3CN[28-30]; $S_2C_2F_4$[31], etc. Nevertheless, in most cases it is still impossible to attribute the luminescence to any definite species (the molecule itself or its fragment).

$\underline{SO_2}$. Below the dissociation limit D of the ground electronic term $|G\alpha\rangle$ ($^1A_1$) (D $\simeq$ 45$\cdot 10^3$ cm^{-1}) in the energy region E $\simeq$ 30$\cdot 10^3$ cm^{-1} there are several triplet and singlet electronic terms $|S\beta\rangle$ ($^3B_1$, $^3A_2$, $^1A_2$, $^1B_1$). In [15] an attempt was made to observe the visible luminescence in SO_2 arising due to IER at IR multiple-photon excitation of the ν_1 mode. The spectroscopic and kinetic characteristics of the luminescence observed coincided with those of UV induced SO_2 luminescence from the $^1B_1$ state. These experiments, however, were performed at comparatively high pressures of SO_2 gas in the cell (P $\sim$ 1 Torr), where the gas was excited by usual CO_2 laser pulse ($\tau_p \simeq$ 100 ns). Therefore, both the MPE and IER were realized under collisional conditions. Later, in [16], the authors managed to excite the SO_2 molecule by a short CO_2 laser pulse (the threshold energy fluence $\Phi \simeq 15$ J/cm^2, $\tau_p \simeq 0.5$ ns) under collisionless conditions (P $\simeq 8 \div 80$ mTorr) and observe the visible luminescence caused by IER. In [17] the SO_2 molecule was excited under collisionless conditions in a two-frequency field by usual CO_2 laser pulses ($\tau_p \simeq 100$ ns) and the process of IER was observed.

$\underline{F_2CO}$. The identification of the CO_2 laser induced emission as belonging to the F_2CO molecule has been drawn by comparing the luminescence spectrum with that of the chemiluminescence in the reaction: $O_2(^1\Delta) + C_2F_4 \longrightarrow F_2CO(S_1) + F_2CO(S_0)$ [18]. The estimation of luminescence intensity shows that $\simeq 10^{-7}$ molecules of the total number in the irradiated volume result in the emission ($\Phi \simeq 50$ J/cm^2, $\tau_p \simeq 250$ ns). Detailed studies of the luminescence of this molecule were not carried out.

$\underline{OsO_4}$. More detailed experimental studies were performed for the visible luminescence in the OsO_4 molecule induced by MPE both in one-frequency and two-frequency IR laser fields with $\Phi \leq 2$ J/cm^2 [8,18-20]. A theoretical model that described experiment well was developed within MPE and IER of the OsO_4 molecule[18]. The studies in a wider range of laser fluences (2 J/cm$^2 < \Phi < 1000$ J/cm^2) show that the luminescence of OsO_4 has three different stages[22]. Each of them has a definite fluence threshold. A theoretical model has been developed, which explains well the three-stage luminescence of the OsO_4 molecule by the sequence of the processes taking place during an IR laser pulse: IER-MPD of the parent molecule (stage I), IER-MPD of the primary fragments (stage II), and IER-MPD of the secondary fragments (stage III). This sequence of processes is apparently rather a general mechanism of emergence of luminescence, and, according to the molecule under IR laser excitation, the molecular or any of the fragmentary emission stage may be absent.

$\underline{CrO_2Cl_2}$. The visible luminescence arising at MPE of the CrO_2Cl_2

MOLECULES AND LASER INDUCED PROCESSES 145

molecule has been the subject of intense studies, too[9,23-26]. Nevertheless, the nature of luminescence in this case is still unclear. There are indirect indications to the molecular character of luminescence at rather low laser fluences[24,25]. It has been shown that at high fluences the luminescence is caused by fragments[26].

$\underline{C_2H_3CN}$. The nature of luminescence is still unclear at present. But the latest experiments with supersonic molecular jets have shown that only two fragments, C_2HCN or C_2CN, can be responsible for the observed luminescence[30]. The authors believe that the most probable mechanism is the sequence of the processes of fragmentation of the C_2H_3CN molecule to C_2HCN and C_2CN and their IER during a CO_2 laser pulse.

For the other molecules given above ($VOCl_3$, $S_2C_2F_4$), as well as for some other cases[13], the collisionless visible luminescence induced by MPE has not been indentified.

4. VISIBLE LUMINESCENCE OF THE OsO_4 MOLECULE

To give an idea of the character of the visible emission induced by IR laser radiation we shall consider below in more detail the case of the OsO_4 molecule[18]. Figure 3 shows a typical time evolution of visible luminescence as the OsO_4 molecule is excited by CO_2 laser pulse. If we choose proper excitation conditions, it is possible to observe easily the collisionless luminescence stage (the excitation of the far "red" wing of the P-branch of the ν_3 OsO_4 mode, the P(40) CO_2 laser line). When excitation is at the centre of the P-branch (the P(10) CO_2 laser line), collisional luminescence occurs in addition to collisionless one. This behaviour is connected with the characteristic property of the MPE which is considered comprehensively in [18,20]. Here we shall discuss in detail the nature of collisionless visible luminescence and particularly the problem of its identification. For this pur-

146 LASERS IN ATOMIC, MOLECULAR AND NUCLEAR PHYSICS

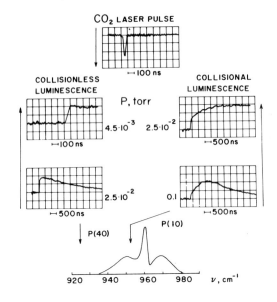

FIGURE 3 The time evolution of the visible luminescence of OsO_4 induced by CO_2 laser radiation[18]. $\Phi = 0.7$ J/cm^2. The IR absorption spectrum of OsO_4 (ν_3 mode) and the excitation frequencies used (bottom).

pose rather a general approach was developed in[21,22]. It is based on separating all possible successive processes at MPE with respect to the energy required for their realization. Really, if IER takes place, this is a process with the lowest energy E_o. This means that, first of all, we must observe IER of the parent molecule. The next process, molecular dissociation, requires the energy D (usually $E_o <$ D, see section 2). To accomplish the process of IER of the primary fragment, we must produce it and then excite to the onset of the upper electronic term E_o^f, that is, expend the energy $E \geq (D + E_o^f)$, and so on for the next fragments. Thus, each of the assumed processes - IER - MPD of molecule, IER - MPD of primary fragment, IER - MPD of secondary fragment - needs a higher laser fluence for its realization. The next important factor is that in this sequence of processes MPD must always quench the luminescence

MOLECULES AND LASER INDUCED PROCESSES   147

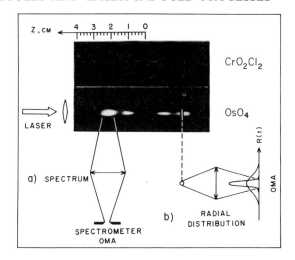

FIGURE 4   Typical photographs of the visible luminescence induced by CO_2 laser radiation focused into a cell with OsO_4 and CrO_2Cl_2 [22]. The focal length of the lens f = 10 cm, the laser pulse duration τ_p = 50 ns, the OsO_4 pressure in the cell P = 80 mTorr, CrO_2Cl_2 - P = 160 mTorr. The used methods of luminescence analysis: a) the study of spectra for each stage; b) the study of the time-of-flight distribution of luminescent species.

caused by IER. So, if this sequence is assumed to be possible for MPE, the luminescence of the parent molecule with increasing laser fluence will appear and then disappear. Then luminescence of primary fragments will take place, and this emission will be quenched again by their dissociation, and so on.

Direct analysis of the spatial distribution of the luminescence produced by focusing the CO_2 laser beam into the cell with the gas is a convenient experimental approach of obtaining such dependences over a wide range of Φ. In this case it is possible to realize laser fluence scale along the focused beam of $\overline{CO_2}$ laser in the cell, with the maximum fluence in the focus. This longitudinal spatial luminescence distribution can be recorded with an optical multichannel analyzer (OMA) and thus the total dependence of

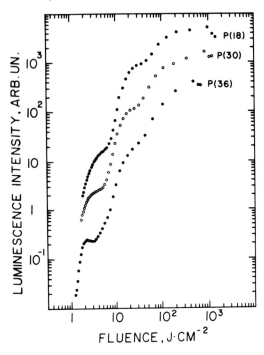

FIGURE 5 The dependence of the luminescence intensity (at the peak of emission pulse) on the laser fluence at the P(18), P(30), and P(36) CO_2 laser lines[22]. The experimental conditions: τ_p = 50 ns, P = 80 mTorr, the gate duration Δ = 400 ns.

luminescence intensity on laser fluence can be obtained after proper processing. Such measurements can be taken during a single CO_2 laser pulse. For qualitative estimation of the presence of different luminescence stages the OMA diode array can be replaced by an ordinary photographic plate. Such photos of the visible luminescence induced by CO_2 laser radiation are presented in Figure 4 for the OsO_4 and CrO_2Cl_2 molecules to explain the used experimental techniques. One can clearly see light and dark spots conditioned by the creation and further quenching of luminescence with increasing laser fluence as the focus z = 0 is approached. The light spot between two dark areas corresponds to a separate luminescence

MOLECULES AND LASER INDUCED PROCESSES 149

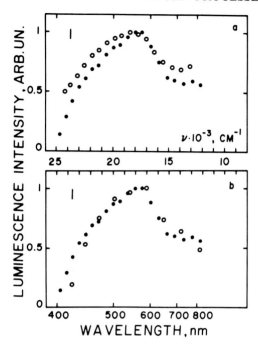

FIGURE 6 The visible luminescence spectrum of stage I obtained at $\Phi = 6$ J/cm^2 for the emission pulse peak (dark circules), (the delay between the emission pulse peak and the gate $\Delta t = 0$) and for its "tail" ($\Delta t = 3$ µs) (open circules) - (a). The luminescence spectrum for the pulse peak ($\Delta t = 0$) with $\Phi = 6$ J/cm^2 (dark circules) and $\Phi = 3.7$ J/cm^2 (open circules) - (b)[22]. The OsO$_4$ molecule was excited at the P(14) CO$_2$ laser line, $\tau_p = 50$ ns, P = 80 mTorr, $\Delta = 250$ ns.

stage. The change of the photographic plate by a more sensitive diode array makes it possible to observe three different luminescence stages and to attain dependences of visible luminescence intensity on laser fluence. Some of such dependences obtained at different CO$_2$ laser lines are given in Figure 5. The luminescence spectra of each stage were obtained. The time-of-flight distribution of luminescent species were measured, to find their velocities.

Figures 6, 7, and 8 show visible luminescence spectra of OsO$_4$

150 LASERS IN ATOMIC, MOLECULAR AND NUCLEAR PHYSICS

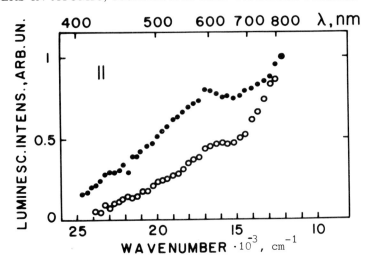

FIGURE 7 The luminescence spectrum of stage II at $\Phi = 23$ J/cm^2 for the luminescence pulse peak (dark circles) and for its tail ($\Delta t = 3$ μs) (open circles). The conditions of experiment are similar to those in Fig. 6.

induced by CO_2 laser radiation for each of stages I, II, III (in order of their appearance with fluence increase; Figures 6, 7, and 8 respectively). Each stage in Figures 6a, 7, and 8a have two spectra characterizing the peak of the luminescence pulse ($\Delta t = 0$, dark circles) and its "tail" ($\Delta t = 3$ μs, open circles). All these spectra are mormalized to the maximum luminescence intensity in the spectral range under study (400 ÷ 800 nm).

Stage I. This stage is characterized by a wide luminescence spectrum with its maximum near $1.8 \cdot 10^4$ cm^{-1} ($\Phi = 6$ J/cm^2). For stage I the spectrum slightly changes in time in the course of luminescence quenching (open circles, Figure 6a). The spectrum at the luminescence pulse "tail" is broadened a little. The spectrum variation with increasing laser fluence Φ within stage I is shown in Figure 6b. As Φ increases, the luminescence spectrum becomes broadened. With its further increase the luminescence spectrum changes stepwise which fully corresponds to the threshold appear-

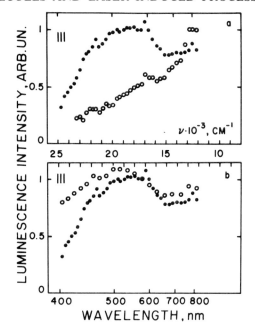

FIGURE 8 The luminescence spectrum of stage III with Φ = 250 J/cm^2 for the pulse peak (dark circules) and its "tail" (Δt = 3 μs) (open circules) - (a). The luminescence spectrum for the pulse peak with Φ = 250 J/cm^2 (dark circules) and Φ = 1000 J/cm^2 (open circules) - (b). The conditions of experiment are similar to those given in Fig. 6 [22].

ence of luminescence at stage II.

Stage II. This stage is characterised by a luminescence spectrum that is red-shifted and differs essentially from the analogous spectrum for stage I. For example, with Φ = 23 J/cm^2 the spectrum maximum lies beyond the spectral range under study and only its short-wave wing can be apparently observed. The time evolution of luminescence spectrum in the process of quenching is quite different for this stage, too (open circules, Figure 7). Indeed, the spectrum wing observed is red-shifted when the delay Δt between the laser pulse and the gate is increased. It is rather difficult to conclude on the character of evolution of the observ-

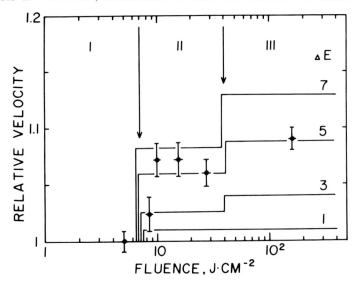

FIGURE 9 The dependence of the relative variations in velocity of luminescent species on laser fluence. All the values are normalized to the species velocity at $\Phi = 5$ J/cm^2 (stage I). The solid straight lines denote the calculated velocities attained within the model[22] at different $\Delta E/\hbar\omega = 1, 3, 5,$ and 7 [22].

ed spectrum with fluence within stage II because there is no maximum in the studied spectral range.

Stage III. With further increase in laser fluence an absorption band appears in the region of $1.8 \cdot 10^3$ cm^{-1} (Figure 8a, dark circules) which corresponds to the luminescence of stage III ($\Phi = 250$ J/cm^2) The luminescence quenching rate for this stage is larger than for stage II. As a result, at the "tail" the luminescence spectrum becomes similar to that of stage II (open circules, Figure 7). The spectrum evolution with increasing Φ is similar to that of stage I. As Φ increases, the luminescence spectrum of stage III is broadened, its maximum blue-shifting.

Thus, the luminescence spectra considered, along with their time evolution, quite unambiguously characterize each luminescence stage.

Analysis of the velocity of expanding of luminescence zone with the use of the same OMA diode array (Figure 4b) allowed measuring the velocities of luminescent species[22] (another method for measurement of such velocities is considered in[20]).

In Figure 9 the dots show the relative change of the luminescent species velocity as the laser fluence increases. All the measured velocities are normalized to that of the luminescent species formed at $\Phi = 5$ J/cm^2 (stage I). When Φ corresponds to stage II, there is a stepwise increase of velocity corresponding to the threshold appearance of this stage (Figure 5). There is not any considerable increase of velocity at stage III. In more detail these results are considered in[22].

The excess of vibrational energy over the dissociation limit at MPD of the OsO_4 molecule (ΔE) is partially converted to the recoil energy of fragments thus causing their most probable velocities to be increased as compared to thermal ones. For example, according to the calculations the value $\Delta E/\hbar\omega = 5$ (where ω is laser frequency) agrees well with the observed increase of velocities[22] (Figure 9, solid lines).

Thus, the whole set of experimental data: 1) the number of potential luminescence stages, 2) the variation of luminescence intensity with Φ, 3) its spectral evolution as Φ increases within each stage and during the transition to the next stage, 4) the velocities of luminescent species at each stage — is explained by the following sequence of processes: IER — MPD of the OsO_4 molecule -- IER — MPD of the primary fragment (OsO_3) -- IER — MPD of the secondary fragment (OsO_2)[22].

In conclusion it should be noted that for a number of molecules the process of IER explains well the visible luminescence induced by IR laser radiation (SO_2, OsO_4). For other molecules it is rather difficult to use IER for interpreting the visible lumi-

154 LASERS IN ATOMIC, MOLECULAR AND NUCLEAR PHYSICS

nescence since the luminescent species is unknown. In the nearest future this problem seems to be solved for many other molecules as well. Another difficulty here is that the information on low-lying forbidden electronic states of molecules and particularly their fragments is practically absent. It is this information needed for detail studying the IER process itself.

REFERENCES

1. E. S. MEDVEDEV, V. I. OSHEROV, Theory of Radiationless Transitions in Polyatomic Molecules (Nauka, Moscow, 1983).
2. Radiationless Processes in Molecules and Condensed Phases, edited by F. K. Fong (Springer, Berlin, Heidelberg, New York, 1976).
3. D. F. HELLER, G. A. WEST, Chem. Phys. Lett., 69, 419 (1980)
4. A. V. EVSEEV, V. M. KRIVTSUN, Yu. A. KURITZIN, A. A. MAKAROV, A. A. PURETZKY, E. A. RYABOV, E. P. SNEGIREV, V. V. TYAKHT, JETP, 87, 111 (1984) (in Russian)
5. J. E. DOVE, H. HIPPLER, J. TROE, J. Chem. Phys., 82, 1907 (1985)
6. A. NITZAN, J. JORTNER, Chem. Phys. Lett., 60, 1 (1978)
7. A. NITZAN, J. JORTNER, J. Chem. Phys., 71, 3524 (1979)
8. R. V. AMBARTZUMIAN, G. N. MAKAROV, A. A. PURETZKY, JETP Lett., 28, 696 (1978)
9. Z. KARNY, A. GUPTA, R. N. ZARE, S. T. LIN, J. NIEMAN, A. M. RONN, Chem. Phys., 37, 15 (1979)
10. J. W. HUDGENS, J. L. DURANT, Jr., D. J. BOGAN, R. A. COVELESKIE, J. Chem. Phys., 70, 5906 (1979)
11. V. N. BAGRATASHVILI, V. S. LETOKHOV, A. A. MAKAROV, E. A. RYABOV, Laser Chem., 1, 211 (1983) and following issues.
12. N. R. ISENOR, M. C. RICHARDSON, Opt. Comm., 3, 360 (1971)
13. H. REISLER, C. WITTIG, in Photoselective Chemistry, Advances in Chemical Physics Ser., Vol. 46, edited by J. Jortner, R. Levine, and S. Rice (Wiley, New York, 1981), pp. 679.
14. P. J. ROBINSON, K. A. HOLBROOK, Unimolecular Reactions (Wiley-Interscience, London, 1972)
15. G. L. WOLK, R. E. WESTON, Jr., G. W. FLYNN, J. Chem. Phys., 73, 1649 (1980)
16. T. B. SIMPSON, N. BLOEMBERGEN, Chem. Phys. Lett., 100, 325 (1983)
17. A. A. PURETZKY, The paper presented at the conference "Recent Advances in Molecular Reaction Dynamics", June 10-14, 1985, Aussois, France.

18. A. A. MAKAROV, G. N. MAKAROV, A. A. PURETZKY, V. V. TYAKHT, Appl. Phys., 23, 391 (1980)
19. R. V. AMBARTZUMIAN, G. N. MAKAROV, A. A. PURETZKY, Appl. Phys., 22, 77 (1980)
20. R. V. AMBARTZUMIAN, G. N. MAKAROV, A. A. PURETZKY, Opt. Comm., 27, 79 (1978)
21. A. A. PURETZKY, H. SCHRODER, in European Conference on Atomic Physics, April 6-10, 1981, Ruprecht-Karls Universität, Heidelberg, Book of Abstracts, edited by J. Kowalski, G. zu Putlitz, H. G. Weber, v. 5A, Part I, p. 288.
22. K. L. KOMPA, H. LAMPRECHT, H. SCHRODER, A. A. PURETZKY, V. V. TYAKHT (submitted to J. Chem. Phys.).
23. J. NIEMAN. A. M. RONN, Opt. Engineering, 19, 39 (1980)
24. I. BURAK, J. Y. TSAO, Chem. Phys. Lett., 77, 536 (1981)
25. J. Y. TSAO, N. BLOEMBERGEN, I. BURAK, J. Chem. Phys, 75, 1 (1981)
26. T. A. WATSON, M. MANGIR, C. WITTIG, M. R. LEVY, J. Phys. Chem., 85, 754 (1981)
27. R. C. SAUSA, A. M. RONN, J. Chem. Phys., 81, 1716 (1984)
28. M. H. YU, H. REISLER, M. MANGIR, C. WITTIG, Chem. Phys. Lett., 62, 439 (1979)
29. M. H. YU, M. R. LEVY, C. WITTIG, J. Chem. Phys., 72, 3789 (1980)
30. T. A. WATSON, M. MANGIR, C. WITTIG, M. R. LEVY, J. Chem. Phys., 75, 5311 (1981)
31. C. N. PLUM, P. L. HOUSTON, Chem. Phys., 45, 159 (1980)

2.3 Laser Induced Chemical Reactions in Combustion and Industrial Processes

J. WOLFRUM

Physikalisch-Chemisches Institut der Puprecht-Karls-Universität, D-6900 Heidelberg, Im Neuenheimer Feld 253, West Germany

and

Max-Planck-Institut für Strömungsforschung, D-3400 Göttingen, Böttingerstrasse 6–8, West Germany

The rapid development of powerful UV-laser sources allows the investigation of macroscopic and microscopic details of elementary chemical reactions important in combustion processes. Experimental results on the effect of selective translational and vibrational excitation of reactants in elementary combustion reactions using laser photolysis and time-resolved atomic line resonance absorption, laser-induced fluorescence and CARS spectroscopy are compared with the results of theoretical studies on ab initio potential energy surfaces and thermal rate parameters. Thermal elimination of hydrogen chloride from 1,2-dichloroethane and 1,1,1-chlorodifluoroethane is a main industrial route to some important monomer compounds. Inducing thic radical chain reactions by UV-exciplex laser radiation offers the advantage that a monomolecular process with low activation energy becomes the rate determining step. This allows lower process temperatures with decreasing energy expense and avoiding the high temperature formation of by products.

INTRODUCTION

It has been known, since the first use of fire by mankind[1], that the rates of chemical reactions depend strongly of the energy of the reactants. Traditionally, the energy dependence of the chemical reaction rate is studied under conditions in which the rate of reaction is slow compared to that of collisional energy transfer. Under these conditions, the energy of the reactants is characterized by a temperature. The temperature variation of the reaction rate can then often be expressed with sufficient accuracy by the Arrhenius equation. The Arrhenius parameters obtained in this way, however, contain no direct information on how the various degrees of freedom of the reacting molecules and in the "activated complex" contribute to the potential pathways of product formation in the chemical reaction. Investigations on the chemical reactivity under a wide range of conditions such as pressure variation and specific excitation of the reactants give important insights into the microscopic dynamics of the chemical reaction. This information on one hand can be compared with the results of theoretical predictions using potential energy surfaces for chemical reactions obtained by ab initio methods, and is also of basic interest to improve the kinetic data used in detailed chemical kinetic modelling of combustion processes. The experimental possibilities to study elementary chemical reactions in detail have expanded quite dramatically in recent years as a result of the development of various laser sources. The coherence, collimation, monochromaticity, polarization, tunability and short pulse duration of laser light sources now available in the infrared, visible, and ultraviolet region allow the preparation and detection of chemically reacting molecules with an unprecedented degree of selectivity.

The precent paper cannot give a comprehensive collection of experimental and theoretical results in this field. Instead, a number of typical examples is presented. The first part gives specific examples of experimental and theoretical investigations with selective translational and vibrational exctation of the reactants. The second part describes studies on the formation of vinylchloride

MOLECULES AND LASER INDUCED PROCESSES   159

and vinylidenfluoride by UV laser-induced chain reactions.

THE EFFECT OF TRANSLATIONAL EXCITATION

Despite the large number of elementary steps taking place even in the oxidation of simple hydrocarbons, important parameters of the combustion process are controlled by relatively few elementary reactions. As shown in Figure 1 in the case of the oxidation of methane, ethane and butane the flame velocities calculated by detailed chemical kinetic modelling are relatively insensitive to reactions specific for the oxidation of these molecules^2. However, there is a strong influence on the calculated flame velocity by a number of unspecific reactions such as

$$H + O_2 \rightleftharpoons OH + O \quad (1)$$
$$CO + OH \rightleftharpoons CO_2 + H \quad (2)$$
$$OH + H_2 \rightleftharpoons H_2O + H \quad (3)$$

At least in one direction considerable collision energies are necessary to surmount the reaction barrier in these elementary steps. The dynamics of such high barrier reactions can be studied in microscopic details by combining translationally hot atom and radical formation by laser photolysis (forming reactants with initially high and monoenergetic collision energies) with time and state resolved product detection by laser-induced fluorescence spectroscopy$^{3-5}$. The experimental apparatus is shown in Figure 2. Fig. 3 gives nascent rotational state distributions for the reactions (1), (-2), and (-3) at relative collision energies around 250 kJ/mol. Despite of comparable total reaction energies, the nascent OH (v=0) rotational distributions are quite different. The distribution is extremely hot from the $H + O_2$ reaction, broadly peaked at rotational quantum numbers around K = 11 in the $H + CO_2$ case and restricted to low values with only 3% of the total available energy partitioned into rotation for the $H + H_2O$ system. The obsered rotational

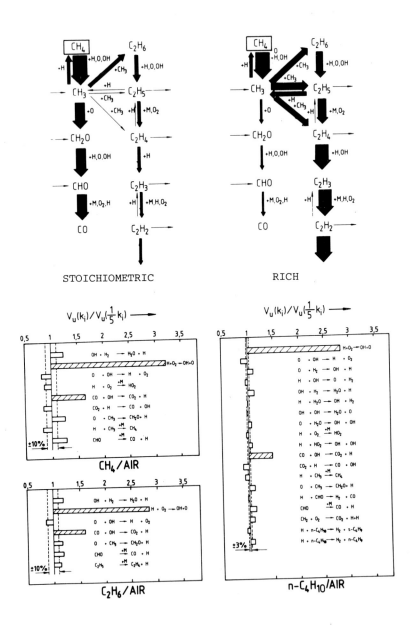

FIGURE 1 Molar chemical fluxes in atmospheric CH_4-air flames and Sensitivity of calculated free flame velocity (V_u) on the reduction of elementary rate coefficients by a factor of five[2].

MOLECULES AND LASER INDUCED PROCESSES 161

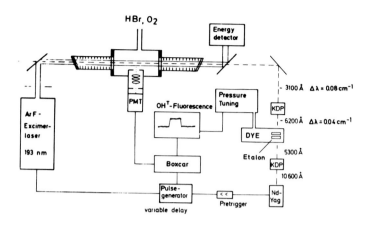

FIGURE 2 Schematic of experimental arrangement for the study of radical reactions with substantial threshold energies.

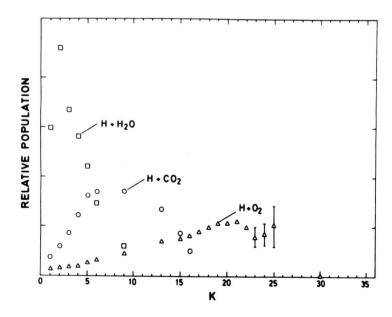

FIGURE 3 Nascent rotational state distribution in OH (V=0) produced in the reactions (1), (-2), and (-3) at relative collision energies of 250 kJ/mol. K is rotational quantum number.

energy distributions give interesting microscopic details on the molecular dynamics of these elementary steps. Spin-orbit and orbital-rotation interactions in the OH radical cause fine-structure splittings for each rotational level. Each of these fine structure levels can be probed by different rotational sub-bands. As shown in Table I, the two OH spin states $^2\Pi_{1/2}$ and $^2\Pi_{3/2}$ are, within experimental error, equally populated. However, the Λ-doublet fine structure states show a clear preference for the lower energy Π^+ component. The experimental result shows that the break-up of the reaction complex generates forces in a plane containing the bond to be broken. OH rotates in that plane and J_{OH} is perpendicular to it and to the broken bond. This picture is consistent only with a preferential planar exit channel in these reactions. This could also be directly demonstrated using polarized photolysis and analysis laser beams[6].

TABLE I Measured absolute reaction cross sections, vibrational and fine structure partitioning for rotational levels of OH produced in the reactions (1), (-2), and (-3)

	$H + H_2O$	$H + CO_2$	$H + O_2$
$E_{coll}^{c.m}$ [kcal/mol]	58.2	60.1	60.7
σ [Å^2]	0.24±0.1	0.37±0.1	0.42±0.2
$\dfrac{\sigma(v=1)}{\sigma(v=0)}$	≤0.1	≤0.1	0.47±0.15
$\dfrac{\sigma(^2\Pi_{3/2})}{\sigma(^2\Pi_{1/2})} \dfrac{K}{K+1}$	1.1±0.2	1.2±0.2	1.2±0.2
$\dfrac{\sigma(\Pi^+)}{\sigma(\Pi^-)}$	3.2±1.0	3.0±1.0	5.9±1.0

Reaction (1) is known to take place adiabatically on the ground state potential surface of the HO_2 ($^2A''$)-radical. Trajectory calculations[7] on an ab initio surface (Melius-Blint[8]) are in agreement with calculated OH rotational distributions from the phase space theory[9] for low relative translational energy. With increasing relative translational energy the OH rotational distribution becomes is considerably hotter than the statistical one, and no long living HO_2- complex excists during the reaction. As given in Table I the experimental total reaction cross section at E_T = 254 kJ/mol is 0.42±0.2 $[Å]^2$ for reaction (1). The theoretical cross section obtained under this condition by quasi classical quantum mechanical threshold calculations[10] at the Melius-Blint surface is 0.38 $[Å]^2$. These numbers cannot be compared directly because the multiplicity of the $^2A"$-surface at infinite $H-O_2$ separation is not taken into account. Miller[10] uses a multiplicity factor of 1/3. This would yield a theoretical cross section of 0.13 $[Å]^2$ which is significantly outside the experimental range. Calculated rate coefficients using this theoretical cross section[10] are in agreement with shock tube measurements of the rate of reaction (1) by Schott[11] and Chiang and Skinner[11]. However, as shown in Figure 4, very recent shock tube experiments by Just and Frank[12] using time resolved atomic resonance line absorption coincide with the extrapolated values recommended by Baulch et al.[13]. Even higher rate data for k_1 are reported by Bowman[14]. Also in the recombination pathway

$$H + O_2 \longrightarrow HO_2^* \xrightarrow{+M} HO_2 \qquad (1a)$$

new measurements of the high prssure limit recombination rate coefficient[15] are higher than calculated values[16]. The observed discrepancies could be attributed to a reduction of calculated reactive cross sections due to a "rigid" character and a barrier of 8 kJ/mol of the Melius-Blint surface for dissociation of the HO_2 in reaction (-1a)[15]. Calculations by Dunning et al.[17] reduce this barrier to less than 1.7 kJ/mol. Also for reaction (-1a) the

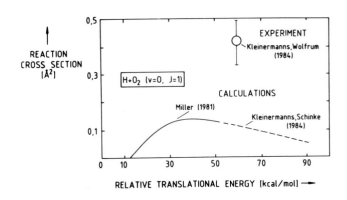

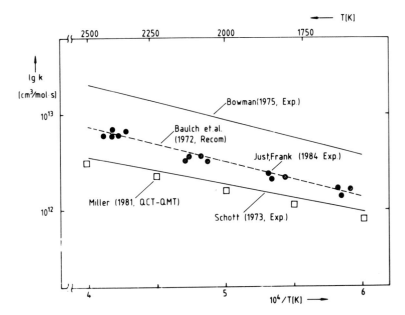

FIGURE 4 Measured and calculated reaction cross sections and thermal rate coefficients for reaction $H + O_2 \rightarrow OH + O$

Melius-Blint surface apparently overestimates the long range O-OH attracion[18] while the Quack-Troe interpolation scheme[19] leads to better agreement with the experimental values at low temperatures[20].

In summary, it is encouraging to see that recent microscopic and macroscopic experimental data start to give a more converging picture on this reaction central for all combustion modellin calculations. However, more work should be done on the potential energy surface used for this system. Further experiments should be directed to additional thermal rate data measurements in the high temperature range and measurements of absolute reaction cross sections at different collision energies.

THE EFFECT OF VIBRATIONAL EXCITATION

The exchange reaction between the hydrogen atom and the hydrogen molecule provides the simplest case where for neutral species the effect of vibrational exctation on the kinetic process of bond breaking under the influence of new bond formation can be studied experimentally and theoretically. As shown in Figure 5, single vibrational quantum excitation of the H_2-molecule exceeds the Arrhenius activation energy (E_a), the threshold energy (E_o) as well as the potential energy barrier height (E_c) of the reaction $D + H_2$. More than half a century ago, London[21] was the first who pointed out that the potential energy of the H_3 system can be expressed in terms of three coulombic interaction integrals and three exchange integrals. As shown in Figure 5, this simle valence bond calcula-

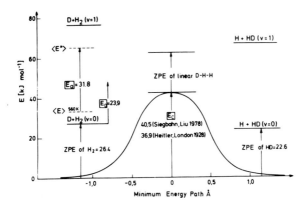

FIGURE 5 Characteristic energies for the $D + H_2$ (v) reaction

tion leads to a potential energy barrier E_c = 36.9 kJ/mol, not far from the result obtained with modern high speed electronic computing devices[22]. However, a more precise treatment of the London method including contributions resulting from everlap and three center terms destroys the good, accidental agreement with modern results. In order of 1500 thoroughlyly selected ab initio points were calculated by Siegbahn and Liu[22] and fitted to an analytical function which represents the potential energy hypersurface mathematically in a smooth and easily handable way by Truhlar and Horowitz[23]. Experiments on the effect of vibrational excitation of the rate of the hydrogen atom exchange reaction were hampered by the difficulties in preparing and measuring known concentrations of vibrationally excited hydrogen molecules. The methods used include thermal generation of H_2(v=1) combined with a hydrogen maser technique[24], energy transfer from laser excited HF(v=1) combined with Lyman-α resonance absorption[25], microwave descharge generation of H_2(v=1) combined with ESR spectroscopy[26], and Lyman-α resonance absorption[27]. Figure 6 shows results of direct measurements of the temperature dependence of the reaction

$$D + H_2(v=1) \longrightarrow HD(v=0,1) + H \qquad (1)$$

using CARS spectroscopy for the detection of H_2(v) - molecules[27]. A distinct temperature dependence is observed in the experiments which compares well with the theoretical calculations performed recently using the SLTH ab initio surface and different methods for treating the dynamics of the chemical reaction.

In calculations using quantised periodic orbits a barrier of 18.3 kJ/mol has been found in the entrance channel of the v=1 adiabatic surface far from the saddle point[28,29] in good agreement with recent exact quantum calculations[30,32,33] and quasi-classical trajectory studies[31]. There is also good agreement of theory and experiment in the case of $D + H_2(v=0)$ and $H + D_2(v=0)$ reactions with

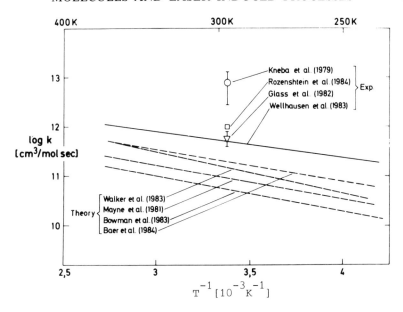

FIGURE 6 Comparison of experimental and calculated rate constants for the reaction $D + H_2(v=1)$

high translational energy of the reactants[34-37]. This indicates that the SLTH surface[23] is correct even at high energies where only a few ab-initio points wre calculated. On the other hand, there are still significant discrepancies for the absolute values of the $H + H_2(v=1)$ rate coefficients between theory and experiments as well as between the different theoretical treatments. The disagreement between theory and experiment for this reaction which should be the best-understood of all chemical processes surely provides a continuing theoretical and experimental challenge.

UV-LASER INDUCED RADICAL-CHAIN REACTIONS

This part descibes application of RGH lasers to investigate the mechanism and kinetics of the photo-initiated chain reaction in 1,2-dichloroethane - (DCE) to form vinyl chloride (VC), and in 1,1,1-chlorodifluoroethane (CDFE) to form vinylidene fluoride (VDF). The reactions are of considerable industrial importance, being the main route to production of vinyl chloride and vinylidene fluoride monomer feedstock for the manufacture of polymers.

Thermal dehydrochlorination of DCE, first observed by H. Biltz in 1902[38], is started by the highly endothermic C-Cl bond rupture in the substrate molecule, followed by the chain consisting of hydrogen abstraction from the substrate by Cl atoms and the unimolecule decomposition of the resulting 1,2-dichloroethyl radicals.

$$CH_2ClCH_2Cl \longrightarrow CH_2ClCH_2 + Cl \quad \Delta H = 334 \text{ kJ/mol} \quad (1)$$

$$Cl + CH_2ClCH_2Cl \longrightarrow CH_2ClCHCl + HCl \quad \Delta H = -24 \text{ kJ/mol} \quad (2)$$

$$CH_2ClCHCl \longrightarrow CH_2=CHCl + Cl \quad \Delta H = 95 \text{ kJ/mol} \quad (3)$$

After the first proposition of this mechanism by Barton and Howlett[39], several authors tried to elucidate the kinetics of the system. Strong effects due to chain promotors, for example O_2 and Cl_2, and chain inhibitors such as propylene and vinyl chloride itself were found. Wall effects turned out to be important for initiation as well as termination in the pyrolytic reaction. Holbrook et al.[40] found that the surface to volume ration and the condition of the reactor surface has a strong influence on the conversion rates. In calorimetric measurements Kapralova and Semenov[41] showed that the reaction is constricted to the region near to the walls.

The most common way used up to now to interpret the experimental reaction order and conversion rates is to describe the system including reactions (1)-(3) and the termination step

$$Cl + C_2H_3Cl_2 \longrightarrow \text{products} \quad (4)$$

by a number of differential equations using the steady state assumption for the radicals Cl and $C_2H_3Cl_2$. This results in an expression for the DCE consumption which is first order in DCE:

$$d[DCE]/dt = \sqrt{k_1 k_2 k_3/k_4} \, [DCE] \quad (5)$$

MOLECULES AND LASER INDUCED PROCESSES   169

The method, first described in the work of Rice and Herzfeld[42], ignores, however, some crucial aspects: the initiation is heterogeneous and can therefore not simply be described by reaction (1). For the termination step one should consider additional channels, e.g. the reverse of reaction (3), the recombination of the dichloroethyl-radicals, and processes involving the surface of reactor. Chlorinated ethanes should show a reactive behaviour in radical chains similar to DCE, provided that there is hydrogen bonded in the position of the chlorine-carrying carbon atom. In contrast to this assumption several experimentators investigating the pyrolysis of these compounds found typical non-radical behaviour of some species when adding inhibitors like propene to chloroethane, 1,1-dichloroethane, and 1,1,1-chlorodifluoroethane. The last is industrially imporatant because it is the precursor of the monomer vinylidene fluoride. The main reason for the observed preference for the four-center elimination compared to the radical chain route could be the formation of δ-radicals like CH_3-CF_2 which are unable to decompose into chain propagation radicals and are therefore acting as recombination partners[43].

The best way for improving the study of the basic chain behaviour is to start the chain reaction homogeneously under wall-free conditions. RGH laser light sources are ideally suited for this purpose. Their defined wavelength and beam geometry together with the short pulse length allow a controlled start of the chain by photolysis. Illuminating a zone in the middle of the reactor ensures that also the termination steps are homogeneous in the observed region.

It is commonly accepted that the primary step in the photolysis of alkyl chlorides in the spectral region of their continuum absorption is the dissociation of the C-Cl bond with a quantum yield near unity[44]. The chemistry of the photochemically induced

170 LASERS IN ATOMIC, MOLECULAR AND NUCLEAR PHYSICS

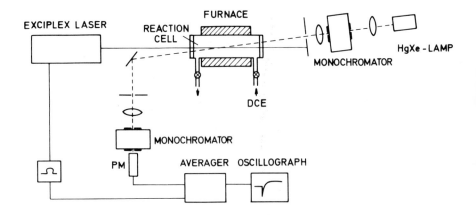

FIGURE 7 Experimental arrangement for the time-resolved study of laser-induced redical chain reactions.

chain reaction should be identical to the thermal reaction. Some recent publications[45-47] on the photolysis of $1,1,1-C_2H_3Cl_3$ and $1,1-C_2H_4Cl_2$ suggest that additional steps such as the α - or β - elimination of HCl or Cl_2 should play a role, but in these measurements secondary reactions were not strictly excluded. In our calculations we took the quantum yield for reaction (1) to be 0.9±0.1. At sufficiently high temperatures the initial chloroethyl radical will decompose in a short time range compared to the total reaction to yield an additional chlorine atom and ethylene.

Figure 7 shows a schematic of the experimental apparatus for the time resolved measurements. The monitor beam passes near axially along the cell so that the space near the windows is excluded from the observation volume. The absorption signals of the products are detected by the photomuliplier, amplified, digitized in a fast transient recorder (Biomation, 8100) and summed by a signal averager (Tracor, NS-575 A) to enhance S/N. Figure 8 demonstrates a typical signal for the time resolved formation of VC after a laser shot.

From the UV-absorption cross sections σ, the laser pulse

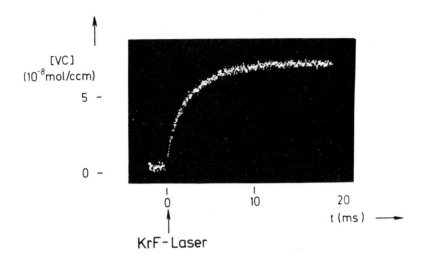

FIGURE 8   UV-absorption signal (T = 570 K, p = 300 Torr, E_L = 188 mJ) showing the laser-induced formation of vinilchloride

E_L, and the amount of product formed, one can calculate the spatially and temporally integrated quantum yield

$$\phi = E_L (\lambda/hc) \, (1 - \exp(-\sigma [DCE]d))/N_{VC} \qquad (7)$$

([DCE] = concentration of the substrate in molecules/cm^3, N_{VC} = Number of product molecules)

In order to study this effect a series of experiments was carried out varying the laser energy from 5 to 300 mJ at fixed photolysis wavelength, temperature, and DCE pressure (λ = 248 nm, p = 300 Torr, T = 570 K). In the double logarithmic plot of quantum yields versus initial radical concentration (Figure 9) one obtains a straight line with a slope of -0.52. The quantum yield decreases inversely proportional to the square root of start radical concentration.

This result, can be explained by the Rice-Herzfeld mechanism keeping in mind the objections stated in the introduction and the fact that it is formulated for continuous radical formatuon. When a lamp with the radiation density I_o is used for photolysis, the

172 LASERS IN ATOMIC, MOLECULAR AND NUCLEAR PHYSICS

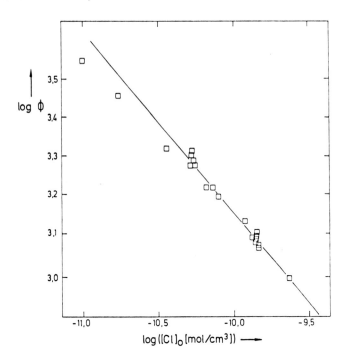

FIGURE 9 Effect of initial radical cincentration of the quantum yield of the laser-induced chain reaction

radical formation at weak absorption may be described by the equation

$$d[Cl]/dt = I_o \cdot \phi_o \cdot \sigma \cdot (\lambda/hc) \cdot [DCE] = k_1' \, DCE \qquad (8)$$

(ϕ = Quantum yield of the primary photolysis step)

With eq. (5) one gets the differential quantum yield

$$\phi = (d[VC]/dt)/(d[Cl]/dt) = \sqrt{k_2 k_3 / k_1' k_4} \qquad (9)$$

which is in agreement with the inverse square root dependence found in the experiments. For the apparent activation energy of ϕ one gets

$$-R(d \ln \phi/dt) = 1/2 \, (E_2 + E_3)/(E_1, + E_5) \qquad (10)$$

One would expect from this relation that the quantum yield activation energy is roughly half of the activation energy of step (3)

MOLECULES AND LASER INDUCED PROCESSES 173

because the contributions of the other steps are negligible. In the next section it will be shown that $E_3 = 83$ kJ/mol.

An additional series of experiments was performed to study the temperature dependence of the quantum yield at a constant conversion rate per laser shot. This has the advantage of working with constant effects like adiabatic cooling or vinyl chloride inhibition. At the highest temperature studied (720 K) and a low laser energy of 1 mJ/cm^2, we obtained quantum yields of more than 10^4.

Two methods were employed to interpret the measured concentration of the coupled differential equations describing the time behaviour of the different species. Starting with a value for the rate determining step (3), whose rate constant has been determined by several authors by indirect methods we found that the calculated chain lengths were always higher than the experimental ones.

Therefore the time resolved experimental signals as well as the quantum yields were simulated by adjusting k_3 and introducing additional inhibition reactions. The addition of the 1,2-dichloroethyl radical to vinyl chloride forming an "inert radical" represents a good explanation for the self-inhibiting effect of the product.

$$C_2H_3Cl_2 + C_2H_3Cl \longrightarrow C_4H_6Cl_3 \qquad (11)$$

Under the experimental conditions, reaction (3) is the rate determining step of the chain, i.e. $k_3 \ll k_2[C_2H_4Cl_2]$ at pressures of some hundred Torrs. After the short laser pulse (30 ns) an equilibrium between the carrier radicals $C_2H_3Cl_2$ and Cl is quickly established ($\tau = 1/(k_2[C_2H_4Cl_2]) < 100$ ns) for which the equation

$$[C_2H_3Cl_2]/[Cl] = k_2[C_2H_4Cl_2]/[Cl] \qquad (12)$$

is valid. After this short period the sum of radical concentrations in nearly equal to the start concentration of Cl atoms produced by the laser pulse

$[C_2H_3Cl_2]_\tau \approx [Cl]_0$ (13)

At the time τ the formation of vinyl chloride can be described by

$(d[VC]/dt)_\tau = k_3[C_2H_3Cl_2] \approx k_3[Cl]_0$ (14)

If the concentration time profiles are simple exponential with a time constant k_f

$[VC] = [VC]_\infty (1 - \exp(-k_f t))$ (15)

$d[VC]/dt = [VC]_\infty \cdot k_f \cdot \exp(-k_f t)$ (16)

and at $t = 0$ $\quad (d[VC]/dt)_{t=0} = [VC]_\infty k_f$ (17)

$k_3 = ([VC]_\infty/[Cl]_0) \cdot k_f = L \cdot k_f$ (18)

The microscopic duration of a chain cycle (determined by the slowest step) is reflected in the macroscopic total reaction time with the chain length L acting as an extension coefficient. The advantage of the analytical method is that it is not necessary to know the termination reactions in detail. Any inhibition reaction shortens the total reaction time but also the chain length so that the value of k_3 remains constant. The algorithm was tested experimentally by a series of runs with fixed temperature and pressure and varying irradiation intensity. The experiments showed that the decrease of the chain length with higher initial radical concentration is paralleled by the increase of reaction time so that the arithmetic product of both remains constant over the whole range of laser energy. Figure 10 compiles data for the change of the unimolecular rate constant k_3 when an inert gas (N_2) was added to 60 Torr of DCE up to a total pressure of 600 Torr. The reaction rate is strongly pressure dependent in this region. In the double logarithmic plot $\log k_3 / \log p$ one obtains a straight line with a slope of 0.5.

The Arrhenius diagram of the measurements at 300 Torr DCE

MOLECULES AND LASER INDUCED PROCESSES

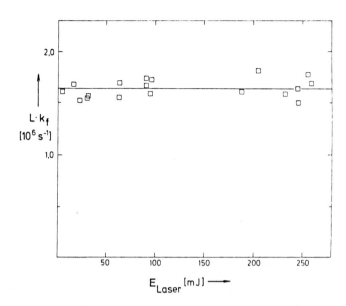

FIGURE 10 Determination of k_3 from overall reaction data

between 520 and 720 K shows a decline of activation energy with temperature. There are two reasons for this effect. First, when moving to higher temperatures at constant pressures, the unimolecular reaction shifts more into the second order region where the activation energy is lower. Secondly, at higher temperatures, the rate of k_3 increases so that this step is no longer strictly rate determining. Therefore the low temperature values are taken to determine the Arrhenius parameters. In the region 520-570 K, one gets the expression

$$k_3 = (6.5 \pm 2) \cdot 10^{13} \cdot \exp(-83 \pm 3 \text{ kJ/mol/RT})$$

Photolysing CDFE with an ArF*-Laser (193 nm) at pressures around 40 Torr we found only small quantum yildes (0.3 at 673 K, 5.4 at 738 K). Two effects might influence the photolysis yield: firstly, the primary photolytic step may be dominated by the molecular elimination of HCl as measured at 147 nm and 123.6 nm by Ichimura et al.[48]. Secondly, the decomposition rate of the intermediate radical

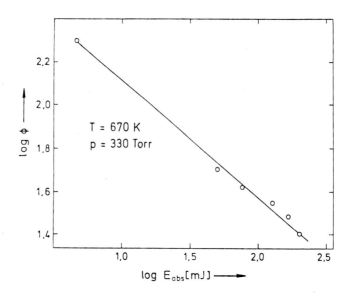

FIGURE 11 Effect of initial radical concentration of Φ in VDF formation

$$CF_2Cl-CH_2 + M \longrightarrow CH_2=CF_2 + Cl + M$$

is assumed to have a higher activation energy relative to the $CH_2Cl-CHCl$ radical.

Additional experiments were carried out using CCl_4 as absorber and changing the laser wavelength to the KrF* line at 248 nm. CCl_4 is known to be a suitable initiator of the radical reaction of $1,1,1-C_2H_3ClF_2$. Figure 11 shows the measured quantum yields for VDF formation as a function of absorbed laser energy which was controlled by the amount of CCl_4 added to the sustrate. At comparable initial radical concentrations the quantum yields are at least an order of magnitude higher than for the ArF* photolysis of the pure compound. A significant increse in the chain length ($\phi > 10^5$) for the laser induced formation of VC is observed using Cl_2 as absorber and changing the laser wavelength to the XeCl line at 308 nm (see Figure 12).

MOLECULES AND LASER INDUCED PROCESSES 177

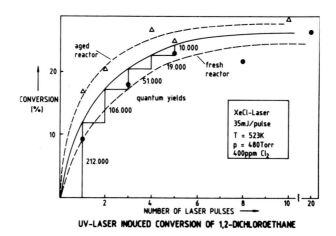

FIGURE 12 Effect of Cl_2 addition on the laser-induced quantum yield of VC-formation

ACKNOWLEDGEMENTS

The financial support by the Deutsche Forschungsgemeinschaft (Sonderforschungsbereich 123, Stochastische, mathematische Modelle), is gratefully acknowledged.

REFERENCES

1. F. J. WEINBERG, Fifteenth Symposium (International) on Combustion (The Combustion Institute, 1975), p. 1.
2. J. WARNATZ, Ber. Bunsenges. Phys. Chem., 87, 1008 (1983)˙
3. K. KLEINERMANNS, J. WOLFRUM, J. Chem. Phys., 80, 1446 (1984)
4. K. KLEINERMANNS, J. WOLFRUM, Appl. Phys., B34, 5 (1984)
5. K. KLEINERMANNS, J. WOLFRUM, Chem. Phys. Lett., 104, 157 (1984)
6. K. KLEINERMANNS, E. LINNEBACH, Appl. Phys. B (to be published, 1984)
7. K. KLEINERMANNS, R. SCHINKE, J. Chem. Phys., 80, 1440 (1984)
8. C. F. MELIUS, R. J. BLINT, Chem. Phys. Lett., 64, 183 (1979)
9. P. PECHUKAS, J. C, LIGHT, C. RANKIN, J. Chem. Phys., 44, 794 (1966)
10. J. A. MILLER, J. Chem. Phys., 74,5120 (1981)
11. G. L. SCHOTT, Combust. Flame, 21, 357 (1973)
12, Th. JUST, G. FRANK, (to be published)

13. D. L. BAULCH, D. D. DRYSDALE, D. G. HORNE, Fourteenth Symposium (International) on Combustion (The Combustion Institute, 1973), p. 107.
14. C. T. BOWMAN, Fifteenth Symposium (International) on Combustion (The Combustion Institute, 1975), p. 869.
15. C. COBOS, H. HIPPLER, J. TROE, J. Chem. Phys., (in press, 1984)
16. J. A. MILLER, N. J. BROWN, J. Phys. Chem., 86, 772 (1982)
17. T. H. DUNNING jr., S. P. WALCH, M. M. GOODGAME, J. Chem. Phys., 74, 3482 (1981)
18. S. N. RAI D. G. TRUHLAR, J. Chem. Phys., 79, 6046 (1983)
19. M. QUACK, J. TROE, J. Chem. Soc. Spec. Period. Report Gas Kinet. Energy Transfer, 2, 175 (1977)
20. N. COHEN, K. R. WESTBERG, Chemical Kinetic Data Sheets for High Temperature Chemical Reactions, Aerospace Report No. ATR 82(7888)-3, El Segundo (1982)
21. F. C. LONDON, Probleme der Modernen Physik (Hirzel, Leipzig, 1928), p. 104.
22. P. SIEGBAHN, B. LIU, J. Chem. Phys., 68, 2457 (1978)
23. D. G. TRUHLAR, C. J. HOROWITZ, J. Chem. Phys., 68, 2460 (1978)
24. E. B. GORDON, B. I. IWANOV, A. P. PERMINOV, V. G. BALALAEV, A. N. PONOMAREV, V. V. FITATOV, Chem. Phys. Lett., 58, 425 (1978); V. B. ROZENSHTEIN, Yu. M. GERSHENZON, A. V. IVANOV, S. I. KUCHERJAVII, Chem. Phys. Lett., 105, 423 (1984)
25. M. KNEBA, U. WELLHAUSEN, J. WOFRUM, Ber. Bunsenges. Phys. Chem., 83, 940 (1979)
26. G. P. GLASS, N. B. CHATURVEDI, J. Chem. Phys., 77, 3478 (1982)
27. U. WELLHAUSEN, Dissertation Universität Göttingen (1984) U. WELLHAUSEN, J. WOLFRUM, Phys. Rev. Lett. (to be publ. 1984)
28. E. POLLAK, Chem. Phys. Lett., 80, 45 (1981)
29. E. POLLAK, R. E. WYATT, J. Chem. Phys., 78, 4464 (1983)
30. R. B. WALKER, E. F. HAYES, J. Phys. Chem., 87, 1255 (1983)
31. H. R. MAYNE, J. P. TOENNIES, J. Chem. Phys., 75, 1794 (1981)
32. J. M. BOWMAN, U. T. LEE, R. B. WALKER, J. Chem. Phys., 79, 3742 (1983)
33. N. ABU SALBI, D. J. KOURI, Y. SHIMA, M. BAER, Chem. Phys. Letters (to be published, 1984)
34. D. D. GERRITY, J. J. VALENTINI, J. Chem. Phys. (to be publ. 1984)
35. C. T. RETTNER, E. E. MARINERO, R. N. ZARE, J. Chem. Phys. (to be publ. 1984)
36. R. GOTTING, H. R. MAYNE, J. P. TOENNIES, J. Chem. Phys. (to be publ. 1984)
37. N. C. BLAIS, D. G. TRUHLAR, Chem. Phys. Lett., 102, 120 (1983)
38. H. BILTZ, Ber., 35, 3524 (1902)
39. D. H. R. BARTON, K. E. HOWLETT, J. Chem. Soc., 148, 155 (1949)

40. K. A. HOLBROOK, R. W. WALKER, W. R. WATSON,
    J. Chem. Soc. (B), 577 (1971)
41. G. A. KAPRALOVA, N. N. SEMENOV, Russ. J. Phys. Chem., 37, 35 and 156 (1963)
42. F. O. RICE, K. F. HERZFELD, J. A. C. S., 56, 284 (1934)
43. G. J. MARTENS, M. GODFROID, R. DECELLE, J. VERBEYST,
    Int. J. Chem. Kin., 4, 645 (1972)
44. J. G. CALVERTS, J. N. PITTS, Photochemistry
    (Wiley, New York, 1967)
45. T. FUJIMOTO, J. H. M. WIJNEN, J. Chem. Phys., 56, 4032 (1972)
46. T. S. YUAN, M. J. J. WIJNEN, Ber. Bunsenges. Phys. Chem., 81, 310 (1977)
47. S. HAUTECLOQUE, J. Photochem., 12, 187 (1980)
48. T. ICHIMURA, A. W. KIRK, E. TSCHUIKOW-ROUX,
    Int. J. Chem. Kin., 9, 696 (1977)
    T. ICHIMURA, A. W. KIRK, E. TSCHUIKOW-ROUX,
    J. Phys. Chem., 81, 2040 (1977)

2.4 Lasers and Reactive Collisions: The Cs*-H$_2$ Reaction

R. VETTER

Laboratoire Aimé Cotton, C.N.R.S. II, Bâtiment 505, 91405 Orsay Cedex, France

The detailed study of the collision between two reacting molecules is the necessary step for the understanding of elementary processes of reaction. Modification, breaking and creation of chemical bonds are determined by a number of parameters, internal energy, kinetic energy, relative orientation of reagents ..., whose control is essential before, during and after the collision. This is the aim of state-to-state chemistry to follow as closely as possible the evolution of these parameters and to evaluate the individual cross-sections.

During the last two decades, the search for an "ideal" experiment where the whole set of parameters could be determined has made considerable progress with the advent of new sophisticated techniques. Among them, the technique of supersonic molecular beams offers unique possibilities, in particular with the good definition of internal and external energies of reagents which can be achieved. Use of two srossed beams of fixed geometry leads to product analysis after a unique collision has taken place, i.e., the analysic is free of complications due to successive encounters; variation of the kinetic energy allows for the determination of activation barriers; angular analysis of the product dictribution yields information about the geometry of the reactive collision.

Although extremely poweful, this technique is generally

restricted to atoms and molecules in their ground state. Furthermore, chemical systems of interest for a useful comparison between theory and experiment are still relative to simple ones: this is due to the fact that the methods of Quantum Chemistry are able to provide accurate potential energy surfaces for systmes implying a small number of atoms only, typically up to three or four. Moreover, the dynamics of collision of these surfaces is still in its infancy since, even for simple systems, one is able to evaluate state-to-state cross-sections for a limited number of circumstances only, a collinear approach of non-rotating reagents for example.

In this context, use of tunable lasers is obviously of great interest for state-to-state chemistry. To schematize, state-selective excitation of reagents (electronic, vibrational, rotational) is easily accomplished and the subsequent effect on reactivity can be observed; second, large amounts of energy can be deposited into the system, hence the possible study of endoergic reactions; third, polarization of laser beams allows for the study of the orientation dependence of the cross-section; fourth, photodissociation products and unstable species (radicals) can be obtained in the zone of collisions by use of intense (pulsed) laser beams. Finally, the technique of laser-induced fluorescence offers two advantages in the detection of nascent products: a quasi-infinite energy resolution and a high sensitivity.

It comes therefore that the combined use of molecular beams and laser beams approaches the conditions required for an "ideal" state-to-state experiment, can multiply the number of experimental situations of interest and lead to the observation of new features. Up to now, there exists only a few experiments of this kind, but numerous data have benn gained already by use of less versatile and less sophisticated experiments. Let us mention briefly several results of major importance (a non-exhaustive list!)[1-7].

The Vibrational Enhancement

Perhaps the most striking effect was observed in the lightly endoergic reaction: HCl + Br $\rightarrow$ HBr + Cl, where the vibrational excitation of HCl molecules with a pulsed HCl laser beam in the infrared yields an increace of the reaction rate by a factor 10^{11} from HCl(v=0) to HCl(v=2)^8. It can be explained simply by considering that the vibrational excitation is used to overcome the activation barrier of the reaction. This experiment was conducted in "bulk" phase but was extended later on to other species in beam + gas arrangements; study of the reactions: HCl + K, HF + Ca, Sr, Ba led to some generalization of the efficiency of vibrational/translational excitation of reagents$^{9-13}$. For instance, theory confirmed by experiment predicts that for an exoergic reaction, vibrational energy is most effective at promoting reaction when the potential energy surface shows a late barrier (mostly an attractive surface) whereas translational energy is most effective in the case of an early barrier. In many cases, the vitrational distribution of products is out of equilibrium.

Similar experiments were performed with a rotational excitation of reagents, but with less spectacular effects. It was shown for example that for endoergic reactions of the type: HX + Na $\longrightarrow$ NaX + H (X = F, Cl) the rate constant shows a minimum for intermediate values of the rotational quantum number^{14}.

Electronic Excitation of Reagents

Electronic exctation by tunable lasers (pulsed or C.W.) is widely used in bulk experiments for a number of simple or heavy molecules^4; it is not the case in beam arrangements although, compared to others (electronic bombardment, photodissociation), this technique offers the advantage of selectivity and simplicity. Let us mention the following cases: $I_2^* + F_2$ [15] ; $Mg^* + H_2$, D_2 [16,17] ; $Ca^* + HCl$, Cl_2,

$CCl_4{}^{18-20}$ and more recently $Cs^* + H_2{}^{21}$.

Perhaps the most appealing example is the one of $Ca^* + HCl \longrightarrow CCl + H$, for which laser excitation of Ca atoms in a beam + gas arrangement directly demonstrates the influence of reagent's orientation. During the reaction, CaCl molecules form preferentially in (A $^2\Pi$) or (B $^2\Sigma^+$) states, depending on the approach direction of Ca p orbitals which is determined by the laser beam polarization: a parallel approach with respect to HCl favors the formation of CaCl in the B state. Other effects are observed with Cl_2. Interpretation suggests that in an electron-jump model, the symmetry of reagents must be conserved.

$Cs^* + H_2 \longrightarrow CsH + H$ is a case where laser excitation of one species is used to compensate the large endoergiticity of the reaction; first observed in a bulk experiment[22], the reaction is now studied in a crossed-beam arrangement, with laser excitation of Cs atoms and laser detection of CsH products. Study of this reaction is the main subject of this Lecture.

The $Cs^* - H_2$ Reaction

The reaction of Cs atoms with molecular hydrogen in their ground state is known to be highly endoergic: $\Delta H = 2.7$ eV, the difference between the energy of dissociation of H_2 and that of CsH. The spectrum of CsH could be studied years ago, production of CsH molecules being ensured by heating Cs and H_2 in high-pressure cells[23,24]. In 1975, however, W. Happer and coll. discoverd that by sending an Ar^+ laser beam at $\lambda = 4579$ Å or $\lambda = 4545$ Å in such cell, quantities of particulates formed in the region of the beam ("laser snow"[22]). This phenomenon was interpreted by initial laser-induced formation of CsH molecules since, Cs atoms being excited to one of the two levels $7P_{1/2,3/2}$, the endoergiticity of the reaction is just compensated (Figure 1); crystallisation of CsH molecules can occur further on. Other experimants conducted in gas cells with tunable

MOLECULES AND LASER INDUCED PROCESSES 185

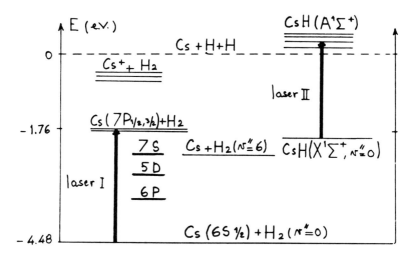

FIGURE 1 Potential energy diagram of the system

lasers confirmed this interpretation[25-27].

These experiments and energy considerations also put into evidence one other aspect of the problem. When Cs atoms are excited to the 7p state, the amount of energy which is available above threshold is drastically small: 0.0016 eV and 0.024 eV for $7P_{1/2}$ and $7P_{3/2}$ respectively. Furthermore, the initial system ($Cs^* + H_2$) is at least on the 11th potential energy surface from the ground state, implying that many surface crossings must necessarily occur during the reaction. Under these conditions, is a unique collision able to promote the direct reaction:

$$Cs^* + H_2 \longrightarrow CsH + H$$

with simultaneously electronic deexcitation and bond breaking, or is it necessary to invoke more complicated processes whose energetics is more favorable? For instance, one could think to the following two-collision processes[28]:

$$\begin{cases} Cs^* + H_2 \longrightarrow Cs + H_2(\dagger) \\ Cs + H_2(\dagger) \longrightarrow CsH + H \end{cases}$$

and

$$Cs^* + H_2 \longrightarrow [CsH_2]^{\ddagger *}$$
$$[CsH_2]^{\ddagger *} + Cs \longrightarrow 2CsH$$

the first one with transfer of electronic excitation in Cs to vibrational energy of H_2, the second with formation of an intermediate excited complex.

The experiments mentioned previously are not able to yield definite answers to this problem since, in gas cells, multiple encounters can occur between atoms and molecules; furthermore the relative kinetic energy is not perfectly determined for each collision. One could attribute the formation of CsH observed in 30's, to multiple collisions between particles of high energy which exist in the tail of the velocity distribution.

Quantum Chemistry on the other hand yilds interesting preliminary predictions in the case of $(Cs + H_2)$[28]. Calculations performed with ab initio methods including relativistic pseudo-potentials, large basis sets, accurate configuration interaction show that the adiabatic ground state potential surface does not present any saddle- point between the entrance valley and the exit plateau, in the case of a collinear approach of the reagents. Therefore Cs* and H_2 could in principle react during the course of a unique collision, but with a small cross-section.

A positive (or negative) answer can come from a crossed-beam experiment: in this case, one is able to observe the effect of a unique collision of well-defined geometry and energy. The facile interfacing with laser beams allows for Cs excitation and CsH detection by use of the technique of laser-induced fluorescence.

Experimental

Figure 2 shows schematically the experiment which has been realized at Laboratoire Aimé Cotton, Orsay. A supersonic beam of molecu-

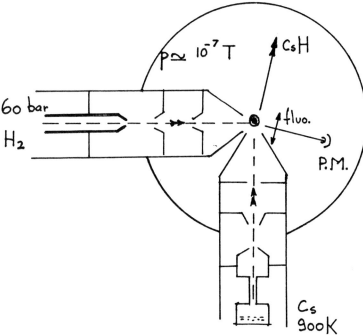

FIGURE 2

lar hydrogen and a supersonic beam of atomic cesium enter a collision chamber and cross at right angle. Perpendicularly to the plane of collision, a first laser beam excites Cs atoms at the crossing and a second laser beam excites CsH molecules to the $(A^1\Sigma^+)$ state from which they fluoresce. The fluorescence light is collected and sent to a photomultiplier through a series of filters; the P.M. singal is analyzed and stored.

The supersonic beam of hydrogen has been built according to technique developed by R. Campargue[29,30]. A high pressure of H_2 (60 bars typically) is established in a first chamber separated from a second one by a nozzle through which the molecule expand. In the second chamber, which is maintained at low pressure by high-velocity pumping, a region of hydrodynamical regime is created, separted from the molecular regime by a shock wave; in the hydrodynamical region, molecules propagating on the axis suffer many collisions and acquire large velocities to the expense of rotational and vibrational energies. A skimmer allows for these molecules to enter a third high-vacuum chamber without perturbation; then they enter the collision chamber through a collimator. This beam offers interesting characteristics since molecules are rotationally and vibrationally cooled (typically up to 1 K and 10 K respectively) and their velocity distribution is narrowed (typically $\Delta V/V \sim 0.1$). Furthermore high dencities of particles are obtained (typically 10^{13} mulecules/cm^3 at the crossing point).

The beam of cesium is obtained by evaporation of the metal held in an alkali-resistant oven heated at 900 K; Cs atoms effuse through a nozzle maintained at 1000 K and two succesive diaphragms which define the geometry of the beam.These diaphragms are cooled at 300 K to avoid secondary emission. Migration of liquid metal and trapping of atoms not directly used in the beam permit the use of moderate quantities of cesium. Typical densities of 10^9 atoms/cm^3 at the crossing point are obtained. By heating more the oven and the nozzle, one should obtain a supersonic regime with the subsequent reduction of the velocity distribution and a larger density of particles.

The two laser beams are provided by C.W. tunable dye lasers; they are mixed on a beam-splitter and they enter the collision chmamber through a Brewster-angle window. Light baffles reduce the stray light inside the collision chamber. The two beams are concentrated at the crossing point so that their waist is adapted to the dimensions of the particle beams, 2 mm - diameter typiclly. The first beam is frequency-locked on the most intense hyperfine component of one resonance line of Cs, at $\lambda = 4555$ Å or $\lambda = 4593$ Å, by means of a servo-controlled system which uses the blue fluorescence signal due to Cs atoms excited at the crossing point. The second beam is frequency-locked on the center of a Doppler-broad-

MOLECULES AND LASER INDUCED PROCESSES 189

ened line of CsH molecules produced in a sealed-off cell outside the collision chamber, by means of the fluorescence signal at this wavelength. By changing CsH transitions in the (A $^1\Sigma^+$ - A $^1\Sigma^+$) system, it is possible to monitor the population distribution of products on each rovibrational level of the X state.

A small fraction (~4%) of the total fluorescence due to CsH molecules produced at the crossing point is collected by an optical system and detected by a low-noise photomultiplier through a series of filters which eliminate stray light and fluorescence light due to excited Cs atoms. Under these conditions the level of the background is typically of 3 photons/second; as one expects very weak signals photon-counting techniques and data storage must be used.

First Results

In search of a signal due to CsH molecules, it has been necessary first to determine accurately the frequency of the relevant (A - X) transitions of the molecule. This has been performed by use of laser-induced fluorescence techniques, by recording simultaneously the spectrum of CsH provided by the sealed-off cell and the spectrum of an I_2 cell[31]. In this manner, 40 lines have been identified, involving the v"=0 and v"=1 levels of the X $^1\Sigma^+$ state[32]; according to energy considerations, the CsH molecules created in the crossed-beam experiment must involve these two levels only (see Frigure 1 and Figure 3).

A weak signal has been detected for several values of the rotational quantum number ($1 \leqslant J" \leqslant 13$) in the v"=0 level[21]. This signal is characteristic of CsH molecules produced at the crossing point of the collision chamber since it vanishes:
i) in the absence of hydrogen;
ii) without excitation of Cs atoms;
iii) by slightly detuning the second laser from line center;

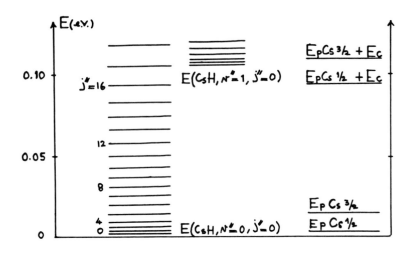

FIGURE 3   Energy diagram of the system. Left: rotational levels of CsH (X $^1\Sigma^+$, v"=0). Center: rotational levels of CsH (X $^1\Sigma^+$, v"=1). Bottom right: potential energy of Cs($7P_{1/2,3/2}$) + H_2). Top right: potential energy + kinetic energy.

for instance, no signal could be detected for detunings as small as $2\cdot 10^{-3}$ cm^{-1}. This indicates that the spectral width of the signal is of the same order, or less, in accordance with the residual Doppler width which is expected in these beam experiments. This test is probably the most sensitive one.

No signal could be detected for transitions involving the v"=1 levels. The surprising result of these measurements however, is in the fact that for a $7P_{1/2}$ excitation of Cs atoms, the signal is larger than for a $7P_{3/2}$ one. Figure 4 shows an example of the data which has been obtained on the (v'=5, J'=7 ← v"=0, J"=6) transition. The signal is multiplied by a factor 6 by simply changing the wavelength of the first laser from λ = 4555 Å to λ = 4593 Å. According to the ratio of excited populations of cesium (n $7P_{3/2}$/ n $7P_{1/2}$ ∼ 1.7) which can be deduced from the measurement of atomic fluorescence, this means that the reactive cross-section is roughly 10 times larger for a $7P_{1/2}$ excitation of Cs atoms.

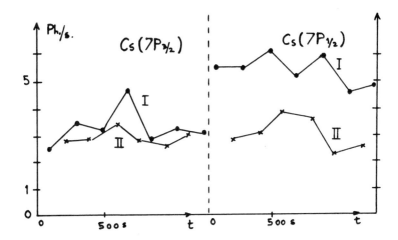

FIGURE 4 Histogram of the signal recorded on the transition
A(v'=5,J'=7) ← X(v"=0,J"=6) in a series of trials
performed under same experimental sonditions except
Cs excitation. Left: Cs($7P_{3/2}$) excitation: curve I
with H_2, curve II without H_2. Integration time: 100 s.
Second laser output power: 170 mW.

A theoretical interpretation of this result has been suggested by J. P. Malrieu by consideration of the relevant potential energy surfaces of the system[21]. Calculations show that a crossing exists at low energy between the (Cs(7P) + H_2) neutral surface and the (Cs$^+$ + H_2^-) ionic one, and that the ground state of CsH is strongly polar, i.e., (Cs$^+$ + H$^-$). Therefore, one can imagine that the system enters the (Cs(7P) + H_2) valley, jumps on the (Cs$^+$ + H_2^-) surface and remains diabatically on the (Cs$^+$ + H$^-$ + H) ionic one when the distance between H atoms increases, despite avoided crossings with the lower neutral surfaces related to Cs(7s),(5d),(6p), and (6s). In this manner, the suggested mechanism of reaction should be a type of harpooning leading to a molecular "zwitterion" and one atom. The interesting feature is that Cs($7P_{1/2}$) with H_2 generates $^2\Sigma^+$ states which have the same symmetry as that of the ionic states, whereas Cs($7P_{3/2}$) generates $^2\Pi$ states which are not relevant for the jump onto the ionic surface, in a collinear approach of the reagents. Similar considerations were already suggested to explain this "fine structure" effect[33,34].

Other data can be extracted from these measurements. First, observation of a signal for J" values up to 13 in the v"=0 level (see Figure 3) indicates that the relative kinetic energy is necessarily transformed into rotational energy during the course of the collision since the only potential energy is not sufficient. For a system at threshold like this one, it is mostly interesting to determine the rotational distribution of products, as can be done by laser-induced fluorescence techniques. Second, by evaluating the density of excited Cs atoms, the efficiency of collection of fluorescence of CsH molecules and their probability of excitation , it is possible to get the value of the reactive cross-section, here 10^{16} cm^2 for a $7P_{1/2}$ excitation of Cs atoms. Due to present experimental uncertainties, this value must be considered as an order of magnitude only, but the same experimental data indicate that a two-collision process would require cross-sections of the order of 10^{-13} cm^2 for each collision to lead to the same signal. Consequently a two-collision process is rather unlikely to occur in this experiment.

Possible Developments

As explained above, determination of the rotational distribution of products is of major importance in such sytem at threshold. However, there are other aspects of the laser excitation of Cs atoms which could be interesting; for instance, excitation to Rydberg levels should lead to different values of the reactive cross-section and to different rotational and vibrational distributions of products. In this case , the optical electron of the atom is less and less bound to the nucleus and the harpooning mechanism which is suggested to occur in the reaction, should become more and more

MOLECULES AND LASER INDUCED PROCESSES 193

efficient. Up to now, there is no systematic result concerning this question, for endoergic and for exoergic reactions.

Second, comparison berween the efficiency of photon excitation versus kinetic energy is made possible by use of the technique of "seeded" beams; addition of a light carrier gas in the beam of cesium should allow, at least partially, for compensation of the endoergiticity of reaction. Also, vibrational excitation of hydrogen could be possible by electronic bombardment; unfortunately, this technique is not yet selective in energy.

Third, the technique of Doppler tuning developed for atom-atom scatering could yield the angular distribution of products. In this technique, the laser beam which induces the fluorescence of CsH molecules is sent into the collision chamber, in the direction of relative velocities at the center of mass of the system. Then, there is a unique relation between the angle of deflection of CsH molecules and the detunign of the laser with respect to line-center: only respond those molecules whose velocity projection on the laser beam obeys the Doppler relation. However, this technique suffers from an important loss in sensitivity due to the angular resolution it-self.

To conclude, the crossed-beam experiment with laser excitation of reagents and laser detection of products which has benn descibed here, appears to have versatile applications in many aspects of reactive collisions. It should be emphasized once more that the necessary need for interpretation implies Quantum Chemistry calculations which are still limited to simple situations. The author is pleased to acknowledge with thanks his colleagues of Laboratoire Aimé Cotton, Orsay, and Laboratoire de Physique Quantique, Toulouse, for participation to the experiment and interpretation of the data.

REFERENCES

1. R. B. BERNSTEIN, R. D. LEVINE, Molecular Reaction Dynamics (Clarendon Press, Oxford, 1974)
2. Atom-Molecule Collision Theory: a guide for the experimentalist, edited by R. B. Bernstein (Plenum-Press, New York, 1979)
3. R. B. BERNSTEIN, Advances in Atomic and Molecular Physics, 15, 167 (1979)
4. Laser Induced Chemical Processes, edited by J. I. Steinfeld (Plenum-Press, New York, 1981)
5. Photo-Selective Chemistry, edited by J. Jorther, R. D. Lev ne, S. A. Rice (J. Wiley & sons, New York, 1981)
6. Advances in Laser Chemistry, edited by A. H. Zewail (Springer-Verlag, Berlin, 1978)
7. Lasers and Chemical Change, edited by A. Ben-Shaul, Y. Haas, K. L. Kompa, R. D. Levine (Springer-Verlag, Berlin, 1981)
8. D. ARNOLDI, K. KAUFMANN, J. WOLFRUM, Phys. Rev. Lett., 34, 1597 (1975)
9. T. J. ODIORNE, P. R. BROOKS, J. V. KASPER, J. Chem. Phys., 55, 1980 (1971)
10. J. G. PRUETT, F. R. GRABINER, P. R. BROOKS, J. Chem. Phys., 60, 3335 (1974)
11. Z. KARNY, R. N. ZARE, J. Chem. Phys., 68, 3360 (1978)
12. A. GUPTA, D. S. PERRY, R. N. ZARE, J. Chem. Phys., 72, 6250 (1980)
13. J. G. PRUETT, R. N. ZARE, J. Chem. Phys., 64, 1774 (1976)
14. B. A. BLACKWELL, J. C. POLANYI, J. J. SLOAN, Chem. Phys., 30, 299 (1978)
15. F. ENGELKE, J. C. WHITEHEAD, R. N. ZARE, Faraday Discuss. Chem. Soc., 62, 222 (1977)
16. W. H. BRECKENRIDGE, H. UMEMOTO, J. Chem. Phys., 75, 4153 (1981)
17. W. H. BRECKENRIDGE, H. UMEMOTO, J. Chem. Phys., 80, 4168 (1984)
18. C. T. RETTNER, R. N. ZARE, J. Chem. Phys., 75, 3636 (1981)
19. C. T. RETTNER, R. N. ZARE, J. Chem. Phys., 77, 2416 (1982)
20. H. J. YUH, P. J. DAGDIGIAN, J. Chem. Phys., 79, 2086 (1983)
21. C. CREPIN, J. L. PICQUE, G. RAHMAT, J. VERGES, R. VETTER, R. X. GADEA, M. PELISSIER, F. SPIEGELMANN, J. P. MALRIEU, Chem. Phys, Lett., to be published (1984)
22. A. TAM, G. MOE, W. HAPPER, Phys. Rev. Lett., 35, 1630 (1975)
23. G. M. ALMY, M. RASSWEILER, Phys. Rev., 53. 890 (1938)
24. I. R. BARTKY, J. Mol. Spectrosc., 21, 25 (1966)
25. J. L. PICQUE, J. VERGES, R. VETTER, J. de Physique Lettre, 41, 305 (1980)
26. B. SAYER, M. FERRAY, J. LOZINGOT, J. BERLANDE, J. Chem. Phys., 75, 3894 (1981)

27. J. P. VISTICOT, M. FERRAY. J. LOZINGOT, B. SAYER,
    J. Chem. Phys., 79, 2893 (1983)
28. F. X. GADEA, G. H. JEUNG, M. PELISSIER, J. P. MALRIEU,
    J. L. PICQUE, G. RAHMAT, J. VERGES, R. VETTER,
    Laser Chemistry, 2, 361 (1983)
29. R. CAMPARGUE, Thèse de Doctorat d'Etat, Paris (1970)
30. R. CAMPARGUE, A. LEBEHOT, J. C. LEMONNIER, Rarefied Gas Dynamics, Progress in Astronautics and Aeronautics 51,1033, (1976)
31. S. GERSTENKORN, P. LUC, Atlas du Spectre d'Absorption de la Molécule d'Iode, Editions du C.N.R.S. (Paris, 1978)
32. C.CREPIN, Thèse de 3ème cycle (Paris-Orsay, 1984)
33. K. BERGMANN, S. R. LEONE, C. B. MOORE, J. Chem. Phys., 63, 4161 (1975)
34. S. HAYASHI, T. M. MAYER, R. B. BERNSTEIN, Chem. Phys. Lett., 53, 419 (1978)

3. LASER STUDY OF CONDENSED MATTER

3.1 Space-Time Holography of Picosecond Pulsed Fields in Highly-Selective Photochromic Media

R. KAARLI, J. KIKAS, A. REBANE and P. SAARI

Institute of Physics of the Estonian SSR Acad. Sci., Riia 142, Tartu 202400, USSR

1. INTRODUCTION

The recording and reproducing of spatial and temporal properties of rapidly-changing optical signals and fields of picosecond duration are a currently central scientific and technological problem in contemporary optics, the solution of which may determine the ways of developing the future ultrahighspeed performance computer technique. The key problem is how to record and then retrieve ultrashort light pulses.

In a series of investigations[1-5], it was shown that by means of spectrally highly selective photosensitive materials, i.e. the media able to memorize the spectral composition of the absorbed light, it is possible to store and reconstruct the time behaviour of optical pulses on a hololgraphic principle.

The experiments performed on photoreactive impurity molecules in frozen matrices, where picosecond light pulses were recorded and highly efficiently (~50%) restored, show how promising such materials are in practical applications.

In the present talk (1) the main idea of holographing the time dependences of the optical signal is explained: it is based on the phenomenon called photochemically accumulated stimulated photon echo; (2) the process of holographing time-development of motion pictures is analysed, and (3) some practical applications are discussed.

2. HOLOGRAPHY OF TIME DEPENDENCES OF THE OPTICAL FIELD IN SPECTRALLY SELECTIVE PHOTOSENSITIVE MATERIALS

In this part the following questions will be answered: (1) How to obtain a spectrally highly selective photosensitive medium; (2) what is the photochemically accumulated stimulated photon echo (PASPE), and (3) how is it possible to perform the holography of optical field time-dependences on the basis of PASPE.

To explain the essence of our method it should be noted that impurity molecules in low-temperature matrices have extremely narrow (10^{-3}-10^{-4} cm^{-1}) lines of purely electronic transition (s.c. optical analogue of the Mössbauer line)[6]. It is also known[7-9] that if such systems undergo photoinduced transformations upon resonance monochromatic excitation, the inhomogeneously broadened spectrum may develop narrow stable gaps whose widths are limited by the homogeneous width of zero-phonon lines (hole burning effect[7-9]) and shich may persist for a number of days, but also for months or even years. Thus, based on photochemical hole burning, it is possible to create a spectrally highly selective photosensitive midium.

If a train of picosecond pulses is introduced into such medium, in the absorption spectrum of the medium a narrow-band transmission grating with the "fringe spacing" $\Delta \nu = 1/\tau$ (τ - interval between the pulses) can be created, i.e. a spectral hologram[1] in the form of the Fourier transform of the temporal shape of the train can be made.

As regards the treatment of the recording, it is important to

---

[1] The calling of such spectral grating as hologram and the described method as a holographic one will be justified below.

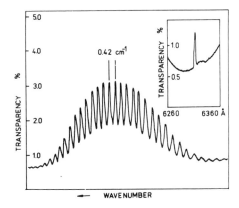

FIGURE 1 The spectral grating (hologram) obtained in sample I after burning by a train of picosecond pulses with 80 ps repetition rate. The envelope of the modulated hole corresponds to a single pulse spectrum; the fine structure is determined by the pulse interval in the train. On the right a general wiew of the hole is depicted, which is recorded with a lower spectral resolution.

keep in mind also that the width of the ingomogeneous spectrum must exceed that of the picosecond pulse spectrum and that the phase relaxation time T_2 of the operating transition must exceed the duration of the train. Transmitting a single attenuated picosecond pulse through such a hologram like linear spectral filter leads to the transformation of the power spectrum of the reading pulse, that in its turn should lead to the appearance of repetitive pulses (interval 80 ps) in the response. The experiment[2] shows that in the response of a single attenuated probe pulse, transmitted through the transparent with a spectral grating (Figure 1), there really appear complementary emission pulses with a repetition rate corresponding to that in the burning train (Figure 2)[2,3].

It is noteworthy that the relative intensity of "echo" pulses is high - 4% of that of the passed-through probing or reading

---

[2] The experimental set-up, data on samples and methods are presented in the appendix.

200 LASERS IN ATOMIC, MOLECULAR AND NUCLEAR PHYSICS

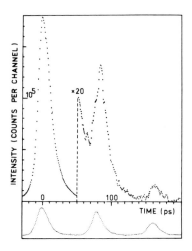

FIGURE 2 Time response of sample I after burning the spectral grating (Figure 1) by a pulse train shown below. First from the left - the probing pulse. For recording a synchroscan streak camera system was used.

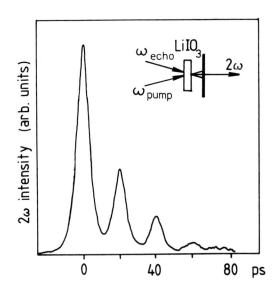

FIGURE 3 Time response of sample II after burning the spectral grating by a pulse train (20 ps intervals) detected by up-conversion. The integral intensity of the restored train is approximately equal to that of passed-through probing pulse.

LASER STUDY OF CONDENSED MATTER 201

pulse. Moreover, after selecting the spectrally selective photochromic medium (sample II) one managed to restore picosecond pulse trains with ~50% efficiency[4] and to observe up to six "echo" pulses (Figure 3).

To explain the nature of this phenomenon and its relation to what has been known earlier, the following must be noted. First, the fine spectral structure of the hole can be regarded as a result of interference: the coherent excitation of molecules, created by the initial reference pulse, interacts with the following ("object" or "code") part of the illumination. Analogously to common (space-domain) holography we store amplitude as well as phase relations between the spectral components of the code and reference pulse. This explains also why the interference-created spectral grating is called a hologram and the proposed method, a holographic one. Second, physically, the complementary pulses in the hologram's time response are a spontaneous coherent emission (free decay singal) of an ensemble of coherently excited dipoles with a specially prepared inhomogeneous distributuion of transition frequencies. It follows that (a) the duration of the recorded and retrieved signal is confined by the phase relaxation time 10^{-9}-10^{-7} s; (b) the phenomenon may be interpreted as a new modification of (stimulated) photon echo or photochemically accumulated stimulated photon echo (PASPE) peculiar to highly-selective photochromic media.

Differences between the phenomena originate from the differences in depopulation mechanisms of the excited state and characteristic restoration times of the absorbers' ground state population (for stimulated echo the excited state lifetime T_1 of molecules is topical for PASPE, the life time of photoproducts $T_p \geq$ 10 hours or maybe years). From here a number of PASPE properties useful in practical applications can be inferred:[3] (1) it is easy to accumulate a high-contrast interference pattern in the medium even in the case of weak pulses by a multiple ($\geq 10^{10}$ times in this experiment) repeating of the recording cycle during a prolonged period; (2) the processes of recording and recalling the run of picosecond events are distinguishable: the formation of PASPE is reduced to a linear filtration of the reading pulse whose intensity application time and even repetition rate do not essentially affect the relative intensity of the response pulses. High efficiency is another advantage of PASPE (see, e.g. Figures 2 and 3) A determining factor here is the modulation depth of the spectral hologram - grating, which in its turn is limited by the homogeneous linewidth ($\sim 1/T_2$) and by saturation in the process of burning.

Assume that the signal (object) pulse is of a complex structure and the reference pulse is shorter than any of the shortest features of signal pulse, then, on recording, owing to interference between the excitation pulses, a hole of a complex form appears in the absorption band, which is determined by the mutual spectra of reference and code pulse. It is the interference with sufficiently short pulse (see above), i.e., whose spectrum in the region of code pulse spectrum can be considered uniform, that allows the phases of the harmonic components of the latter to be fixed.

On transmitting a short reading pulse through such spectral hologram as a linear filter, its response reveals a pulse with a form and structure coinciding with that of the initial signal pulse, i.e. analogously to the previous simple case of a train a pulse of a complex form is retrieved.

---

3) The correlation of stimulated-echo shapes with those of excitation pulses was first pointed out in[10,11], applicabilities of "ordinary" photon echo in holography is discussed in[12-15].

LASER STUDY OF CONDENSED MATTER 203

3. TIME-SPACE HOLOGRAPHY

In the previous chapter the recording and retrieval of only the time dependence of pulses with a plane wavefront was regarded. Now we shall consider the holography of both space and time domain properties of pulsed light fields. In the course of this observation we shall try to answer the following questions:

(1) What requirements are set to the recording process?

(2) What are the peculiarities the spectral hologram reveals on recording the pulses with complex wavefront?

(3) What are the distinguishing features of the retrieval process in comparison with the common space domain holography?

(4) How the changes of the refraction index (phase grating) of the spectral hologram act?

Suppose that on recording the short reference pulse with a plane pulse front and the pulse scattered from the object scene (object pulse that is of a complex spatial and temporal structure) fall onto the spectrally selective photochromic medium at some angle (Figure 4). Such scheme (s.c. off-axis scheme) is well known in ordinary holography and it allows one to avoid a spatial overlapping of the undiffracted wave with restored images. Now, if the reference pulse is delayed with respect ot the object one, then in each medium volume element, besides the spectrum of each pulse, their mutual spectrum is traced in the form of a spectral absorption grating whose fringe spacing is determined by this delay[5]. In different sites of the hologram, in off-exis scheme, the delay between excitation pulses varies and therefore the carrier frequency of the spectral grating also changes from site to site.

It should be noted that in the spectral hologram the spatial modulation of medium transmittance is observed only in the case of narrow monochromatic radiation components. As the period of this

204 LASERS IN ATOMIC, MOLECULAR AND NUCLEAR PHYSICS

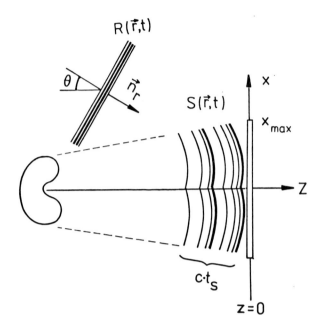

FIGURE 4 The recording scheme of a space-time hologram $R(\vec{r},t)$ is the reference pulse; $S(\vec{r},t)$ the object pulse. The photosensitive medium with the dimension $2x_{max}$ lies in the plane $z = 0$.

spatial modulation is determined by the frequency of the component, the interfringe spacing varies for different components, that results in the smearing of the spatial light-induced grating for white light.

However, if the object pulse will be overlapped by the reference one in time, in the region of pulse overlapping an ordinary interference pattern can be observed and as a result a spatial modulation of hologram in white light transmittance appears.

Proceeding to the process of restoring, one should note that in order to avoid the loss of information the reading pulse should be short in the same sense as the reference one, or the latter should be used for reading.

On restoring by a pulse with a flat wavefront towards the

LASER STUDY OF CONDENSED MATTER

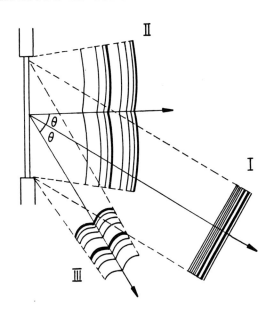

FIGURE 5 The pulses forming behind the hologram in the course of restoring process. I - the undiffracted part of the restoring pulse in $\vec{K}_R$ direction; II - the pulse in $\vec{K}_S$ direction, which forms the virtual image run; III - the pulse in $2\vec{K}_R - \vec{K}_S$ direction, which forms the real image of the events.

reference pulse ($\vec{K}_3 = \vec{K}_R$) there are two possible directions for the image-forming hologram-diffracted waves (see Figure 5): in the direction of $\vec{K}_S$ appears the virtual image of object events and in the direction of $2\vec{K}_R - \vec{K}_S$, the real one.

The most interesting feature of the spectral hologram is its ability to distinguish on restoring the "future" and the "past" in the word-for-word meaning: the virtual image of the events taking place in the object scene after the application of the reference pulse, tends to one direction, while the real image of the object events before the reference pulse, to anoter (Figure 5). In other words, depending on the sequence of the arrival of signal and reference pulses, on restoring on, one of the images is imposed prohibition.

206 LASERS IN ATOMIC, MOLECULAR AND NUCLEAR PHYSICS

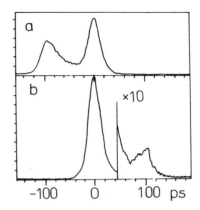

FIGURE 6 Recording and restoring the asymmetric signal pulse in the case when it is first sent to the medium.
a - the asymmetric signal pulse and the delayed reference pulse; b - the restoring pulse and the restored time-reversed pulse; in Figure 5 to the latter corresponds the 3rd beam.

As became clear in the previous part already, the spectral hologram is able to memorize also the arrival time of the object pulse. This means that the replicas of the object scene scattered over such hologram are delayed with respect to the restoring pulse in accordance with the delay on recording.

Besides, the real image is formed by a wave whose front is conjugated with respect to the object wave and which has an inverse time behaviour (see Figure 6). The on-and-off switching of the directions of hologram-scattered waves or the prohibition of one of the images is conditioned by the causal behaviour of the volume element polarization, that is always ensured by the dispersion relations between the absorption coefficient and the refraction index of the medium.

If one disregards the spectral phase grating, which inevitably exists together with the amplitude grating, the restoration of both the virtual and real images is allowed, that contradicts the causality principle. More exactly, the account of the refrac-

LASER STUDY OF CONDENSED MATTER 207

tion index changes in the spectral hologram forbids the restoration of the image which, according to the pulse sequence, should be scattered until the arrival of the reading pulse. This is what leads to the violation of causality in the previous case.

4. SOME APPLICATIONS OF SPACE-TIME HOLOGRAPHY

High efficiency, linearity of recording and reading, and especially the longevity of storing allows the PASPE-based space-time holography to be considered promising for applicational purposes. The results of this work leave no doubt about the feasibility of the holography of space-time events of picosecond and sub-picosecond scale holographyc movie of ultrahighspeed processes. Besides, the PASPE phenomenon itself and the holographic methods based on it have a number of applications. Some of them are considered below.

1. Phase Relaxation Time Measurements

The interference of the reference and object pulse in a selective photosensitive medium are limited by the phase relaxation time T_2 of the excited state of the impurity. The longer the interval between the pulses is, the larger is the number of the excited molecules which have forgotten the phase of the first excitation pulse before the arrival of the second one. This leads to the decrease of the modulation depth of the spectral hologram, that in its turn leads to the decrease of the relative intensity of the echo pulse according to the law $\exp(-4t/T_2)^4$. Thus the relaxation time T_2 of the impurity molecule can be found from the dependence of the relative intensity of the echo signal on the delay t.

2. Parallel Detection of Photochemical Holes

The employment of photochemical burning of persistent holes in the impurity absorption band of low-temperature solutions allows the density of binary information stored in optical memories to be increased at the expense of a spectral coordinate by several orders[18,19]. In principle the number of the holes that can be formed in the inhomogeneously-broadened impurity absorption band by photochemical burning is limited by the ratio of the widths of the ingomogeneous absorption band and the homogeneous nophonon line. For a purely electronic line this ratio may be as much as 10^4 - 10^5, that leads to a limit information packing of 10^{12}-10^{13} bit/cm^2. Unfortunately, a successive recording of holes by monochromatic excitation is a time-consuming procedure. The process of hole burning and detection can be essentially accelerated by using the PASPE phenomenon. This allows one to perform the parallel recording as well as reading of photochemical holes. Thus, by means of PASPE it has been shown[20] that it is possible to burn 1600 holes in the absorption band of object II whereas such amount of holes does not engender any distortion in their form or contrast. In conslusion, this approach allows a parallel information recording and reading over all frequency components of the memorizing medium and the use of ultrahighspeed light pulse modulation for information coding.

3. Synthesis of Light Pulses with a Given Arbitrary Form and Wave Front

By using CW tunable single-frequency lasers it is possible, in principle, to synthesize and fabricate such spectral amplitude-phase holograms which on transmitting a reading pulse allow one to obrain pulses of any form and wavefront. In other words, we can form light pulses of any spatial and temporal properties. For that it is necessary to create the spectral grating-hologram in each volume element (of λ-size magnitude) of the selectively photosensitive medium by means of the laser so that on restoration the scattering from the whole area of the hologram should result in the formation of a pulsed light field with the desired spatial and

LASER STUDY OF CONDENSED MATTER

APPENDIX

Experimental Methods and Samples

As a spectrally selective photosensitive medium solid polymerized solution of H_2-tetra-tret-butyl-porplyrazine (sample I) and octaethylporphin (sample II) in styrol, cooled down to liquid helium temperature, were used. The inhomogeneous width of the operating absorption bands were respectively 500 cm^{-1} and 200 cm^{-1}, the homogeneous width of purely electronic lines, ~ 0.1 cm^{-1} and ~ 0.05 cm^{-1}. The samples were prepared in the form of 3-10 mm-thick platelets with optical density D = 1-3.

Photochemical holograms were burned and probed by picosecond CW dye (R6G) laser, which was synchronously pumped by an argon laser with active mode-locking (Spetra Physics, models 375 and 171). Pulse repetition rate (duration 2-3 ps, spectral width 5 cm^{-1}) was 82 MHz. Spectral measurements were performed by the same dye laser working in the CW mode and having a complementary scanning etalon in the resonator (oscillating linewidth 0.075 cm^{-1}). Transmission spectra were recorded by a double-channel photon counting system.

The hologram time response was studied by a synchroscan streak camera system[20] with time resolution of about 20 ps or by up-conversion of the echo signal in a $LiIO_3$ crystal in a noncollinear scheme by using a gating picosecond pulse with variable delay. The dependence of up-conversion signal on the delay was recorded by a phase-sensitive detector.

The average intensity of laser emission of burning (recording) and probing (image restoring) was 0.1 mW/cm^2 and 0.1 W/cm^2, respectively.

To form the reference and object pulses delayed with respect to one another Fabri-Perot and Michelson-type interferometers were used.

210 LASERS IN ATOMIC, MOLECULAR AND NUCLEAR PHYSICS

temporal characteristics.

4. Wave Front Conjugation

As was shown above, if on recording in the selectively photochromic medium an object pulse is sent first then a time-reversed replica of the object pulse is restored. When including to the treatment 3 space dimensions, on restored by a pulse of a direction reverse to the reference one ($-\vec{K}_R$), a simultaneous conjugation of the wave front occurs.

In conslusion, the restored pulse is a backward-propagating phase conjugated and time reversed exact copy of the signal pulse.

5. CONCLUSION

In the case under consideration the holographic process is generalized to a case of photosensitive materials which are able to memorize not only the spatial distribution of the field intensity but also its spectral compositon. The proposed and experimentally realized method of the holographic recording in the familiar media with photochemically active impurity absorption centers allows an object scene of 10^{-8} to 10^{-13} s duration to be restored in its full time dependence with an efficiency sufficient for practical applications. In essence the space-time holography solves the problem of real recording of moving space domain pictures, if to bear in ming that the cinematographic method of imagewise shooting does not allow the image changes to be restored in the direct sense of the word.

It is also shown that photochemically accumulated photon echo and the space-time holography based on this phenomenon have a number of promising applications.

The authors are indebted to Prof. K. Rebane for fruitful and stimulating discussions.

Two geometries were used: (a) collinear direction of all beams to investigate only spectral-time dependences; (b) the signal and reference beams separated to an angle $\theta \leqslant 1°$ to study both spatial and temporal dependences. For a spatial formation of the signal pulse one shoulder of the Michelson interferometer was supplied with 2m-focal-length mirror. Behind the cryostat, where the samples were introduced, on a special screen it gave an image of a point adjacent to the spot from the passed-through plane reference wave. The angle between the reference and object wave was chosen by detuning of the interferometer.

REFERENCES

1. A. K. REBANE, R. K. KAARLI, P. M. SAARI, Opt. Spectrosc., 55, 405 (1983) USSR
2. A. K. REBANE, R. K. KAARLI, P. M. SAARI, JEPT lett., 38, 383 (1983)
3. A. REBANE, R. KAARLI, P. SAARI, A. ANIJALG, K. TIMPMANN, Optics Commun., 47, 173 (1983)
4. A. REBANE, R. KAARLI, Chem. Phys. Lett., 101, 317 (1983)
5. A. K. REBANE, R. K. KAARLI, Izvestija Akademii Nauk SSSR, serija fizicheskaya, 48, 545 (1984) (in russian)
6. K. K. REBANE, Impurity Spectra of Solids (Plenum Press, New York, 1970)
7. A. A. GOROKHOVSKII, R. K. KAARLI, L. A. REBANE, JEPT Lett., 20, 216 (1974)
8. B. M. KHARLAMOV, R. I. PERSONOV, L. A. BYKOVSKAYA, Optics Commun., 12, 191 (1974)
9. R. I. PERSONOV, Lecture in this edition.
10. V. A. ZUIKOV, V. V. SAMARTSEV, R. G. USMANOV, JEPT Lett., 32, 293 (1980)
11. S. O. ELYUTIN, S. M. ZAKHAROV, E. A. MANYKIN, JEPT, 76, 835 (1979)
12. Electromagnetic Superradiance (Kazan, 1975) (in russian)
13. K. I. SHTYRKOV, V. V. SAMARTSEV, Phys. Stat. Sol., 45, 647 (1978)
14. Yu. N. DENISYUK, Holography and its Prospects (Nauka, Leningrad, 1981), p. 7, (in russian).
15. T. W. MOSSBERG, Optics Lett., 7, 77 (1982)
16. P. M. SAARI, A. K. REBANE, Proceening of Acad. Sci. Estonian SSR, Phys. Mat., 33, 322 (1984)

17. P. M. SOROKO, Principles of Holography and Coherent Optics (Moscow, 1971) (in russian)
18. G. CASTRO, D. HAARER, R. MACFARLANE, H. P. TROMMSDORFF, US Patent 4,103,346 (1978)
19. K. K. REBANE, Laser Study of Inhomogeneous Spectra of Molecules in Solids, Proceedings of Conference "Laser-82", New Orleans, USA, Dec. 1982
20. J. V. KIKAS, R. K. KAARLI, A. K. REBANE, Opt. Spectosc., 56, 387 (1984) (USSR)
21. A. O. ANIJALG, P. M. SAARI, T. B. TAMM, K. E. TIMPMANN, A. M. FREIBERG, Kvant. Elekton., 9, 2449 (1982) (USSR)

3.2 Picosecond Processes in Semiconductors and their Application

V. BRÜCKNER

Friedrich-Schiller-Universität Jene, Department of Physics, DDR-6900 Jene, German Democratic Republic

1. INTRODUCTION

Recent advances in short pulse laser physics suggest new experimental methods for the investigation of semiconductor processes. Highly excited states in the semiconductor can be produced by the absorption of strong laser light, on the other hand laser pulses down to the femtosecond time scale[1] can be used to study the time behaviour of the optically excited carriers. Therefore, using the picosecond technique it is possible to measure semiconductor processes on a picosecond time scale, which cannot be achieved by other techniques.

In this paper the results of transient reflectivity measurements and of time-resolved photoconductivity in semiconductors are summarized which were achieved in our laboratories during the last 2-3 years.

2. BASIC PRINCIPLES

If the photon energy of the excitation laser light $\hbar\omega$ is more than the energy gap of the semiconductor E_g a dense free carrier plasma can be generated. For example, assuming the absorption constant to be $\alpha \simeq 10^4$ cm^{-1} a peak charge carrier concentration in the order of 10^{20} cm^{-3} can be achieved using picosecond solid-state laser sources with about 10^{16} photons/cm^2 ^{2}. A typical value of the intrinsic density of free carriers is less than 10^{10} cm^{-3}. Therefore the optically generated electron-hole density N completely determines both the electrical and the optical properties of the semiconductor. The changes of the optical (i.e. of the complex permittivity ε) and electrical (i.e. conductivity σ) properties can be calculated using the well-known Drude model of a free electron gas[3]

$$\varepsilon = \varepsilon_1(\omega_T) - i\varepsilon_2(\omega_T)$$
$$\varepsilon_1(\omega_T) = \varepsilon_r - Ne^2\tau^2/\varepsilon_o m_{opt}(1 + \omega_T^2\tau^2)$$
$$\varepsilon_2(\omega_T) = Ne^2\tau/\varepsilon_o m_{opt}\omega_T(1 + \omega_T^2\tau^2)$$
(1)

$$\Delta\sigma = eN\mu$$
(2)

where ω_T is the frequency of a probe beam, ε_o is the vacuum dielectric constant, τ is the thermalization time, m_{opt} is the "optical mass" of the carriers[4], and μ is the carrier mobility. In Equations (1) and (2) the influence of the holes was neglected. It should be mentioned that the Drude model is a good approxiamtion in most of the experimental cases. A more correct discribtion is presented in[5] using the dynamical behaviour of the Fermi energy level at different temperatures of hot carriers. From (1) and (2) we can conclude that it is possible to determine the dynamical behaviour of carriers N(t) using the excite-and-probe technique[6] (i.e. time resolved reflectivity of transparency) or/and the optoelectronic switching in semiconductors[7] (i.e. time-resolved conductivity).

In order to describe the temporal and spatial behaviour of the non-equilibrium charge carriers it is necessary to take into consideration various processes, the most important of them are the carrier generation $G(\vec{r},t)$, the linear - $R_L(\vec{r},t)$ and nonlinear - $R_{NL}(N,\vec{r},t)$ recombination, surface recombination $R_S(\vec{r},t)$, carrier diffusion $R_D(\vec{r},t) = \text{div}\,\vec{j}(\vec{r},t)/e$ and carrier thermalization to the band edge $R_{th}(N,\hbar\omega-E_g,\vec{r},t)$. The temporal and spatial behaviour of the charge carriers can be described by the continuity equation[8]

$$\frac{\partial N(r,t)}{\partial t} = G(\vec{r},t) - R_L(\vec{r},t) - R_{NL}(\vec{r},t) - R_S(\vec{r},t) -$$

$$- \text{div}\,\vec{j}(\vec{r},t)/e - R_{th}(N,\hbar\omega-E_g,\vec{r},t) \qquad (3)$$

where $\vec{j}(\vec{r},t)$ is the current density vector.

A more detailed discussion of the influence of these processes to the carrier dynamics is given in[9]. Additionally, the influence of the spatially inhomogeneous carrier concentration[2] and of interferences in thin semiconductor films[10] should be taken into account.

In general, it is possible to separate the influence of these processes experimentally in the following manner:

1. Linear recombination is an exponential decay process which does not depend on the carrier concentration $N_o = N(t=0)$;
2. Diffusion is a non-exponential decay independent of N_o (at least up to $N_o \simeq 10^{19}$ cm^{-3} [11]);
3. Nonlinear (e.g. Auger) recombination is a non-exponential process which strongly depends on N_o;
4. Thermalization is an extremely rapid process (typically in less than 1 ps [12]) which depends on the excess energy of the carriers $\hbar\omega-E_g$.

In all cases it is necessary to take into account the finite excitation pulse duration[9]. The influence of surface recombination can be checked by a grating method[13] or by comlete comparison of time-resolved reflectivity and transparency at different excitation wavelength[14].

3. EXPERIMENTAL METHODS

a) <u>Time Resoved Reflectivity and Transparency</u>

216 LASERS IN ATOMIC, MOLECULAR AND NUCLEAR PHYSICS

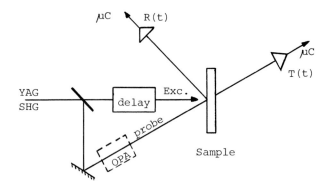

FIGURE 1 Time-resolved reflectivity and transparency

For measurements of the optical properties of the semiconductros the well-known excite-and-probe technique was used. The strong second harmonic radiation of a mode-locked Nd-YAG laser system acts as excitation (Figure 1). This 532 nm, 25 ps pulse incidents perpendicularly onto the sample, its energy can result in a maximum carrier density of about 10^{21} cm^{-3} in silicon using the laser beam of about 2 mm diameter. Experimentally it was found that there is a linear dependence of the reflectivity changes on the excitation energy density from 1 to 20 mJ/cm^2 and no irreversible changes of the semiconductor structure take place[15]. The optically excited change of the refractive index in the semiconductor and, therefore, the relaxation processes can be checked by the reflectivity and/or transparency change for a weak probe pulse (the fundamental wave of the YAG laser or the radiation of a optical parametrical amplifier tunable from 800 to 1500 nm [16]) delayed in time by a variable optical delay line (maximum delay time 1 ns). Of course, the probe pulse measurements should be performed advantageously in the absorption-free spectral range of the semiconductor. For example, at 1064 nm the small absorption constant in amorphous silicon of about 50-100 cm^{-1} [17] and the attenuated probe pulse energy (less than 10^{-9} J) leads to a neglecting change of the refractive index by the

LASER STUDY OF CONDENSED MATTER 217

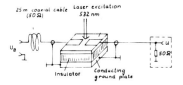

FIGURE 2   Time-resolved photoconductivity

probe beam itself. The angle of incidence of the probe pulse was optimized to be about 68°[6]. To avoid the influence of lateral diffusion processes the probe beam diameter was chosen to be one order of magnitude smaller than the excited part of the sample. The energies of the exciting pulse, of the incident and the reflected (transmitted) probe pulses were measured by a laser energy monitoring system[18]. The data were evaluated by a microcomputer K 1510.

b) Time-resolved Photoconductivity

The time-resolved conductivity experimentes were performed using a sandwich-like device[19]. It consists of a conducting ground plate, an insulator (mostly glass of about 1 mm thickness) and the semiconductor (less than 0.3 mm thickness) (Figure 2). Aluminium contacts were evaporated onto the semiconductor. In order to satisfy the 50 Ω line impedance the stripwidth was chosen to be about 1 mm, the gap between the electrodes depends on the used laser pulse energy (typically between 0.05-1 mm). The contacts of the coaxial cable were pressed onto the strip line, a dc bias up to about 2 kV can be applied to the device. The laser controlled electrical pulses were examined by a fast sampling oscilloscope (S7-13, time resolution of about 200 ps) or by a boxcar integrator (BCI 280, 220 ps time resolution) connected with the microcomputer K 1510.

c) Measurements of the Correlation Function

If the carrier relaxation time in the semiconductor is less than the time resolution of the detection in time-resolved conductivity experiments, these ultrafast relaxation processes can be studied using the optoelectronic autocorrelation technique[20]. Unfortunately, the fast carrier recombination is accoumpanied typically by a small carrier mobility. This results in a small voltage transmission through the correlation measurement device[21] even at the highest excitation energies. This difficulty can be overcome using the correlation measuring technique, where the first optoelectronic switch is changed by an arrangement based on the extinguishment of the electric pulse by its time-delayed reflection at an external charging resistor[19]. Integrating this charge resistor into the gap between the striplines of switch 1[22] electrical pulses as short as the used pulse duration without any limitations of the transmitted voltage can be obtained. The typical arrangement is seen in Figure 3. A part of the gap of switch 1 was covered with an insulating light-tight layer. This non-illuminated part of the gap acts as the charging resistor, the non-covered part acts both as switching element and charged line. Using this arrangement the generation of short ($\leqslant 5$ ps) electric pulses was reported[22] even if the recombination time of the used GaAs sample is comparatively

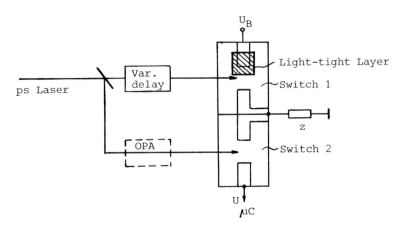

FIGURE 3   Correlation technique using a partly covered gap

LASER STUDY OF CONDENSED MATTER

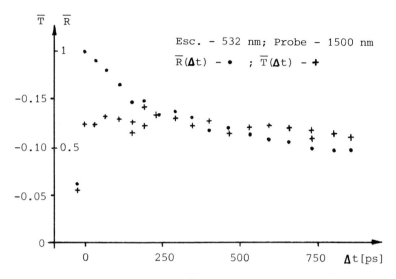

FIGURE 4 Normalized reflectivity $\bar{R} = (R - R_o)/(R_{min} - R_o)$ and transparency $\ln(T/T_o) = \bar{T}$ in monocrystalline Si (R_o, T_o - values without excitation, R_{min} - minimum reflectivity).

long (about 500 ps [19]). It was shown theoretically[23] that a time resolution less than 1 ps can be acieved using partly covered gaps.

4. RESULTS AND DISCUSSION

a) Monocrystalline Silicon

The silicon samples used in our experiments are disks of about 0.2 mm thickness. The acceptor concentration was less than 10^{13} cm^{-3}, i.e. the silicon was of high resistivity (about 2000 Ωcm). We have performed time-resolved reflectivity and transparency experiments using the second harmonic of the YAG laser as the excitation radiation. The excitation energy density was varied between 0.5 and 7 mJ/cm^2. Some typical results are presented in Figure 4. The temporal behaviour of the normalized reflectivity can be explained assuming the diffusion to be dominantly[9], on the other hand in the transparency behaviour dominates the influence of the carrier recombination (including surface recombination). Comparing numerical

220 LASERS IN ATOMIC, MOLECULAR AND NUCLEAR PHYSICS

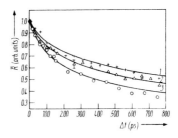

FIGURE 5 Normlized reflectivity dependence on the excitation energy density : + - 1.4 mJ/cm^2, ∆ - 2.9 mJ/cm^2, o - 6.6 mJ/cm^2, solid lines represent the numerical calculations with (1) - D = 18.1, (2) - 30, (3) - 60 cm^{2}s^{-1}.

calculations with the experimental results the best fit was found at an ambipolar diffusion constant D = 18.0 cm^{2}s^{-1} and at a recombination time ≃ 10 ns. Using the fundamental wave of the YAG laser as the excitation is possible to reduce the influence of the surface recombination time[14].

The dependence of the normalized reflectivity on the delay time between excite (532 nm) and probe (1064 nm) pulses at three different excitation levels is shown in Figure 5 [11]. Assuming the linear recombination time to be more than 10 ns, and comparing the numerical results with the experiments the best fit was found at diffusion constants of 18, 30, and 60 cm^{2}s^{-1} at excitation energy densities of 1.4, 2.9, and 6.6 mJ/cm^2, respectively. These experimental results can be compared with the theoretical calculations taking into account many-body effects[24]. As it is seen from Figure 6, the experimental results are in a reasonable agreement with the theoretical ones taking into account the influence of Auger recombination at the highest excitation energy density used (corresponding to a carrier concentration N = 8·10^{19} cm^{-3}).

b) Amorphous Silicon

We have studied ultrafast processes in amorphous Si films prepared

LASER STUDY OF CONDENSED MATTER 221

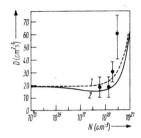

FIGURE 6 Density dependence of the ambipolar diffusivity in noncrystalline Si (the lines represent the calculations of [24]).

by different techniques and/or at different conditions. The samples are a-Si films deposited onto a glass substrate by different techniques:

(i) by plasma decomposition of SiH_4 in a silan-hydrogen mixture at various pressure (GD-a-Si:H) or at different substrate temperatures (e.g. GD-90 at 90°C substrate temperature);

(ii) by ultrahigh vacuum evaporation (ev-a-Si);

(iii) by ion sputtering performed at a substrate temperature of about 170°C (sp-a-Si).

Experimentally it was proved that there is a linear dependence of the reflectivity changes on the excitation energy density in all samples used in time-resolved reflectivity experiments.

Analyzing the experimental results at 532 nm excitation the time behaviour of the non-equilibrium charge carriers can be subdevided in three typical groups:

(i) Two-exponential decay of the carrier concentration independent of the excitation energy density (obrained in GD-a-Si:H prepared at different hydrogen pressure at room temperature, in GD-90 and in ev-a-Si);

(ii) One-exponential decay of the carrier concentration (obtained in sp-a-Si);

(iii) One-exponential decay; dependence of the time behaviour on

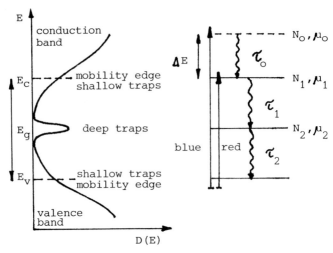

FIGURE 7 Model of the relaxation processes in amorphous silicon

the excitation energy density (obtained in GD-200).

These experimental results can be discussed using the following 4-level-model (Figure 7). The strong absorption of laser photons leads to the generation of non-equilibrium charge carriers in extended states ($\hbar\omega > E_g$). Within a time τ_o short with respect to our excitation pulse duration these carriers relax to the shallow traps at the the mobility gap[12]. The recombination of the thermalized carriers can be described by a bimolecular process[25] (recombination rate γ) and/or relaxation from the shallow traps to deep localized states (τ_1) and to the valence band (τ_2).

Results of the fitting procedure, i.e. τ_1, τ_2, and γ, in GD-a-Si:H films, deposited under systematic variation of the silanhydrogen pressure, are presented in Table I.

TABLE I

Pressure [Pa]	N_S [10^{18} cm^{-3}]	c_H [%]	τ_1 [ps]	τ_2 [ps]
25	0.75	25	85	2.5
80	0.25	35	35	0.75
150	1.4	26	50	1.25

LASER STUDY OF CONDENSED MATTER   223

Taking into account the measurements of some film parameters (spin desity N_s, hydrogen concentration c_H) we can conclude that the recombination is obiously determined by the structure of the amorphous film[26]. For example, nearly the same hydrogen concentration was obtained at 25 Pa and at 150 Pa, but the spin density differs by a factor of about two. This points to a "better" structure of the film produced at 25 Pa, which causes an increasing recombination time τ_1. A similar behaviour of the recombination time in moncrystalline silicon was described in[28]. In our experiments we did not obtain any dependence of the time behaviour of the carriers on the excitation energy density. Therefore, taking into account our measuring error the bimolecular recombination rate can be estimated to be less than 10^{-12} cm^3/s^{-1}.

TABLE II

Sample	T_s [°C]	N_s [10^{17} cm^{-3}]	E_g [eV]	τ_1 [ps]	τ_2 [ns]	γ [$cm^3 s^{-1}$]
GD-20	20	10	1.81	32	0.75	$<10^{-12}$
GD-90	90	5	2.05	75	1.4	$<10^{-12}$
GD-200	200	1	1.62	165	-	$3.3 \cdot 10^{-10}$

The temporal behaviour of photoexcited charge carriers in GD-a-Si prepared at different substrate temperatures is presented in Table II[29]. It can be seen that the carrier recombination time τ_1 increases with decreasing defect density. In the samples GD-20 and GD-90 no bimolercular recombination was found within the measuring error, i.e. $\gamma < 10^{-12}$ $cm^3 s^{-1}$. A bimolecular recombination process was found only in the sample GD-200. In this sample the smallest optical gap E_g was obtained and, therefore, the probability of band-to-band recombination increases in agreement with our experimental results.

Let us compare the results of time-resolved reflectivity measurements in amorphous silicon films prepared by different techniques[27]. As it is seen from Table III two recombination times were obtained in GD-a-Si:H and ev-a-Si. Unlike this time behaviour we obtained only a single exponential decay with a time constant of about 2.6 ns in sp-a-Si.

TABLE III

Technique	N_S [10^{18} cm^{-1}]	τ_1 [ps]	τ_2 [ns]
GD	0.75	85	2.5
ev	10	50	3.0
sp	10	25	2.6

The decreasing values of τ_1 with increasing spin density (Table III) give hint to the assumption that the recombination time τ_1 in our sp-a-Si films is less than the excitation pulse width (25 ps). In this case τ_1 cannot be calculated from these reflectivity experiments because the charge carriers relax already during the excitation process into the intermediate state. Another situation we can found in photoconductivity experiments, where the mobility of localized charge carriers in the intermediate state is much less than in the conduction band. Therefore, changes in the photoconductivity are caused predominantly by the carrier concentration in the conduction band, i.e. τ_1 recombination gives rise to the time behaviour in time-resolved photoconductivity experiments.

Using two identical sp-a-Si samples the autocorrelation function was measured[27]. Assuming the laser to be of Gaussian shape the electrical pulse duration was found to be about 25 ps, i.e. optical excitation and electrical response have an identical pulse duration. Therefore, the relaxation time τ_1 in sp-a-Si is much less than the laser pulse duration in agreement with other measurements in a-Si[30].

c) Silicon-on-Sapphire (SOS)

We have studied fast carrier processes in SOS by time-resolved

reflectivity and conductivity measurements[31]. Starting from a SOS disk of 0.5 μm thichness different preparations were performed:
(i)   no additional preparations, i.e. "pure" SOS;
(ii) Ar- or Si-ion-beam damaged SOS, where the dose was varied between 10^{14} and 10^{16} cm^{-2};
(iii) hydrogenation of the ion-beam damaged SOS.

In the basic material we have found a recombination time of about 270 ps at 0.8 mJ/cm^2 excitation. This behaviour can be explained by the influence of structural defects in the interface. In ion-beam damaged unhydrogenated SOS a ultrafast recombination process (< 25 ps) was found acompained by a slow decay (> 1 ns). These results are in agreement with the 4-level model described above, assuming τ_0 and τ_1 to be less than the excitation pulse duration τ_p. Note that the maximum photoconductivity in the undamaged SOS was about 500 (1500) times more than in silicon (argon) ion-beam damaged samples at the same excitation level. That may be the reason why it was difficult to observe any switched voltage using a ps dye laser of low (10^{-10} J) pulse energy.

In hydrogenated ion-beam damaged SOS a carrier relaxation process shorter than the pulse duration of the excitation (532 nm, 25 ps) was obtained by correlation measurements. Using the optical parametrical amplifier it was possible to change the excess energy ($E_{exc} = \hbar\omega - E_g$) of the excited carriers. Using "red" pulses ($E_{exc} \simeq 0.1$ eV) an exponential decay of about 80 ps together with a slow one (about 1 ns) were found. Using "blue" pulses ($E_{exc} \simeq 0.6$ eV) a very fast (< 5 ps) relaxation together with the 1 ns relaxation were obtained. It was shown by numerical calculations that it is possible to explain both results assuming the relaxation times to be $\tau_0 < 5$ ps, $\tau_1 = 80$ ps, $\tau_2 = 1$ ns. The conductivity changes can be described by

$$\Delta\sigma = e(\mu_0 N_0 + \mu_1 N_1 + \mu_2 N_2) \tag{3}$$

226 LASERS IN ATOMIC, MOLECULAR AND NUCLEAR PHYSICS

where μ is the carrier mobility. Near to the mobility edge we have a mobility change of some orders of magnitude. That means, a small value of N_o has a strong influence on the conductivity. To explain our experiments we have to assume the mobility μ_o to be at least 10 times more than μ_1 [22]. It should be mentioned that the time-resolved methods presented here demonstrate the possibility to study the dependence of the carrier mobility on their excess energy $\mu(E_{exc})$. Further experiments on this matter are under consideration [31].

5. APPLICATIONS

During the last time high-frequency microelectronic elements in the 10-100 GHz range were developed. Because many electronic and optoelectronic devices utilize the laser - induced creation of non-equilibrium charge carriers, studies of ultrafast optoelectronic switching[7] are of importance due to its picosecond precision and simplicity of operation. Ultrafast switching devices have been developed to switch kilovolt amplitudes with ps risetime[32], to study ultrafast carrier processes[21,33], to detect picosecond pulses[30], to switch a streak camere with low jitter[34], or to measure the electronic response of a field-effect transistor[35]. In our laboratory optoelectronic switching experiments in CdS[36], GaAs[19], a-Si[27], and SOS[31] including correlation function measurements were performed.

There are some experimental difficulties to determine the autocorrelation function, which are caused by the requirement to work at saturation-free conditions, i.e. at $U/U_B \ll 1$. Therefore, it is fortunable to use a switch giving short pulses of high amplitude. In this case the experimental arrangement is similar to the well-known sampling technique. In our experiments we used a GaAs switch with a partly covered gap[22,23] which is able to generate electrical pulses shorter than 5 ps. This method can be used

LASER STUDY OF CONDENSED MATTER 227

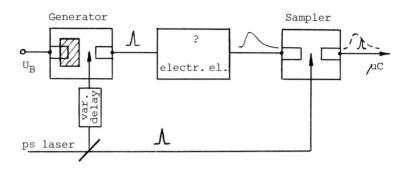

FIGURE 8 Scheme of the arrangement for response time measurements

to study the response time of electronic elements (Figure 8). By means of switch 1 (GaAs with a partly covered gap) an ultrashort electrical pulse is generated. Mostly the action of the electronic element leads to an increase in the pulse duration and to changes in the pulse shape, which can be checked by the "sampling" switch 2 (a-Si). Using this arrangement the response time in a transistor KT 391 A was found to be in the order of 50 ps^{37}.

Acknowledgement

I wish to thank my co-workers H. Bergner, F. Kerstan, M. Supianek, and B. Schröder. Furthermore, I am grateful to W. Nowick (Karl-Marx-Stadt) and Professor M. Schubert for helpful cooperation.

REFERENCES

1. B. WILHELMI, Femtosecond Light Pulses, Paper presented to the 3rd Vilnius School on Laser Application.
2. H. BERGNER, V. BRUCKNER, B. SCHRODER, Opt. & Quant. Electr., 14, 245 (1982)
3. K. KREHER, Festkörperphysik (Akad.-Verlag, Berlin, 1973)
4. V. L. BONCH-BRUEVICH, S. G. KALASHNIKOV, Halbleiterphysik (Verlag der Wiss., Berlin, 1982)
5. C. B. SHANK, D. H. AUSTON, E. P. IPPEN, O. TESCHKE, Solid State Commun., 26, 567 (1978)
6. H. BERGNER, V. BRUCKNER, B. SCHRODER, Kvantovaya Elektronika, 10, 1150 (1983)

7. D. H. AUSTON, Appl. Phys. Lett., 26, 101 (1975)
8. P. S. KIREJEW, Physik der Halbleiter (Akad.-Verlag, Berlin, 1974)
9. H. BERGNER, V. BRUCKNER, Opt. & Quant. Electr., 15, 477 (1983)
10. H. BERGNER, V. BRUCKNER, Optica Acta, 31 (1984) (in press)
11. H. BERGNER, V. BRUCKNER, Phys. Stat. Sol. (a), 79, K.85 (1983)
12. J. TAUC, Festkörperprobleme, 22, 85 (1982)
13. A. S. PISKARSKAS, in Primenenie laserov v atomnoi, molekularnoi i jadernoi fizike (Nauka, Moskva, 1979)
14. H. BERGNER, V. BRUCKNER, M. SUPIANEK, (unpublished)
15. H. BERGNER, Thesis (Friedrich-Schiller-Universität, Jena, 1983)
16. Y. K. VAITKUS, in Primenenie laserov v atomnoi, molekularnoi i jadernoi fizike (Nauka, Moskva, 1979)
17. N. S. MOTT, E. A. DAVIS, Electronic Processes in Non-Crystalline Materials (Clarendon Press, Oxford 1979)
18. K. RUHLE, W. TRIEBEL, Feingerätetechnik, 29, 166 (1980)
19. V. BRUCKNER, F. KERSTAN, Kvantovaya Elektronika, 11, 1344 (1984)
20. D. H. AUSTON, A. M. JOHNSON, P. R. SMITH, J. C. BEAN, Appl. Phys. Lett., 37, 371 (1980)
21. V. BRUCKNER, F. KERSTAN, Exp. Techn. der Physik, 32, 139 (1984)
22. V. BRUCKNER, F. KERSTAN, Electr. Lett. (in press)
23. F. KERSTAN, M. SCHUBERT, Opt. & Quant. Electr., (in press)
24. J. F. YOUNG, H. M. VAN DRIEL, Phys. Rev., B26, 2147 (1982)
25. J. M. HVAM, M. H. BRODSKY, J. de Physique, 42, C4-551 (1981)
26. H. BERGNER, V. BRUCHNER, D. DIETRICH, F. KERSTAN, W. NOWICK, Phys. Stat. Sol. (b), 117, 197 (1983)
27. H. BERGNER, V. BRUCKNER, F. KERSTAN, W. NOWICK, Phys. Stat. Sol. (b), 117, 603 (1983)
28. A. E. BAKANOWSKI, J. H. FORSTER, Bell Syst. Techn. J., 39 87 (1960)
29. H. BERGNER, V. BRUCKNER, D. DIETRICH, W. NOWICK, Phys. Stat. Sol. (b), 120, 655 (1983)
30. D. H. AUSTON, P. LAVALLARD, N. SOL, D. KAPLAN, Appl. Phys. Lett., 36, 66 (1980)
31. H. BERGNER, V. BRUCKNER, F. KERSTAN, W. NOWICK, Phys. Stat. Sol., (submitted)
32. G. MOUROU, W. KNOX, S. WILLIAMSON, Laser Focus, 18, 96 (1982)
33. V. K. MATHUR, S. ROGERS, Appl. Phys. Lett., 31, 765 (1977)
34. W. KNOX, G. MOUROU, Opt. Commun., 37, 203, (1981)
35. D. H. AUSTON, P. R. SMITH, Laser Focus, 18, 89 (1982)
36. V. BRUCKNER, F. KERSTAN, Electr. Lett., 18, 885 (1982); Opt. Commun., 45, 187 (1983)
37. V. BRUCKNER, F. KERSTAN, (unpublished)

3.3 Optical Bistability of the Light Induced Orientation Effects in the Liquid Crystals

S. M. ARAKELIAN, A. S. KARAIAN and Ju. S. CHILINGARIAN

Yerevan State University, Yerevan, USSR

1. STRONG OPTICAL BISTABILITY EFFECTS IN THE LIQUID CRYSTALS

Dispersive Optical Bistability Systems. Optical bistability (OB) or multistability is the general property of nonlinear optical system for a given set of input conditions[1]. These states are determined by amplitude as well as polarization parameters of the light. The state of the bistable system actually assumes depends on the direction in which it changes the input conditions - i.e. it has a memory. The hysteretic curve on the dependence, e.g. of the output light intensity (I out) from input intensity (I in), can be obtained by adjusting the feedback. Note that it a necessarily to distingwish between optical bistability and optical hysteresis[13.]. With these definitions a bistable system will always exhibit hysteresis, but for pulsed excitation the observation if hysteresis does not necessarily imply that the system with a characteristic time t could exhibit hysteresis for excitation with pulses with a duration of the order of t, but for cw excitation there would only be one possible output for a given input.

Among the various nonlinerar optical effects which results in OB the effects based on optical-field-induced refractive indices Δn^{NL} (so called "dispersive nonlinear systems") have attracted a great deal of attention. An application is often most appealing if it is operative with cw laser beams. This would require a nonlinear medium with an extremely large nonlinearity. Among fluids, liquid crystals (LC's) appear to have by far the largest Δn^{NL} because of the extremely strong optical-field-induced collective molecular reorientation in the media[2]. In come cases Δn^{NL} as large as 0.1 can be obtained with a laser intensity of < 100 W/cm^2.

The physics of OB system depends critically on the various time constants[1].

In the simplest case the sytem will have three important time constants: the first, τ_p, the characteristic time of input intensity variation, the second, τ_M, characterizing the speed of response of the nolinear material (via the microscopic nonlinearity) and the third, τ_F, describing the overall response time on the macroscopic feedback. In a system with a resonator, the resonator round trip time, τ_{RT}, is also relevant (the quantity τ_{RT} is different from τ_F as it may require many round trips of light in the cavity to establish a stable feedback).

In this hand the unique for nonlinear media possibility to vary in the large scale the time relaxation parameters which occurs in the LC's (from tens seconds to microseconds and even to nanoseconds)[3] allow investigate the OB at steady-state, non-steady-state and transient regimes.

In contrast with traditional nonlinear media in LC's can arise, at last, specific light-induced orientational effects: laser-induced phase transitions (Ist or IInd order) which results in qualitatively new types of OB regimes[4,5].

Nonlinear Fabry-Perot Resonators. Nonlinear resonators, in particular Fabry-Perot resonators (FPR), are the classic OB-system[6]. In this type of OB, the feedback is provided by the multiple reflections inside a FPR, and the form of the microscopic optical nonlinearity on optical intensity. In a sample FPR consisting of two

LASER STUDY OF CONDENSED MATTER   231

mirrors separated by a distance L and filled with a medium of refractive index N, the transmission depends on the resonator tuning.

OB in nonlinear FPR has a direct physical explanation[1]. If the system is initially off resonance, increasing I_{input} pulls it toward resonance because of the nonlinear refractive index. But moving toward resonance gives a further increase in I_{inside} due to resonant magnification, this pulling the system even further towards resonance. Under the right conditions this process becomes regenerative and provides a switching into the near on-resonance state. Once there is it can be held on resonance with less I_{input} because of the complete resonant magnification of I_{inside}.

The results of a detailed theoretical and experimental study of the dynamic behaviour of a nonlinear FPR filled with a isotropic phase of nematic LC (NLC)-nonlinear Kerr type medium, are given in[7]. The three modes of operation, namely, power limiter, differential gain, and optical bistability with hysteresis are considered. It is shown that the quasi-steady-state operation requires not only $\tilde{\tau}_M \ll \tilde{\tau}_{RT}$ but also $\tilde{\tau}_p$ several hundred times larger than $\tilde{\tau}_{RT}$. The experimental results covering a wide range from the extremely transient to the quasi-steady-state case are all in excellent agreement with theory.

The nonlinear FPR in which the nonlinear medium is a thin nematic films was investigated in[8]. Because of the large induced phase shift even at relatively low laser intensities, the higher-order bestable loops was easily observed. An intersting phenomenon occurs when the laser intensity I_{in} becomes higher than a certain value: it is found that output will then break into oscillation. The oscillation was periodic and lasts indefinitely. This is the result of two opposing mechanisms with very different response times contributing to Δn^{NL}. In[8] the two mechanisms are first, the molecular reorientation, which is strong and slow with a time constant of the order of seconds, and second, the laser heating, which is much weaker but faster with a time constat of ~ 0.01 s. (At put up the NLC cell inside of Ar^+ - laser FPR the

the similar auto-oscillation regime was obtained in[9].)

Nonlinear Oscillator. The second case of OB will be illustrated with the behavior of an assembly of Duffing nonlinear (anharmonic) oscillators driven by an intense electromagnetic field. Thys type of OB have a large number of nonlinear optical phenomena: four-wave mixing[10], stimulated scattering[11] et al.

At microscopic description this problem is very well known for nonlinear optics, that is the anharmonic oscillations of the atoms and molecules. The detail analysis of various OB regimes which arise here are given in the theory of nonlinear oscillations. In optics this problem was discussed in[12] for the first time it seems.

The feedback in nonlinear oscillator appears through the resonance frequency and amplitude of oscillations coupling which results in non-isochronism property. These nonlinear oscillations are very interesting things for LC's dynamics and can occur if the external fields including light fields applied to the sample. The strong nonlinear optical effects in LC's can be described by assuming that the behavior of the material system in the presence of an intense electromagnetic field can be modelled by that of an assembly of driven anharmonic oscillators. In another hand it is already obvious, that new types of OB can arise for laser radiation and LC's interaction if this analogy does work.

Mirrorless Optical Bistability. Phenomena on Surface. OB can appears for optical phenomena on interface between two dielectric materials (at the total internal reflection[13] and at Brewsterian angle) as well as between dielectric and metal (resonance of surface plasmons)[14] or/and probably other media. One (or both) of these materials has an intensity-dependent refractive indes. Under appropriate conditions the reflectivity of these interfaces would not only be intensity dependent, but would exhibit optical hysteresis and OB. For LC'sas a nonlinear medium these effects is developped under cw laser beams[15].

Distributed-feedback Systems. Optical systems with distributed feedback can also be with mirrorless OB. The periodic structures can induce here by laser radiation (dynamic self-diffraction effects including phase conjugation[16]). These wave mixing processes can be easily observed in a NLC film where the diffraction of the probe beam from the laser induced grating can be of multiple orders[2,3]. In particular, up to 8 orders of diffraction can be discerned by eye at a beam intensity of ~ 100 W/cm^2 [17], and an output phase-conjugation efficiency is of 1 order at a laser intensity of $\sim 10^3$ W/cm^2 [18].

In periodic media light induced distortions of its pitch for waves in the Bragg regime result of to that the reflection coefficient is a multivalued function of the input intensity. This leads to an intrinsic OB. In this hand the cholesteric and probably smectic LC's are presented the most interesting. The resulting distortion of the cholesteric helix under laser field and the increase in its period means that a light wave which at low intensity satisfies the Bragg condition for the structure may not suffer Bragg reflections at higher intensities[19]. A transition from high reflection to high transmission occurs as the incident intensity is increased beyond a critical value. In the high-transmission state, the light field extends throughout the length of the medium and is able to maintain the molecules distribution in the perturbed condition until the intensity is reduced to a level considerably lower than the critical switch-on intensity. This explains the hysteresis both for the amplitude and polarization characteristics.

A critical internsity for bistable reflection is quite large-on the order of 1 MW/cm^2 for the typical cholesteric LC's and incident wave-length of 1 μm [19]. More less intensity is required if hybrid schemes are using when light as well as others external fields are applied.

Cross-section Effects. OB regimes can be associated with aperture effects-self-focusing of light[20]. In these cases the changing of the light beam cross section leads (via feedback, including some optical elements-mirrors and diaphragms) to the hysteresis. Especially in LC's these aperture changings are the most strong and when the intensity is sufficiently high (on order of 100 W/cm^2) the output displays a multiple diffraction ring pattern, which can be easily understood from the spatial self-phase modulation effect[21].

Nonhomogeneous Media. Two states of the optical system-with high level of loss and transmission can arise in nonlinear light scattering experiments if nonhomogeneous LC's are under optical orientation conditions. As from the very begining the light wave enters the strong scattering medium the clearing region spreads out into the medium[22]. The nonlinear scattering and incident intensities coupling is determined by cubic equation[23]. That results in OB via the hysteresis[24]. The necessary intensities for LC's are $10^3 - 10^5$ W/cm^2. These phenomena have quite common character and determin by nonlinear clearance of the medium which includes various processes ot the dissipation (absorbtion, evaporation etc.)[22]. In LC's these effects are the most intersting at existence of the heat, hydrohynamics, diffusion, and others processes.*

The specific of these phenomena in LC's are connected with the threshold process. In particular the molecular fluctuations and light scattering quickly grow on the threshold of reorientation. That produces the supplementary loss channel of the system input energy. Over the threshold these loss disappear. Therefore, a new type of OB arise: the consecutive field decreases and increases results in hysteresis.

---

* The spread out of the switching waves for the spatial arise here[25].

LASER STUDY OF CONDENSED MATTER   235

Conclusions. Thus many nonlinear optical effects in LC's have OB (or multistability) properties which can be investigated in cw laser fields. The most important thing here is the resonance curve or another singular point existense, nearby of which the optical system propeties are strongly changing. Than via nonlinearity of the medium, if there is certain initial shift from this point, the strong feedback appears and transition between various stable states of optical system can occur. The feedback for all examinated above systems was determined by nature of self-action effects: light-induced refractive index changing leads to redistribution of intensity and phase of going waves inside of the medium that back again acts on spatial distributions of refractive index.

In next sections we will discuss with more details a few of lignt-induced orientational effects in LC's in the viewpoint of common schemes of OB. Concrete experimental evidences of these OB effects requires the quantitative estimations of LC's parameters and of the experimental conditions for every separate case.

2. LIGHT-INDUSED REORIENTATION IN THE LC'S AND ANDLOGY WITH NONLINEAR DRIVEN OSCILLATOR

Let us consider the NLC layer (by the thickness of d) having the everage direction of molecular orientation $\vec{n}_o$ along the Z-axis. We assume the NLC is placed in external light field E ‖ X, X ⊥ Z which leads to molecular reorientation. Rejection angle θ characterizes this changing of the initial orientation (in the XZ plane). The boundary conditions are hard and $\theta(0) = \theta(d) = 0$.

An equation which describes the molecular motion perturbed by laser field is given by

$$J\frac{\partial^2 \theta}{\partial t^2} = K_{33}\frac{\partial^2 \theta}{\partial z^2} - j_1 \frac{\partial \theta}{\partial t} + \frac{\varepsilon_a \varepsilon_\perp}{\varepsilon_\| c}^{1/2} g(\theta) I_z \qquad (1)$$

where J is the molecular reorientation inertia moment (on unity of volume), K_{33} is an elastic constant, j_1 is viscosity, $\varepsilon_a = \varepsilon_\| - \varepsilon_\perp > 0$

is the optical dielectric susceptibility anisotropy ($\varepsilon_{\shortparallel}$, $\varepsilon_{\perp}$ are the basic values along and perpendicular to $\vec{n}_o$, respectively), $g(\theta) = (tg\,\theta/\cos\theta) \cdot [1 + \frac{\varepsilon_{\perp}}{\varepsilon_{\shortparallel}} tg^2\theta\,]^{-3/2}$, I_z is the magnitude of the Poynting vector z-component which is the constant, c is the light velocity.

Equation (1) differs from known one for NLC [3], because of the inertial member, and describes the oscillating regime[26]. Usually nonperiodic damping of orientation deformation are proposed for LC but collective molecular oscillations can occur in initial stage of reorientation process.

Assuming the existence of the most energetic profitable deformation of NLC $\theta = \theta_m(t)\sin(\pi z/d)$ which is satisfy of the given boundary conditions, and taking into account that at $\theta \ll 1$ $g(\theta) \approx \theta + [(5\,\varepsilon_a - 4\varepsilon_{\perp})/6\,\varepsilon_{\shortparallel}]\,\theta^3$ we have instead of (1) a new equation

$$\frac{\partial^2\theta}{\partial t^2} + \alpha\frac{\partial\theta}{\partial t} + R_o\theta + \beta\theta^3 = 0 \qquad (2)$$

where $\alpha = j_1/J$, $R_o = [K_{33}(\pi/d)^2 - (\varepsilon_a \varepsilon_{\perp}^{1/2}/\varepsilon_{\shortparallel}\,c)I_z]/J$,
$\beta = (\varepsilon_a\,\varepsilon_{\perp}^{1/2}/6cJ\,\varepsilon_{\shortparallel}^2)(4\varepsilon_{\perp} - 5\,\varepsilon_a)I_z$

Nonlinear equation (2) is classical; if $\alpha = 0$ it is Duffing equation[27]. If $\beta > 0$ and $R_o > 0$ all solutions of (2) are periodic. These solutions differ in principle from linear equation($\beta = 0$) ones - the oscillations frequence of (2), which is determing by $R_o^{1/2}$, depends on amplitude. But the motions qualitatively reminded harmonic, correspond to small energies W of system, if $\beta < 0$ and $R_o > 0$. The critical meaning W_c exists with W increasing, i.e. if $W > W_c$ periodic motions can not arise.

The Wc value is $W_c = R_o^2/4|\beta|$; it is achieves when $\theta = \pm(-\frac{R_o}{\beta})^{1/2}$. Unlimited mitions can appear and of $W < W_c$ but initial regection of molecular orientation have to be large enough.

If $R_o < 0$ but $\beta > o$ there are two stable equilibrium states:

LASER STUDY OF CONDENSED MATTER 237

$\theta = \pm(-R_o/\beta)^{1/2}$ (because two possible molecular reorientation directions exsist) and one unstable state: $\theta = 0$. Periodic motion takes place in this case only.

Thus $R_o = 0$ condition determines the light intensity threshold meaning ($I_{z,th}$) which produces the NLC deformation ($\theta \neq 0$). The $I_{z,th}$ is order 100 W/cm^2 for typical NLC, i.e. comparely sligt-strength laser fields reorient of NLC.

OB arise here if one more external field (variable and weak) besides light one is applied to NLC along Z. Really, driven by external field $\vec{F}$ nonlinear oscillations (2) lead to the hysteresis on dependence of oscillation amplitude $|\theta_{m_o}|$ versus frequence $R^{1/2}$ of the field $F = F_o \sin R^{1/2} t$, $F_o \gg 0$, $\vec{F}_o \perp \vec{E}$ (e.g., if external field is the magnetic one (strength $\vec{H}'$, $H' \ll H$) for that near the threshold of reorientation $F \approx -(RH'/H)$ (for $R^{1/2} \gg j_1/J$)), which have to be add in the right part of (2)^{27}. In this case $\theta_m = \theta_{m_o} \sin R^{1/2} t$, $\theta_{m_o} \gtrsim 0$, the $|\theta_{m_o}(R)|$ dependence has the resonance (if $\alpha \ll R^{1/2}$) and $R = R_o + \frac{3}{4}\beta\theta_{m_o}^2 - F_o/\theta_{m_o}$ ($\alpha = 0$ for simplisity). If R would be fixed the hysteresis arises at the increasing and decreasing consecutive of F_o as a result of resonant curve continious deformation. The resonant existence condition, $\alpha \ll R^{1/2}$, defines that NLC parameters meaning, which provide the oscillation regime in fact.

The hysteresis arising exactly on the reorientation threshold is of the most interest. Then we can speak about first order field induced phase transition (compare with[4]). In this case the physical cause of OB appearing is connecting with reorientation indefinite of molecules in initial moment: both rejection directions are equivalent. If we achieve the threshold intensity from strength field side one of the possible state is already realized in the system so the field increasing and decreasing cases are not equivalent.

If $\vec{F_o} \parallel \vec{E}$ or F_o is not small enough it may be shown that we will

have analogous with a nonlinear Mattew equation instead of (2).
The parametric unstability development in this problem ia analized
for example in[27].

The equation (2) may be generalized by taking into account
the fluctuations of the NLC molecules. In this cade the random
force f(t) should be introduced to the right part of (2), i.e. the
problem is the thermal molecular fluctuations development in presence of light field. The solutions of equation (2), if f(t) is the
gauss and δ-correlated in the time function, are well known[28].
Its statistic behaviour is determined by sign of R_o. If $R_o = 0$ the
transition from regime with maximum at $\theta = 0$ for steady-state probability destribution of $\theta(t)$, $P(\theta)$, to the regime with $P(\theta)$
maximums at $\theta = \pm (R_o/\beta)^{1/2}$ occur. In this case the fluctuations
sharply increase as well as the scattering light intensity for
$R_o = 0$. Thus the supplementary loss chanal of the system input energy is introduced. When the reorientation threshold is excited
these loss are disappear. Therefore when field decreases arise
because of the light scattering.

The analysis of OB at dynamics self-diffraction with phase-conjugation effects can also be given by equation (1) if the material system is driven by the light field $\vec{E}$, consisting of two
counterpropagating plane waves, and if the field $\vec{F}$ consists of
strong $(\vec{E})$ and weak waves; the last one has an oblique direction
of propagation to $\vec{E}$. The analysis of modified by this way the (2)
equation can be done by standard method of meanfield approxomation[10]. The output phase conjugation efficiency is of 1 order at cw
laser internsity of $\sim 10^3$ W/cm^2 [3].

3. SELF-INDUCED NONLINEAR SCATTERING OF THE LIGHT AT OPTICAL ORIENTATION CONDITIONS OF THE MOLECULES IN A NLC

The large optical nonlinearity of NLC is not a occurrence by is a
result of it very important property to strongly scatter of the

LASER STUDY OF CONDENSED MATTER 239

light. This coupling is a development of the general one which arranges by the fluctuationar-dissipative theorem[23]. The phenomenon of molecules optical orientation, which is determined by this nonlinearity and which leads to the clearence of initialy non-homogeneous medium[29], can be used for observation of the OB. Two states of the optical system in this case are: the first one is the state with strong light scattering and the second one is the state with strong light transparence.

The feedback arises here because of the self-accordance of the problem: when the light enters into the medium it scatters as well as reorients of the molecules; but this last effect changes the intensity of the scattering light and loss is decreasing[22]. In contrast with division of 2 the local nonlinear oscillations in every point of the medium are taking into account.

The precise description of this process requires to solve an equation of clearance wave propagation with the extinction coefficient which is determined by the light intensity.

The light-induced order parameter changing as well as elastic properties of NLC is the important point here. But qualitative picture can be understand by the assuming that the behavior of this material system in the presence of an intense electromagnetic field can be modelled by that of an assembly of anisotropic isolated domains which interact with laser field via induced dipole moments.

To concrete the problem let us now consider that linear polarized (along X axis) laser radiation propagates in a nonoriented NLC with the thickness along Z axis. The scattering light is observed in ZY plane(along Y axis) with polarization in the same plane.

The expression which can be derive in this geometry for the nonlinear changing of the scattering intensity, ΔI_{SC}^{NL} is simply given (for steady-state regime) by [23]

$$\Delta I_{SC}^{NL} \equiv (I_{SC}^{NL} - I_{SC}^{L})/I_{SC}^{L} = \frac{1}{2}[15L_2(q) - 15L_4(q) - 2] \qquad (3)$$

where I_{SC}^{NL} and I_{SC}^{L} are the light scattering intensities with and without optical orientation effects consequently; $L_2(q)$, $L_4(q)$ are the generalized Langevin functions of the second and forth orders; $q = I_{in}(\alpha_\| - \alpha_\perp)/2K_B Tc$ is the orientational parameter of the light-induced dipoles, I_{in} is the light intensity inside of the midium, $\alpha_\| - \alpha_\perp > 0$ is the anisotropy of the optical dielectric constant, K_B is the Boltzman constant, T is the temperature, c is the light velocity.

Near from optical saturation ($q \geqslant 10$) the functions $L_2(q)$ and $L_4(q)$ can be expand into a series in ascending powers of $1/q$ [23] and expression of (3) we may rewrite in the form of cubic equation

$$\Delta I_{SC}^{NL} = -1 + \frac{15}{\alpha I_{in}} - \frac{60b}{\alpha^2 I_{in}^2} - \frac{120d}{\alpha^3 I_{in}^3} \qquad (4)$$

where $\alpha = (\alpha_\| - \alpha_\perp)/2K_B T$; b, d are numerical parameters.

Assuming that nonlinear midium can be devided by consequence of thin layers, when ΔI_{SC}^{NL} for each layer describes by (4), and passing to the integral (along Z) we obtain the coupling between outgoing and input intencities which is not univalued. This leads to the hysteresis at $b > 0$, $d < 0$, $b^2 > 4|d|$. The solutions can be obtained by the graphic method (compare with [6]) or by the numerical calculations.

Thus OB regimes arise in non-homogeneous NLC's at optical orientation condition of the molecules. Typical laser intensities when optical orientation of the molecules occurs in NLC's is about 10^3 W/cm^2 [29].

4. CONCLUSION

The main developings of OB effects in LC's which are possible within the third-order optical orientational susceptibility have been

LASER STUDY OF CONDENSED MATTER 241

described in this lecture. It was demonstrated that majotity of these light induced effects can be analysed on the basis of the general OB principles.

But we studied comparatively simple problems. Of great interest it seems is both theoretically and experimentally study of polarization instability effects in LC's, in particular the OB of nonlinear optical activity. This very informative phenomenon which has already been used in the modern laser spectroscopy requires the search for a media with large nonlinear constants and LC's are one of the best examples of this materials.

In another hand the speciticity of LC's leads to new types of OB: that incluses also and hydrodynamics effects which have principle meaning for the common problems of turbulence and dynamic stochastic processes.

The most important seems to be the developing of a detailed quantitative theory of these phenomena in LC's as well as of the mentioned above effects. The all of instability regimes and in particular the transition to the chaos can be appear via the numerical analysis. Multiplex optical instability can arise in laser active media. Very important as yet are the non-steady-state processes of OB. Note, that sufficiently strong optical nonlinearity of LC's can determine by the reorientation in optical field of the molecule fragments; the time constants, characterizing the speed of response of the nonlinear material, are here in the order of 10^{-12} s. It means that LC's have great interest not only from the viewpoint of the study of physical principle of OB and it practical demonstration but also of many possible application in a variety of subjects and at creation of a new class of optical processing of information. The practical applications are possible and for large time response of LC's if hybrid schemes (the laser as well as static)fields are using.

Among the experimental study of OB in LC's the surface schemes in reflection are the most interest for the present time. This type of OB attractive from a practical point of view because of the relative case with which small devices, perhaps compatible with integrated optics, can be fabricated.

REFERENCES

1. D. A. B. MILLER, Laser Focus, No. 4, 79 (1982)
2. R. M. HERMAN, R. I. SERINCO, Phys. Rev., A19, 1757 (1979)
3. S. M. ARAKELIAN, Yu. S. CHILINGARIAN, Nonlinear Optics of Liquid Crystals (Nauka, Moscow, 1984), p. 360
4. H. L. ONG, Phys. Rev., A28, 2393 (1983)
5. S. M. ARAKELIAN, Z. E. ARUSHANIAN, A. S. KARAIAN, Yu. S. CHILINGARIAN, Izvestia AN SSSR: Fiz, 47, 2453 (1983)
6. F. S. FELBER, J. H. MARBURGER, Appl. Phys. Lett., 28, 782 (1976)
7. T. BISCHOFBERGER, Y. R. SHEN, Appl. Phys., 32, 156 (1978)
8. S. D. DURBIN, M. M. CHEUNG, S. M. ARAKELIAN, Y. R. SHEN, J. Phys. (Fr), 44, Colloq., 161 (1983); Izvestia AN SSSR: Fiz., 47 2463 (1983)
9. S. M. ARAKELIAN, A. Yu. DALLAKIAN, A. S. KARAIAN, Yu. S. CHILINGARIAN, Lasers Optics conf. paper (Leningrad, 1984)
10. C. FLYTZANIS, G. P. AGRAWAL, C. L. TANG, in Lasers and Applications, edited by Guimaraes et al. (1981), p. 317
11. G. P. JOTIAN, L. L. MINASSIAN, Opt. Comm., 49, 117 (1984)
12. C. FLYTZANIS, C. L. TANG, Phys. Rev. Lett., 45, 441 (1980)
13. A. E. KAPLAN, J. Exp. and Teor. Fiz., 72, 1710 (1977); P. W. SMITH, J.-P. HERMAN, W. J. TOMLINSON, P. J. MALONEY, Opt. Lett., 5, 323 (1980)
14. G. M. WYSIN, H. J. SIMON, R. T. DECK, Opt. Lett., 6, 30 (1981) I. I. CHEN, G. M. CARTER, Sol. St. Comm., 45, 277 (1983); Appl. Phys. Lett., 41, 307 (1982)
15. S. M. ARAKELIAN, L. M. ASLANIAN, G. L. GRIGORIAN, S. T. NERSISIAN, Y. S. CHILIGNARIAN, Izvestia AN SSSR, No. 4 (1985)
16. N. V. KUHTAREV, V. N. STARKOV, Pis'ma v J. Techn, Fiz., 7, 692 (1981); Opt. and Spectrosk., 56, 569 (1984); M. CRONIN-GOLOMB, J. O. WHITE, B. FISCHER, A. YARIV, Opt. Lett., 7, 313 (1982)
17. S. D. DURBIN, S. M. ARAKELIAN, Y. R. SHEN. Opt. Lett., 7, 145 (1982)
18. S. M. ARAKELIAN, S. D. DURBIN, Y. R. SHEN, Pis'ma v Jour. Tech. Fiz., 8, 1353 (1982)

19. H. G. WINFUL, Phys. Rev. Lett., 49, 1179 (1982)
20. A. E. KAPLAN, Opt. Lett, 6, 360, (1981); J. E. BJIRKHLOM, P. W. SMITH, W. J. TOMLINSON, A. E. KAPLAN, Opt, Lett., 6, 345 (1981)
21. A. S. ZOLOT'KO, V. F. KITAEVA, at. al., Pis'ma v Journ. Exper. and Teor. Fiz., 32, 170 (1982); S. D. DURBIN, S. M. ARAKELIAN, Y. R. SHEN, Opt. Lett., 6, 411 (1981)
22. M. B. VINOGRADOVA, O. V. RUDENKO, A. P. SUCHORUKOV, Waves Theory(Nauka, Moscow, 1979), p. 384
23. S. KIELICH, Molecular Nonlinear Optics: Polska/Trans., edited by I. L. Fabelinskii (Nauka, Moscow, 1981)
24. G. B. ALTSHULER, V. S. ERMOLAEV, Doclad AN SSSR, 268, 844 (1983); 273, 597
25. N. N. ROSANOV, Journ Exper. and Teor. Fiz., 80, 96 (1981); Pis'ma v Journ. Techn. Fiz., 7, 351 (1981); Opt. and Spectrosk., 48, 108 (1980)
26. J. L. FERGASON, G. H. BROWN, J. Am Oil. Chem. Soc., 45, 120 (1968)
27. O. BLAKYER, Nonlinear Systems Analysis: Trans. Engl., edited by R. V. Khokhlov (Mir, Moscow, 1969); N. N. MOISEEV, Nonlinear Mechanics Asimptotic Methods (Nauka, Moscow, 1981)
28. V. I. KLYATSKIN, Dynamic Systems with Fluctuated Parameters; Statistic Description (Nauka, Moscow, 1981)
29. A. M. ARAKELIAN, L. E. ARUSHANIAN, et al., Journ. Technich. Fiz., 52, 943, (1982)

3.4 Identification of Relaxation Mechanisms in the Nonlinear Spectroscopy of Semiconductors

V. M. PETNIKOVA, S. A. PLESHANOV and V. V. SHUVALOV

Department of Physics, Moscow State University, 117234, Moscow, USSR

A new technique has been developed for identifying and
measuring the intraband relaxation times in semiconductors
in the femtosecond time domain. The technique is based on
the polarization nonlinear spectroscopy of electronic
resonances. An experimental investigation was carried out
on ε-GaSe crystal and the following parameters were
obtained for the first time:
 intraband relaxation energy $T_3=(36\pm4)$ fs
 orientation of the electron quasi-momentum $T_4=(69\pm13)$ fs
 transverse relaxation $T_2=(23.1\pm1.6)$ fs.
The carrier recombination time was measured to be
$T_1=(380\pm20)$ ps and the coefficient of the carrier spatial
diffusion $D_r=(7.5\pm1.4)$ cm^2/s.

1. INTRODUCTION

Progress in the nonlinear spectroscopy of condensed media has

stimulated the energance of a whole number of new experimental

techniques[1-8,11].

The method of the probe beam has been traditionally applied

for slow times measuring. This technique makes use of the light

pulses that are shorter than relaxation times. To study fast proc-

esses, however, the parametric interaction of fields with differ-

ent frequencies (biharmonic pumping), which is quasi-stationary with regard to the measured times, seems to be more promising[5-7,11]. At the same time it is, as a rule, imposible to isolate in a particular experiment one mechanism out of many relaxation mechanisms with different nature, while experimental data are often compared without sufficient reason to the simplest models of a particular isolated process[1-3,5-8.10].

The objective of the given paper was to obtain a more comprehensive model of the nonlinear susceptibility, that would take account of several types of relaxation mechanisms at once. Such a model may be instrumental in somparing the role of each mechanism in a particular experiment, in designing and conducting a run of experiments at separating and identifying these mechanisms.

The resonance parametric interactions are studied on the cubic nonlinearity of single-photon electron transitions. Once a substance is exposed to three fields with amplitudes $\vec{E}_{1,2,3}$, frequencies $\omega_{1,2,3}$ that are close to the resonance frequency of the transition ω_o, and wave vectors $\vec{K}_{1,2,3}$, a fourth wave $\vec{E}_4$ is generated with frequency $\omega_4 = \omega_1 + \omega_2 - \omega_3$ in the direction of $\vec{K}_4 = \vec{K}_1 + \vec{K}_2 - \vec{K}_3$. The advantage of this technique is the absence of background in this direction.

2. THE MODEL OF THE MEDIUM NONLINEAR SUSCEPTIBILITY

The nonlinear response of a semiconductor is determined by various types of nonuniformities and by relaxation mechanisms inherent to them. This is, above all, a spectral nonuniform widening of electronic transition (band structure). The action of field upon the substance exhibits polarization nonuniformity in the distribution of the orientations of the electron quasi-momentum. The spatial nonuniformoty is due to the interference of interacting fields, that disturbs an equilibrium and uniform distribution of excitation in space.

The phenomenological approach adopted in this work is based

LASER STUDY OF CONDENSED MATTER

on the kinetic equations (KE) for the density matrix[9] and on the model notions of the relaxation processes. These will be further considered as markovian processes, while fluctuations from the equilibrium state will be assumed to be small. This assumption is justified in case of long enough light pulses, high equilibrium population of the top level and weak excitation. In this case, relaxation coefficients are independent of time and satisfy the principle of the detailed equlibrium.

The internal state of the model medium is characterized by the quantum electron number (i = 1,2 for the valence and conduction bands), electron energy (ε_i) which is estimated from the bottom of the corresponding band, orientation of the electron quasi-momentum, or parralel to it dipole moment[13] $\vec{\mu}(\vec{\Theta} = (\Theta, \varphi))$ and spatial coordinate ($\vec{r}$). The three latter value may be considered as continuous and are determined by the distribution functions f(a), a = ($\vec{\Theta}, \varepsilon_i, \vec{r}$).

The above listed relaxation mechanisms are assumed to be independent and combination of three possible types of the kernel of the collision integral is used for their description

I $\varphi(a,a') = f_o(a)$;   II $\varphi(a,a') = \delta(a-a')$;

III the relaxation process of diffusion type.

The relaxation prosesses in semiconductor may be divided into two groups - intraband and interband ones. It will be further assumed that transition from conduction band into valence band may take place both in accordance with the energy selection rules and without these rules. The rate of interband transition is denoted γ_1' and γ_1'' correspondingly, The rate of reverse transition is assumed to be much slower.

The following processes are allowed for intraband transitions:
1. Inelastic relaxation proceeding both with the rotation of dipole momentum and without it. The relaxation rates are γ_3'' amd γ_3'.

2. Elastic relaxation proceeding both at a rate γ_4' and with the diffusion coefficient D_θ.

3. All the above mentioned processes were considered to be localized in space. Apart from these, an elastic relaxation with preservation of the orientation of quasi-momentum and spatial carrier diffusion (D_r) is also considered.

The rate of disturbance of the polarization phase γ_2 is determined by the sum of all relaxation processes rates. Equilibrium distribution functions are assumed to be rectangular and equal for both zones. The medium is assumed to be isotropic ($f_o(\vec{\theta}) = 1/4\pi$).

In the case of a straight-band semiconductor it is natural to presume that under the action of light optical transitions occur with the preservation of the values $\vec{\theta}$ and $\vec{r}$ and fulfillment of the selection rules by electron number and energy ($\varepsilon_2 = -\varepsilon_1$). Thus the following KE can be written:

$$\frac{\partial \sigma_1}{\partial t} = 2\mathrm{Im}V\sigma_{21}/\hbar + \gamma_1'\sigma_2 + \gamma_1''f_o(\varepsilon)\int\sigma_2 d\varepsilon - \gamma_3'[\sigma_1 - f_o(\varepsilon)\int\sigma_1 d\varepsilon] -$$
$$- \gamma_3''[\sigma_1 - f_o(\varepsilon)f_o(\vec{\theta})\int\sigma_1 d\varepsilon d\vec{\theta}] - \gamma_4'[\sigma_1 - f_o(\vec{\theta})\int\sigma_1 d\vec{\theta}] +$$
$$+ (D_\theta \nabla_\theta^2 + D_r \nabla_r^2)\sigma_1; \qquad (1a)$$

$$\frac{\partial \sigma_2}{\partial t} = -2\mathrm{Im}V\sigma_{21}/\hbar - (\gamma_1' + \gamma_1'')\sigma_2 - \gamma_3'[\sigma_2 - f_o(\varepsilon)\int\sigma_2 d\varepsilon] -$$
$$- \gamma_3''[\sigma_2 - f_o(\varepsilon)f_o(\vec{\theta})\int\sigma_2 d\varepsilon d\vec{\theta}] - \gamma_4'[\sigma_2 - f_o(\vec{\theta})\int\sigma_2 d\vec{\theta}] +$$
$$+ (D_\theta \nabla_\theta^2 + D_r \nabla_r^2)\sigma_2; \qquad (1b)$$

$$\frac{\partial \sigma_{21}}{\partial t} = -iV(\sigma_1 - \sigma_2)/\hbar - \sigma_{21}[\gamma_2 + i(\omega_o + 2\varepsilon/\hbar)] \qquad (1c)$$

Here $\sigma_i = \sigma_{i\varepsilon, i\varepsilon}(\vec{\theta},\vec{r})$, $\sigma_{ij} = \sigma_{i\varepsilon, j\varepsilon}(\vec{\theta},\vec{r})$ - are elements of the density matrix; $V = V_{i\varepsilon, 2\varepsilon} = -(\vec{\mu}/2)\sum_{j=1}^{3}\vec{E}_j \exp(-i\omega_j t + i\vec{k}_j \vec{r}) + \mathrm{c.c.}$ - the interaction hamiltonian in a dipole approximation; $\hbar\omega_o$ - the forbidden band width.

A self-congruent problem of estimating the scattered field was determined by the plotted KE(1) and shortened expressions for field amplitudes[14]. Axis Z is normal to the flat nonlinear layer

LASER STUDY OF CONDENSED MATTER 249

with 1-width. The solution was made in relation ot the frequency and spatial $\vec{x} = (K_x, K_y)$ Fourier-components of the density matrix, using the methos of successive approximations. The method can be used in this case since the excitation is assumed to be weak. Determination of linear polarization allows to estimate the absorption coefficient of the medium in a linear approximation $S_i = 2\pi\omega_i\mu^2 N/3\hbar c\gamma_2^*$. Here N - is the concentration of particles; $\gamma_2^* = 2\Delta\epsilon/\pi\hbar$ - nonuniform width of transition; $2\Delta\epsilon$ - the width of the distribution function $f_o(\epsilon)$. γ_2^* was assumed to be much larger than the relaxation rates and detunings of the optical excitation frequencies $|\omega_i - \omega_j|$, (i,j = 1-4). The boundary problem of spatial diffusion was solved with boundary conditions of the second order corresponding to the absence of the excited particles flow on the boundaries of the nonlinear layer. The solution of standard shortened equations with calculated polarization allowed to estimate the amplitude of the scattered field.

The general expression obtained for $\vec{E}_4$ is a set of Lorents loops with width determined by combinations of rates of induced relaxation processes. A tensor nature of nonlinear susceptibility is described by $Y_{\alpha\beta\xi\eta} = (4\pi\mu^4)^{-1}\int\mu_\alpha\mu_\beta\mu_\xi\mu_\eta d\vec{\Theta}$; $Y_{\alpha\beta} = (4\pi\mu^2)^{-1}\int\mu_\alpha\mu_\beta d\vec{\Theta}$; $\alpha,\beta,\xi,\eta = x,y,z$.

Expressions for $\vec{E}_4$ can be substantially simplifyed through a particular selection of the interacting waves polarizations. In isotropic media there is a certain angle between the polarization of waves $\vec{e}_1 = \vec{e}_2$, $(\vec{e}_3 \hat{}\vec{e}_{1,2}) = \psi_o = \text{arctg}\sqrt{2}^{-4}$, at which the field amplitude $\vec{E}_4$ is independent of times responsible for the processes of orientational relaxation. The means of simplifying the obtained expressions, aimed at identifying the relaxation processes are discussed below.

3. PRINCIPLES OF IDENTIFICATION OF RELAXATION PROCESSES

Two substantially different groups - fast and slow relaxation mechanisms - are characteristic of semiconductor materials:

$$\gamma_2, \gamma_3', \gamma_3'', \tilde{\gamma}_4 \gg \gamma_1', \gamma_1'', \gamma_{5j}, \gamma_{6n} \qquad (2)$$

$$\tilde{\gamma}_4 = \tilde{\gamma}_4' + D_\Theta/6, \quad \gamma_{5j} = D_r|\vec{æ}_j - \vec{æ}_3|^2, \quad \gamma_{6n} = D_r(\pi n/l)^2; \quad j = 1,2; \quad n = 0,1,2,\ldots$$

Times shorter than 1 ps are typical of fast processes, while with slow ones times are more than 100 ps[1,2,4,7,10,13]. Among experimental techniques for measuring both fast and slow relaxation times, a combination of probe pulse and biharmonic pumping may be regarded as an optimal technique. It is necessary to use two frequency tunable pulse radiation sources with pulses duration $\mathcal{T}_p \simeq 10$ ps, which is intermediate for (2).

The measurements for slow times group are based on the probe pulse technique. Fields $\vec{E}_{1,3}$ superimposed in time with the polarization angle equal to ψ_o produce in the medium in question a diffraction lattice. It is probed by pulse $\vec{E}_2$ delayed by $\mathcal{T} \sim \gamma_{slow}^{-1} \gg \mathcal{T}_p$ relatively $\vec{E}_{1,3}$ fields. The expression for scattered field is now substantially simplifyed:

$$E_{4\alpha}^{(\psi_o)}(\mathcal{T}) \sim \sum_{n=0}^{\infty} C_{\alpha 21}^{(n)} \exp[-(\gamma_1 + \gamma_{51} + \gamma_{6n})\mathcal{T}], \qquad (3)$$

$$C_{\alpha ij}^{(n)} = [-\pi \omega_4 \mu^4 N \exp(-S_4 l) / \operatorname{ch}^3 \gamma_2^* l (1 + \delta_{on})] \psi(K_{iz} - K_{4z}, S_i - S_4) \cdot$$

$$\cdot \psi(K_{jz} - K_{3z}, S_j + S_3) \sum_{\beta, \xi, \nu} Y_{d\beta \xi \nu} e_{i\alpha} e_{j\beta} e_{3\nu}; \quad i,j = 1,2$$

$$\psi(x,y) = (ix - y)[1 - (-1)^n \exp(ix-y)]/[(ix-y)^2 - (\pi n/l)^2];$$

$$\gamma_1 = \gamma_1' + \gamma_1''$$

where δ_{on} - the Kronecker's symbol.

The processes of longitudinal relaxation and transverse spatial diffusion are separated by changing the incident angle of interacting waves $\vec{E}_{1,3}$, that changes the period of spatial lattice.

With fast times group $\gamma_{2,3,4}^{-1}$ the measurements are made by biharmonic pumping which is provided by frequency tunable field $\vec{E}_3$ and coincident fields $\vec{E}_1 = \vec{E}_2$. Depending on their polarization the scattered field is equal to

LASER STUDY OF CONDENSED MATTER   251

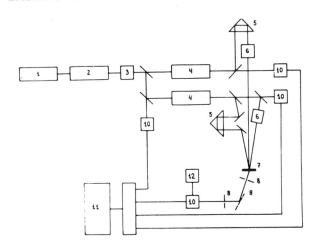

FIGURE 1 The diagram of experimental installation. 1 - main
         oscillator; 2 - amplifier; 3 - frequency doubler;
         4 - dye laser; 5 - delay line; 6 - rotators of pola-
         rization; 7 - sample; 8 - diafragms; 9 - monochromator;
         10 - radiation receptors; 11 - electronic computer;
         12 - photomultiplier power supply.

$$(\vec{e}_{1,2} \vec{\cdot} \vec{e}_3) = \pi/2; \ E_{4d}^{(\pi/2)} \sim [(\gamma_2 - i\Delta\omega_1)^{-1} + (2\gamma_3'/\gamma_2^*)(\gamma_4 - i\Delta\omega_1)^{-1}] /$$
$$/(\gamma_3' + \tilde{\gamma}_4 - i\Delta\omega_1);$$
$$(\vec{e}_{1,2} \vec{\cdot} \vec{e}_3) = \psi_o; \ E_{4d}^{(\psi_o)} \sim \sum_{n=0}^{\infty} C_{d\,u}^{(n)} / (\gamma_3 - i\Delta\omega_1)[(\gamma_2 - i\overset{\Delta}{\omega}_1)^{-1} +$$
$$+ (2\gamma_3/\gamma_2^*)(\gamma_1 + \gamma_{51} + \gamma_{6n} - i\Delta\omega_1)^{-1}] \quad (4)$$

$\gamma_3 = \gamma_3' + \gamma_3''; \ \gamma_4 = \tilde{\gamma}_4 - \gamma_3''; \ \Delta\omega_1 = \omega_1 - \omega_3.$

The technique of measuring the relaxation times that remain unknown will be discussed later.

4. EXPERIMENTAL SET-UP

Investigations were carried out with one of the most promising
semiconductors in quantum electronics[15-20] - monocrystalic films of
$\mathcal{E}$ - GaSe (15-20 μm) with optical axis perpendicular to the layer.

Block diagram of the experimental installation for studying
relaxation processes is shown in Figure 1. The duration of generation pulses of lasers on organic dye was determined using the system of coupled resonators[22]. The cell with the dye was placed

252 LASERS IN ATOMIC, MOLECULAR AND NUCLEAR PHYSICS

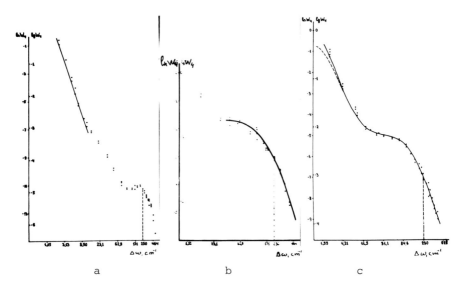

FIGURE 2 The method of biharmonic pumping. The dependence of the scattered field energy on the frequency detuning:
a - parallel waves polarizations
b - crossed polarizations
c - the angle between polarizations is equal to ψ_o.

inside the main resonator matched with the length of the main oscillator. The cell's walls formed a low - quality active nonlinear Fabry-Perot interferometer. At the cell width 2 mm DYE generated 20 ps pulses with limited spectrum width in the 590-650 nm range. Conversion efficiency in tunable radiation was 15%, peak power - 50 kW.

Figure 1 demonstrates the formation of beams used for measurements. The dynamic range of measurements covered 7 orders of signal values and was obtained through a change in the photomultiplier due to discrete calibrated switching of supply voltage[12]. The registration system worked in the mode of establishing digital "gates".

Thus, the installation described provided all the necessary degrees of freedom for experimental realization of the measurement procedure.

5. BIHARMONIC PUMPING

The experimental dependence of the scattered field energy W_4 on the central frequency separation of interacting pulses $\Delta\omega = \omega_{10} - \omega_{30}$ is given in Figure 2 for various polarizations of interacting waves. A typical feature of these curves is the presence of three essentially different sections - central peak, a well pronounced smooth segment near the inflection point with further sharp decline.

Comparison of theoretical and experimental results, as well as determination of the fast relaxation processes times can be made only through thoroughly stepwise analysis of various sections in all curves. Only this procedure makes it possible to separate one relaxation mechanism after another at each section and eliminates the need for a large number of adjusting parameters. First of all one should take the integral over pumping spectra in[4]:

$$E_j(\omega_j) \sim 1/[1 - i(\omega_j - \omega_{jo})/\delta\omega], \quad j = 1,2,3; \quad \omega_{10} = \omega_{20},$$

which corresponds to an exponential pulse shape.

A. <u>Parallel Polarization</u>. In the range of low detunings, there occurs an "anomalous" growth of W_4, while the relative rise of the central peak is $\sim 10^5$. In the range of $\Delta\omega \simeq 200\text{-}250 \text{ cm}^{-1}$ a characteristic rise is seen on the curve. The rise can be explained by scattering on the optical phonons LO_1 with frequency $\sim 250 \text{ cm}^{-1}$ [20]. Thus, the polariton scattering previously not taken into account is quite important here. Its acoustical branch accounts for the path of the curve in the range of low detunings. The discrete acoustical modes, however, are not resolved[12], since the pumping spectrum width is quite large.

In the first approximation the effect of polariton scattering can be considered independently. It means that Mandelshtam and Brillouin cubic nonlinearity should be added to the following expression[12,14]:

$$\chi_{\alpha\beta\xi\eta} \sim (\delta_{\alpha\beta}\delta_{\xi\eta} + \delta_{\alpha\xi}\delta_{\beta\eta})/(\Omega^2 - \Delta\omega^2 - 2i\gamma\Delta\omega) \tag{5}$$

where α,β,ξ,η = x,y,z, Ω is frequensy and γ^{-1} is life time of acoustical phonon. In parallel pumping light polarizations the contribution of this nonlinearity is maximal and predominates in the range of zero detunings. The presence of such slow nonlinearity made it possible to measure the function of field coherence[21].

The initial section of the surve (Figure 2a) is in a good agreement with a theoretical decline obtained for Mandelshtamm and Brollouin nonlinearity (5) at pumping spectrum width $\delta\omega$ = 1.08 ± 0.10 cm^{-1}.

B. <u>Crossed Polarizarions.</u> Due to selection rules (5) the processes of polariton scattering exert no influence of the measurement results. Some rise in the curve in the range of law detunings is determined by an error of $\sim 2°$ in the adjustment of the rotation angle in the polarization plane.

In $\Delta\omega \gg \delta\omega$ frequency range, the theoretical curve should be described by expression (4), where the second member can be neglected and in accordance with (2) one can write:

$$\gamma_2 \approx \tilde{\gamma}_4 + \gamma_3' \tag{6}$$

Thus, in this case the superimposing of theoretical and experimental dependencies $W_4(\Delta\omega)$ by only one adjustment parameter γ_2 (Figure 2b) allows to estimate unambiguously the rate of transverse relaxation in the sample γ_2 = 230 ± 10 cm^{-1}.

The superimposing of theoretical surves with experimantal points was carried out using the least squares method. The error in the value of adjustment parameter corresponds to a two fold increase in the sum of deviation squares.

C. <u>Polarizations Eliminating the Times of Elastic Orientation relaxations.</u> A stepwise superimposing of calculated curve with experimental relationship allows to determine in sequence the rate of inelastic interband relaxation γ_3 and γ_2^*.

For samples of gallium selenide 15-20 μm in width and the angle between the directions of pumping light propagation 9.2°, the coefficients $C_{\alpha\|}^{(n)}$ in (4) decrease so rapidly that it is sufficient ot use only the first of them. In this way, the curve slope after inflection is determined by the first term in (4) comprising the previoudly found rate γ_2 and the sole unknown value γ_3. The calculated curve (Figure 2b) in the $\Delta\omega > 50$ cm^{-1} range corresponds to $\gamma_3 = 150\pm15$ cm^{-1}. The rate of elastic orientation relaxation in accordance to (6) is equal to $\gamma_4 = \gamma_2 - \gamma_3 = 80\pm15$ cm^{-1}. The expressins $\gamma_3/\gamma_2 = 0.65\pm0.05$ and $\gamma_4/\gamma_2 = 0.35\pm0.05$ are determined more accurately.

After the integration over the pumping spectrum, the theoretical expression for W_4 in the low detunings range contains again only one unknown value $2\gamma_3/\gamma_2^*$. The calculated curve (Figure 2b, dash-line) corresponds to $2\gamma_3/\gamma_2^* = (4.2\pm0.2)10^{-2}$ and estimates $\gamma_2^* = (6.7\pm0.5)10^3$ cm^{-1}. The full line in Figure 2c corresponds to the additive contribution of Mandelshtamm and Brillouin nonlinearity (5) with a relative weight $(8\pm2)10^{-4}$.

6. THE PROBE BEAM METHOD

In measuring the slow relaxation times group (2), the scattered field is determined by expression (3), where it is still enough to leave only the first term of the sum with n = 0, while $\gamma_{60} = 0$.

Figure 3a shows the experimental dependence of the scattered field energy on time delay between pulses $W_4(\tau)$. The angle between the pumping beams $\vec{E}_{1,3}$ was $(9.2\pm0.2)°$.

The scattering on the lattice of long-lived acoustical phonons and accumulation along train led to formation of background, the level of which can be determined by the value W_4 in the $\tau < < -\tau_p < 0$ range. The subsequent "deduction" allows to calculate the effective relaxation rate of electronic nonlinearity.

256 LASERS IN ATOMIC, MOLECULAR AND NUCLEAR PHYSICS

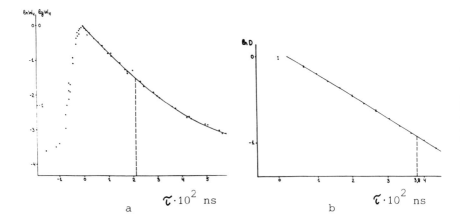

FIGURE 3 The probe beam method
a - dependence of the scattered field energy on time delay between light pulses;
b - dependence of the sample optical density on time delay

$$\gamma_{eff} = \gamma_1 + \gamma_5 = (4{,}69\pm0.21)10^9 \text{ s}^{-1} \qquad (7)$$

We have also carried out the direct determination of the rate γ_1 according to classical technique of measuring the relationship between the sample optical density and the delay time of probing pulse with respect ot the pumping light pulse^4 (Figure 3b). The rate of longitudinal relaxation was $\gamma_1 = (2.63\pm0.14)10^9 \text{ s}^{-1}$. It follows from (7) that the rate of spatial diffusion $\gamma_5 = (2.06 \pm 0.35)10^9 \text{ s}^{-1}$, which corresponds to the value of spatial diffusion coefficient (3) $D_r = (7.5\pm1.4) \text{ cm}^2/\text{s}$.

7. CONCLUSIONS

It should be noted that in gross limiting cases the theoretical results obtained in this work turn into relationships obtained earlier by a number of authors[1-7]. At the same time they account for some observed "anomalies" in the behaviour of nonlinear responce, clarify the interpretation of the measured values and establish the limits of the models validity.

LASER STUDY OF CONDENSED MATTER

Within the frameworks of phenomenological theory the optimal combination of biharmonic pumping and probe beam made it possible to measure and identify the times of the following relaxation prosesses in

ε - GaSe:
1. transverse relaxation $\quad T_2 = 23.1 \pm 1.6$ fs
2. inelastic intraband relaxation $\quad T_3 = 36 \pm 4$ fs
3. elastic intraband relaxation $\quad T_4 = 69 \pm 13$ fs
4. interband relaxation $\quad T_1 = 380 \pm 20$ ps
5. transverse spatial diffusion $\quad T_5 = 490 \pm 80$ ps

which corresponds to the value of the spatial diffusion coefficient $D_r = 7.5 \pm 1.4$ cm^2/s.

The experimental studies performed at room temperature on thin monocrystallic films of ε-GaSe confirm the applicability of theoretical analysis based on the phenomenological description of relaxation processes, while their interpretation does not require the theory of nonmarkovian relaxations.

REFERENCES

1. A. L. SMIRL, Dynamics of High-Density Transient Electron-Hole Plasmas in Germanium (Preprint Academic Press, New Yord, 1983)
2. T. E. BOGGES, A. L. SMIRL, B. S. WHERRET, Opt. Comm., 42, 128 (1982)
3. T. YAJIMA, Y. TAIRA, J. Phys. Soc. Jap., 47, 1620 (1979)
4. H. E. LESSING, A. VON JENA, Chem. Phys. Lett., 42, 213 (1976)
5. T. YAJIMA, H. SOUMA, Phys. Rev., A-17, 309, 324 (1978)
6. J. J. SONG, J. H. LEE, M. D. LEVENSON, Phys. Rev., A-17, 1439 (1978)
7. S. Y. YUEN, P. A. WOLFF, Appl. Phys. Lett., 40, 457 (1982)
8. M. A. VASIL'EVA, V. I. MALYSHEV, A. V. MASALOV, Kratkie Sobshenia po Fizike FIAN, No. 1, 35 (1980)
9. P. A. APANSEVICH, Osnovy Teoree-Vzaimodeistvia Sveta s Veshestvom (Nauka i Tehnika, Minsk, 1977)
10. V. L. VINETSKY, N. V. KUKHTAREV, M. S. SOSKIN, Kvantovay Electronica, 4, 420 (1977)
11. V. M. PETNIKOVA, S. A. PLESHANOV, V. V. SHUVALOV, Preprint MGU, Fizicheskiy Facultet, No. 18 (1984)
12. N. I. KOROTEEV, M. F. TERNOVSKAY, Kvantovay Electronica, 9, 1967 (1982)
13. A. I. ANSELM, Vvedenie v Teorij Poluprovodnikov (Nauka, Moskva, 1978)

14. S. A. AKHMANOV, N. I. KOROTEEV, Metody Nelineinoy Optiky v Spektroskopee Rassenia Sveta (Nauka, Moskva, 1978)
15. Z. S. MEDVEDEVA, Halcogenidy Elementov 3B Podgrupy Periodicheskoi Sistemy (Nauka, Moskva, 1968)
16. V. V. SOBOLEV, Zony i Exitony Halcogenidov Gallia, India i Tallia (Shtiincea, Kishinev, 1982)
17. S. S. YAO, J. BUCHERT, R. R. ALFANO, Phys. Rev., B25, 6534 (1982)
18. J. L. STAEHLI, A. FROVA, Physica (Utrecht), 99B, 299 (1980)
19. V. ANGELLI, C. MANFREDOTTI, R. MURRI, L. VASANELLI, Phys. Rev., B17, 3221 (1978)
20. R. BALTRAMEJUNAS, G. GUSEINOV, V. VARKEVICIUS, V. NIUNKA, J. VAITKUS, J. VISCAKAS, Opt. Comm., 11, 274 (1974)
21. R. BALTRAMEJUNS, J. VAITKUS, R. DANELIUS, M. PETRAUSKAS, A. PISKARSKAS, Kvantovay Electronica, 9, 1921 (1982)
22. V. M. PETNIKOVA, S. A. PLESHANOV, V. V. SHUVALOV, Kvantovay Electronica, 11, 1668 (1984)

3.5 Photochemical and Photophysical Hole Burning in Electronic Spectra of Complex Organic Molecules

R. I. PERSONOV and B. M. KHARLAMOV

Institute of Spectroscopy, Academy of Sciences USSR, 142092 Troitzk, Moscow Region, USSR

1. INTRODUCTION

The last 10-15 years were marked by a considerable progress in the understanding of the spectral band broadening nature in electron-vibrational spectra of complex organic compounds in solutions. It was established that in many cases, at sufficiently low temperatures, wide spectral bands (typical for solutions) are broadened mainly inhomogeneously and conceal a great number of narrow zero phonon lines (ZPL) [1,2]. It was also established that in case of the so-called quasi-line spectra of some organic impurity crystals the inhomogeneous line broadening proves to be decisive. By now methods have been developed which are capable of eliminating this inhomogeneous broadening and revealing the line structure in fluorescence, phosphorescence, and absorption spectra of complex molecules in arbitrary solvents by means of selective laser excitation (see, e.g. reviews [3,4] and references therein). These methods are often called "Site Selection Spectroscopy"*. In the emission spectra inhomogeneous broadening can be eliminated upon selective excitation of one type centres with an absorption line at laser

---
*) Upon monochromatic irradiation one excites selectively the molecules with identical energy of electron-vibration transition. Then, strictly speaking, it would be more correct to say "Energy Selection Spectroscopy".

frequency (sometimes this method is called "Fluorescence Line Narrowing Spectroscopy"). The second method enables to reveal a fine structure in absorption spectra. It is based upon selective photochemical or photophysical transformation of centres absorbing laser radiation. After transformations of such kind narrow stable holes can be observed in broad absorption bands at the electonic-vibrational transition frequencies of "burnt" molecules. At present this method is known as "Hole-Burning Spectroscopy". By now all the main peculiarities and characteristics of line spectra, obtained via selective methods, have been studied. The said methods opened up new possibilities for fine spectroscopic investigations of complex organic molecules and their interaction, and for many new practical applications.

The present review deals only with one of the above selective methods, i.e. the hole burning method.

2. PRINCIPLES OF THE METHOD

2.1. Essence of the Method

In order to eliminate inhomogeneous broadening in absorption spectra one must selectively effect the object so that it will be possible to separate a homogeneous component of the spectrum. In gaseous spectroscopy of atoms and simple molecules (sometimes even of solids) the saturation of absorption upon intensive monochromatic pumping is widely used. In our case we deal with effects of another type. Here the hole-burning is carried out by means of selective low-temperature photoreactions in a solid solution[5,6] (see also reviews [3,4,7-9] and references therein). If the photoreaction takes place, the absorption spectrum of the phototransformed molecules is shifted. The concentration of molecules absorbing at the "burning" frequency diminishes. This results in the emergence of a narrow hole in the inhomogeneously broadened absorption band. The

LASER STUDY OF CONDENSED MATTER 261

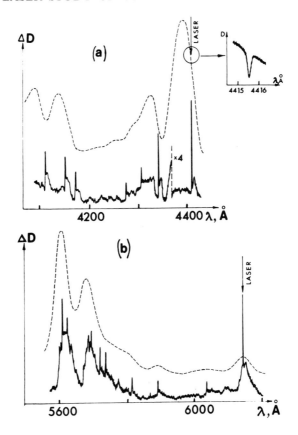

FIGURE 1 "Hole-burning" spectra (solid lines) at 4.2 K: (a) perylene in ethanol (burning time t = 5 min, P = 5 mW/cm^2), the inset shows the hole in absorption band burnt at laser frequency; (b) porphin in polystyrene (t = 1 min, P = 5 mW/cm^2). Dashed lines indicate the absorption spectra of the same samples.

life-time of such a hole is determined by the rate of the back reaction (if it is reversible) and turns out to be very long. At low temperature in the darkness these stable holes can conserve for dozens and hundreds of hours. As an example Figure 1a (top right) demonstrates one of the first results in stable-hole-burning.

2.2. Hole-Burning Spectra

It is clear, that the hole-burning with a monochromatic source results in the emergence of a hole in the density distribution function $n(\nu)$ of guest molecules along the purely electronic transition frequency. However, holes are to appear not only at the burning frequency but at all the vibronic transition frequencies of burnt molecules, as well. The difference of the sample absorption spectra before and after burning is called the "hole-burning spectrum"[10]. It is a fine-structure absorption spectrum of "burnt" molecules. Figure 1 exhibits two examples of the hole-burning spectra: for perylene in ethanol and for porphin in polystyrene. One can see that along with the 0-0 line (the hole at laser frequency) these spectra contain a great number of vibronic lines. The hole-burning spectrum, in particular, permits to identify the vibration frequencies of the molecule in the excited electronic state.

2.3. Burning Mechanisms

After first publications[5,6] hole-burning has been observed on many objects. At present it is widely employed to investigate homogeneous spectra characteristics of organic molecules in matrices[7-9]. In these procedures one deals with most various hole-burning mechanisms that can be divided into two types.

The first type is characterized by various photochemical reactions (photochemical hole-burning - PHB): photodissociation[11], proton phototransfer[12], electron phototransfer[13], etc. Specific mechanism of PHB in case of metal-free porphyrins is widely known, in particular. It is connected with the transfer of two protons in the centre of the porphin ring[14,15] (Figure 2a). Such modification in the proton pair position in the molecule fixed in a solid matrix results in its spectrum displacement by dozens or hundreds of wavenumbers. Figure 2b shows another example of the photoreaction observed in solid solutions of quinizarin[12]. In this case, upon irradiation, there is a break (or formation) of a hydrogen

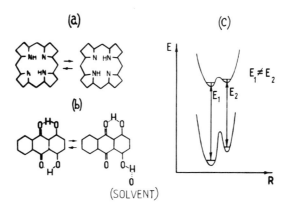

FIGURE 2 Some photoreactions in low temperature solutions: (a) proton tautomerization in free-base porphyrins; (b) intramolecular hydrogen-bond break (intermolecular hydrogen-bond formation) in quinizarin and some other molecules; (c) two-well potentials for ground and excited states of guest molecule.

bond. In both examples the reaction is reversible (i.e. upon further irradiation in the absorption band of the photoproduct the initial spectrum is restored).

Of considerable interest is the second type of hole-burning processes - the so-called nonphotochemical hole-burning (NPHB) first observed in[1,2,5]. This term is used for photoprocesses which take place in low-temperature solutions of photostable (under ordinary conditions) molecules[8]. NPHB is apparently non-specific and can be observed in polar glassy solutions (e.g. in alcohol solutions) of many compounds and in some polymer matrices. As a rule, NPHB is not observed in non-polar crystalline matrices (e.g. in crystalline n-paraffins). By now this mechanism has not yet been established in full detail. One of the possible NPHB models (the so-called reorientation model[16]) suggests there be 2 (or more) minima at the potential curves of the guest molecule electronic states in a disordered matrix (Figure 2c)[8]. The displacement of the molecule after excitation into the other minimum results in the frequency change of the purely electronic transition, and a hole emerges at burning frequency. From the qualitative viewpoint this model is in full accord with the NPHB one-quantum character and its universality. However, certain detailed investigations of the NPHB peculiarities on some objects show a picture that does not fit this model[17]. Possibly, the NPHB is not that universal and can have a different nature (in some cases positively photochemical) for various compounds.

2.4. Shape of Hole Profile

Let us analyse the shape of an optical band in the hole-burning spectrum in more detail. It is necessary to take into account that the spectral absorption band of the impurity centre is characterized not only by a zero phonon line (ZPL) but also by a broad phonon wing (PW), placed on the shortwave side from ZPL. The burning is realized not only via narrow ZPLs (i.e. "resonantly") but also via broad PWs (i.e. "nonresonantly"). This results in the additional wide distribution in hole-burning spectrum which is located both on the shortwave and on the longwave side from ZPL and determines the specific shape of the hole (Figure 1).

The shape of the inhomogeneously broadened absorption band is described by the convolution of the absorption band profile of a separate impurity centre - $\varepsilon(\nu - \nu_0)$ with the distribution function of the number of impurity centres along the 0-0 transition frequency - $n(\nu - \nu_1)$. It is clear that the band shape in the hole-burning spectrum - $\Delta\varepsilon(\nu)$ is determined then by the convolution $\varepsilon(\nu - \nu_0)$ with the distribution function difference $\Delta n(\nu_0 - \nu_1)$ before and after burning:

$$\Delta\varepsilon(\nu) = \int \varepsilon(\nu - \nu_0) \Delta n(\nu_0 - \nu_1, \nu_B) d\nu_0 \qquad (1)$$

where ν_B is the burning frequency.

Generally speaking, the shape of $\Delta n(\nu_0 - \nu_1, \nu_B)$ depends on

LASER STUDY OF CONDENSED MATTER

the intensity and the time of laser burning, on the position and shape of the photoproduct absorption band, etc. If we neglect the overlapping of absorption bands of the initial substance and the photoproduct then at small burning irradiation intensity (and at one-quantum burning mechanism) the expression for Δn is relatively simple:

$$\Delta n(\nu_o - \nu_1, \nu_B) = n(\nu_o - \nu_1)[1 - \exp[-kA\varepsilon(\nu_B - \nu_o)]] \quad (2)$$

where $A = It$ is the exposition, and k is the constant. The latter depends upon the photoreaction mechanism and the kinetic parameters of the molecule. At small expositions instead of (2) one can write:

$$\Delta n(\nu_o - \nu_1, \nu_B) \simeq kA\varepsilon(\nu_B - \nu_o)n(\nu_o - \nu_1) \quad (3)$$

When the distribution $n(\nu - \nu_1)$ is considerably wider than the homogeneous absorption band $\varepsilon(\nu - \nu_o)$, one can approximately present $n(\nu - \nu_1) = \text{const}$. In this case it follows from (2) that:

$$\Delta n(\nu_o - \nu_1, \nu_B) \sim \varepsilon(\nu_B - \nu_o) \quad (4)$$

i.e. the distribution Δn repeats the shape $\varepsilon(\nu - \nu_B)$ reflected mirror-like with respect to ν_B. In this case (1) looks as follows:

$$\Delta \varepsilon(\nu) \sim \int \varepsilon(\nu - \nu_o)\varepsilon(\nu_B - \nu_o)d\nu_o \quad (5)$$

The expression (5) is an autocorrelation function and thus, when it is valid, the band shape in the absorption spectrum must be symmetrical with respect to ν_B.

The band shape is schematically exhibited in Figure 3. The molecules burnt via ZPL (i.e. resonantly) form a narrow resonant hole (and a shortwave PW). The molecules burnt via PW (i.e. non-resonantly) yield a wide distribution, mainly on the longwave side and partly on the shortwave side from the narrow hole.

If we present the true absorption band as the summation of two terms: ZPL - $\varepsilon_1(\nu)$ and PW - $\varepsilon_2(\nu)$, then the expression (5) is

266 LASERS IN ATOMIC, MOLECULAR AND NUCLEAR PHYSICS

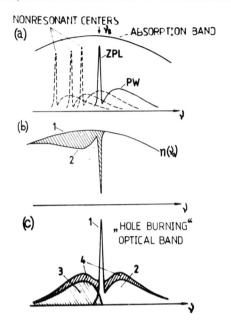

FIGURE 3 Formation of optical band in a hole-burning spectrum:
(a) inhomogeneously broadened absorption band and homogeneous bands of resonantly (solid line) and nonresonantly (dashed lines) absorbing molecules; (b) impurity centres distribution function $n(\nu_o)$ before - (1) and after burning - (2). The hatched area represents a burnt part - $\Delta n(\nu_o)$. The narrow hole is formed by the resonantly burnt molecules, the wide longwave distribution - by the nonresonantly burnt molecules. (c) Optical band in the hole-burning spectrum. The figures 1-4 mark the contributions of different terms in (5).

the summation of four terms. The narrow hole $\Delta\varepsilon_1(\nu)$ is determined by one of these terms:

$$\Delta\varepsilon_1(\nu) \sim \int \varepsilon_1(\nu - \nu_o)\varepsilon_1(\nu_B - \nu_o)d\nu_o \qquad (6)$$

When ZPL has the Lorenzian shape (as usually happens), the narrow hole (6) has the Lorenzian shape, as well. Its width Γ_h is connected with homogeneous ZPL width as follows:

$$\Gamma_h = 2\Gamma_{ZPL} \qquad (7)$$

LASER STUDY OF CONDENSED MATTER   267

This relation is usually employed in the investigation of homogeneous ZPL width via hole-burning method.

With the exposition increase the shape $\Delta n(\nu)$ is not identified any more by (4) and is continuously varying with time. In this case we observe the change in both the shape of the narrow resonant hole and the shape and relative intensity of broad wings.

With burning time increase the narrow hole is broadened. In fact, those molecules are burnt very fast which have their ZPL maxima located very closely to the ν_B frequency. Due to the reduction in these centre concentration their burning rate is gradually slowed down in comparison with the centres the ZPL maxima of which are more displaced from ν_B (by the value of the Γ_{ZPL} order). To take into account the hole broadening related to the said, one usually extrapolates the measured width Γ_h to the limit $t \rightarrow 0$.

As was first noted in[18] there exists one more mechanism of hole broadening which is valid even at small burning time-intervals. This mechanism is due to the population of a metastable triplet state. High population of a triplet state, possible even at moderate intensities of burning radiation, results in the non-linear sample bleaching and in the hole broadening related to it. In this case the hole width depends upon laser intensity and is expressed as follows[18]:

$$\Gamma_h = \Gamma_{ZPL}(1 + \sqrt{1 + p}); \quad p = I\varepsilon_{max}\Phi_T/K_T \tag{8}$$

where I is the burning source intensity; ε_{max} is the absorption coefficient in the ZPL maximum; K_T is the deactivation rate of the triplet state; Φ_T is the output of interconversion. At sufficiently low temperatures, when the ZPL width tends to the radiation limit ($\sim 10^{-3}$-10^{-4} cm^{-1}), ε_{max} reaches the values of 10^{-10}-10^{-11} cm^2. So the saturation can be achieved at the source intensities of the order of 10-100 $\mu W/cm^2$ [19].

With the exposition increase one can observe the deformation of broad wings along with the narrow hole broadening. A longwave wing component, in this case, grows faster (due to a relative increase in the integral contribution of the non-resonantly burnt centres as compared to the resonant centres). As was shown in[19] the saturation of a triplet state results in a similar increase in the longwave part of the wing at sufficient sourse intensity. This effect takes place even at the smallest expositions. It is worth noting that the wing assymetry can be due to the band overlapping of the initial compound and the photoproduct, to peculiarities of the true shape of the $n(\nu - \nu_1)$ function, etc. All these factors should be taken into consideration at the quantitative analysis of the hole-burning spectra.

3. METHOD APPLICATIONS

The hole-burning method opened up new possibilities for fine spectroscopic investigations of complex molecules in a condensed medium. These possibilities are: investigation of the vibrational structure of excited states; determination of homogeneous absorption spectra characteristics and of electronic and vibrational relaxation constants; studying of electric and magnetic field influence upon molecular spectra; determination of the photochemical reaction rates by the hole width; investigation of glass-like state peculiarities; etc. Here we shall briefly consider only some of them.

3.1. Measurement of Homogeneous ZPL Width and its Temperature Dependence

The homogeneous line width and its temperature dependence contain valuable information on the electron-phonon coupling in solid solutions. For the last 7-8 years a big number of investigations on the ZPL temperature broadening of organic molecules in crystalline and glassy matrices have been performed[20-23] (see also reviews[7,8]).

LASER STUDY OF CONDENSED MATTER

For a number of molecules, both in crystalline and in glassy matrices, a radiation limit of the width ($\Gamma \sim 10^{-3}$ cm^{-1}) was obtained. This occurred upon measuring the ZPL width in the low temperature range and upon extrapolation of results to a T $\rightarrow$ 0 limit[21-23]. However, in other cases of molecules in glassy matrices the evaluation of Γ at the T $\rightarrow$ 0 yields appreciably greater values: 0.1 - 1 cm^{-1} [8,9]. These values are 2 or 3 orders greater than the radiation width. The reasons for this are not quite clear. One of the possible reasons, briefly discussed in[9], is the presence of relaxation processes in glasses developing even at extremely low temperatures. These processes chaotically shift the absorption lines of separate guest molecules and result in the so-called spectral diffusion. If typical-for-these-processes periods are less than the observation time, they will give an additional contribution to the ZPL width being measured. But the fact, whether these processes play an important role or not at low temperatures, does not have a reliable experimental proof.

In the course of investigations it turned out that the temperature dependence of $\Gamma(T)$ for molecules in crystalline and glassy matrices is appreciably different. While dealing with crystalline matrices, the temperature broadening appears to be either $\sim T^2$, or $\sim \exp(-E/kT)$. That falls in line with the theoretical assumptions on the interaction of an impurity centre with local or quasilocal vibrations. At the same time, when treating glassy matrices, in the majority of cases there was discovered a specific and universal enough dependence: $\Gamma(T) \sim T^{1.3 \div 1.4}$. This dependence does not fit the ZPL broadening theory so well-developed earlier. In the works of recent years[24-26] this unconventional temperature dependence is explained through the interaction of the impurity centre with the so-called two-level systems (see reviews [8,9] and references therein). At first these two-level systems

(TLS) were introduced to explain the abnormal behaviour of specific heat, thermal conductivity, saturation effects in ultrasonic absorption, etc. at low and extremely low temperatures (see, e.g. [27]). As different from phonons, such TLS are the Fermi systems. Therefore, if we take into account the impurity centre - TLS interaction we can obtain the following temperature dependence: $\Gamma(T) \sim T^n$ with the n values in the interval $1 \leqslant n \leqslant 2$ (see [26] and references therein). Thus, the use of TLS model enables to explain the ZPL broadening observed in molecular spectra in glasses. However, more direct proofs of the TLS influence upon the optical impurity characteristics are, apparently, needed to finally solve the problem. Physical concretization of this phenomenological model is also necessary.

3.2. Stark Effect

The possibilities to investigate the influence of external electric and magnetic fields (Stark and Zeeman effects) upon the spectra of complex molecules are essentially limited by a large width ($\Delta \nu \sim 10^2 - 10^3$ cm^{-1}) of their spectral bands. Thus, for instance, the value of the absorption band shift under field influence is often two or three orders of magnitude less than their width. The situation is considerably improved if we follow the changes in the profiles of narrow holes under field influence. Owing to the small hole width ($\Delta \nu \sim 10^{-1} - 10^{-2}$ cm^{-1}) the sensitivity of measurements increases by several orders.

Investigations of the Stark effect (which serve, in particular, to determine dipole moments and polarizabilities of molecules in excited electronic states) via hole-burning method were started in[28-31]. In order to successfully interpret experimental data it was necessary to analyse the dependence of a hole profile upon the external field magnitude. Such theoretical analysis was performed for several simple cases in[29]. It was supposed that under the

LASER STUDY OF CONDENSED MATTER

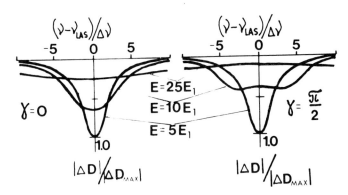

FIGURE 4 Dependence of the hole profile on external electric field (linear Stark effect). Parameters: $\gamma = \angle(\vec{\mu}, \vec{\mu}_{ga})$; $2\Delta\nu$ - homogeneous line width; $E_1 = hc\Delta\nu/(f_e |\Delta\vec{\mu}|)$.[29]

external field influence ZPL absorption maxima of all molecules in solid solution undergo the displacements $\Delta\nu_0$:

$$\Delta\nu_0 = -(f_e \Delta\vec{\mu} \vec{E} + \tfrac{1}{2} f_e^2 \vec{E} \Delta\hat{\alpha} \vec{E})/(hc) \tag{9}$$

where $\Delta\vec{\mu}$ and $\Delta\hat{\alpha}$ are changes of a static dipole moment vector and of tensor of molecular polarizability at electronic transition, and f_e is a local field factor. The first term in (9) corresponds to a linear Stark effect and the second - to a quadratic one. Using (9) and (5) it is possible to obtain an expression for the hole profile in the presence of the field (this is done with taking into account orientation factors and averaging over all possible orientations of molecules with respect to the field direction). The result depends on mutual orientation of the transition dipole moment $\vec{\mu}_{ga}$, $\Delta\vec{\mu}$ and polarizability tensor axes. Some results of numerical calculations for the dependence of the hole profile upon the applied field magnitude for the linear Stark effect are demonstrated in Figure 4. It is clearly seen, that the hole, upon the field influence broadens, but remains symmetrical with respect to the initial maximum position. Its shape depends upon the angle γ between $\vec{\mu}_{ga}$ and $\Delta\vec{\mu}$. According to the calculation, in case of the quadratic Stark effect the hole also broadens, but becomes asymmetrical and is shifted along the frequency scale[29]

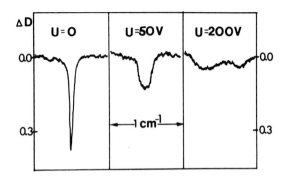

FIGURE 5 Change in the hole profile (linear Stark effect) with the voltage applied to 40 μm thich chlorin in polyvinylbutural film at T = 2 K.[30]

(see next section about the quadratic Zeeman effect).

Figure 5 exhibits the experimental data for the linear Stark effect on the hole burnt in the absorption band of chlorin in polyvinylbutyral. Comparing the experimental data for chlorin with the calculated ones it was possible to fing $|\Delta\mu|$ = 0.2-0.4 Debye.

Along with the spectral measurements of the hole shape depending upon the external field magnitude it is possible to make another type of the Stark hole experiments, i.e. to measure the so-called "field curves"[29,32]. In the latter case we register the change in the optical density $\Delta D(E-E_B)$ (where E_B is a field at which the burning is performed) at a fixed frequency ν (e.g. $\nu = \nu_{las}$) as a function of the applied field E. A "field curve" $\Delta D(E-E_B)$ has the maximum at $E=E_B$, and its width is connected with the homogeneous absorption ZPL width of separate impurity centres. At known Stark molecular parameters the $\Delta D(E-E_B)$ measurements can serve as a relatively simple means for determining homogeneous ZPL width[32]. Such measurements are rather simple, do not need any spectrometer and can be performed by means of one laser with a fixed frequency.

The application of modulation technique seems to be also quite perspective for the investigations of the Stark effect on the holes[33]. We only note that the Stark hole measurements turn out to be a subtle tool in studying impurity molecular crystals with line spectra, as well[34].

3.3. Zeeman Effect

Similar to the Stark spectroscopy the use of narrow spectral holes enable to appreciably increase the sensitivity of Zeeman experiments. In the last 5 years magnetic properties of singlet states of complex molecules have been investigated via this method.

As known, owing to low symmetry of compex molecules the latter, as a rule, do not conserve either the electronic moment or its projection on the molecular axis. Diamagnetism is a most characteristic manifestation of magnetic properties of complex molecules. In some cases, however, when the electronic levels of the corresponding symmetry are closely located we can come across the so-called Van-Fleck paramagnetism. The ZPL shift magnitude is propotional to H^2 and is expressed as follows:

$$\Delta \nu_o = - \frac{1}{2} \vec{H} \Delta \hat{\chi} \vec{H} \tag{10}$$

where $\Delta \hat{\chi}$ is the magnetic susceptibility change of the molecule at electronic transition ($\Delta \hat{\chi}$ is a summation of dia- and paramagnetic terms). The magnitudes of the effects expected are very small ($< 10^{-4}$ cm^{-1}/kG^2), and under ordinary conditions are not accessible for observation. The hole-burning method made it possible to perform direct spectroscopic measurements of such shifts. The exceptional narrowness of the holes burnt in the absorption bands of free base porphyrins ($\Delta \nu \sim 10^{-2}$-10^{-3} cm^{-1}) was used to investigate induced paramagnetism in porphin and chlorin in the n-paraffin monocrystals[35,36]. In this case, magnetic shifts ≤ 0.1 cm^{-1} in the fields up to 86 kG were measured.

A much wider variety of objects to be investigated and the simplicity in sample preparation are provided by glassy solvents and polymers, and not by n-paraffin matrices. Two specific points should be born in mind while using such amorphous isotropic matrices. First, as was noted, the hole width for amorphous matrices is

274 LASERS IN ATOMIC, MOLECULAR AND NUCLEAR PHYSICS

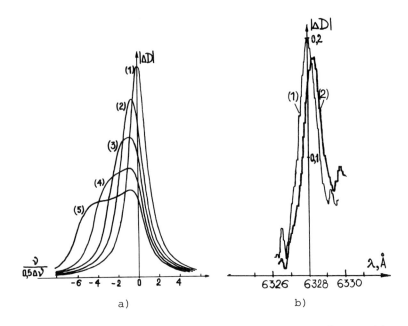

FIGURE 6 a) Dependence of the hole profile on external magnetic field (quadratic Zeeman effect), calculated for the case, when the electronic transition dipole moment is perpendicular to z-axis of molecule and magnetic field direction coincides with incident and registered beam direction[37]. b) Quadratic Zeeman effect on the hole in absorption band of chlorin in polystyrene film ($\lambda_{burning}$ = 6328 Å, P = 1 mW/cm^2, t = 5 min): (1) - hole at H = 0; (2) - hole at H = 400 kG. Experimental conditions are the same as in calculated case (a)[37].

appreciably larger than for crystalline ones. At helium temperatures this width is usually 0.1 - 1 cm^{-1}. Second, owing to the system isotropy it is necessary to average over all possible molecular orientations while theoretically analysing the experimental data. In case of the quadratic Zeeman effect in accordance with (10) the hole is broadened asymmetrically and is shifted (as in the quadratic Stark effect). In order to carry out reliable experimental observations of small magnetic shifts (even employing hole-burning) it is necessary, in such situation, to make use of strong fields produced in pulse set-ups. The first Zeeman effect measurements on the holes in the strong pulse magnetic fields (up to 400 kG) were performed in[37]. Figure 6a displays the calculated dependence of the hole profile upon a field magnitude. Experimental data for chlorin in a polymer film are shown in Figure 6b. These data yielded a magnetic susceptibility value for chlorin: $\chi = (2.6\pm0.3)\cdot10^{-5}$ cm^{-1}/kG^2 (which practically coincides with the measurement data on monocrystals in stationary fields[35]).

In conclusion, it is worth noting that the strong pulse magnetic field technique enables to investigate magnetic properties of high excited electronic states (S_2 and higher), transitions into which yield relatively broad ZPLs ($\Delta\nu \sim$ 1-10 cm^{-1}). Such measurements can assist in solving the problem of the contribution separation of dia- and paramagnetic parts to the measured χ value.

3.4. Vibrational Relaxation. Rates of Photochemical Reactions

Here we shall briefly point to one more type of information which can be obtained upon the hole width measurement. The homogeneous width and the temperature ZPL broadening have been already discussed. This temperature broadening is conditioned by a so-called pure dephasing time T_2^*. (In the considered cases the ZPL width tended to a natural limit determined by a life-time T_1 only at $T \rightarrow 0$.) However, even at $T \neq 0$ one can come across a situation when $T_1 \ll T_2^*$, and the ZPL width is determined by T_1. Then, measuring the ZPL width via hole-burning technique, one can determine T_1. Such situation is realised for vibronic transitions at low temperatures, for vibrational relaxation time in a condensed state lies within a picosecond range. The measurements of vibrational relaxation time via this technique were performed in[38,39].

The second case, when $T_1 < T_2^*$ is realised in some photoactive systems where photoreaction is the main excited state desactivation channel. Here the hole width yields information on the photoreaction rate. This method has been employed for determining the

276 LASERS IN ATOMIC, MOLECULAR AND NUCLEAR PHYSICS

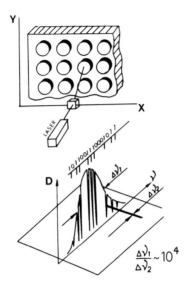

FIGURE 7 Three dimensional (x-y-ν) optical storage scheme.

rate of the proton phototransfer[12], and of the electron phototransfer[13]. The possibility to apply this method for investigation of biological systems and initial photosynthesis processes seems also exceptionally interesting.

3.5. Other Applications

It goes without saying that all possible domains for the hole-burning method application have not yet been studied to the end. It might happen that they lie beyond the scope of pure spectroscopy. Of practical interest are the works aiming at the increase in memory volume of optical storage devices (OSD). Diagram in Figure 7 shows the mechanism of such a device proposed in[41]. Using laser hole-burning a third - spectral - OSD coordinate should be added to the two spatial ones. The presence or absence of the hole at a given frequency will correspond to the information in binary code. The spectral coordinate capacity is determined by the ratio of the inhomogeneous absorption band width to the hole width. At low temperatures this ratio can equal 10^3-10^4. This means that owing to the spectral coordinate the OSD capacity can be increased by 3 or 4 orders. Now one has a right to say that it is quite possible to provide a large volume of the OSD memory. However, quite a number of problems should be solved for its technical realisation. These, in particular, are: providing high material sensitivity, providing a necessary speed of response, protecting information from erasing upon reading-out, etc. The problem, whether these difficulties will be overcome and a technically advantageous device of such type will be constructed, rests with the future.

Still another interesting usage of the hole-burning method is its application in the so-called spatial-time holography[42]. Employing a frequency-selective medium for recording (i.e. employing a photoactive substance with a large inhomogeneous and a small homogeneous spectrum broadening) one can register not only a spatial, but also a time signal structure with a high time resolution. Whereas a time memory of usual media is determined by a life-time of the excited states, the hole-burning effect provides a media set with practically unlimited information storage time. These problems are considered in more detail in[43].

REFERENCES

1. R. I. PERSONOV, E. I. AL'SHITS, L. A. BYKOVSKAYA, Pis'ma v Zh. Eksper. i Teor. Fiz., 65, 609 (1972); Optics Commun., 6, 169 (1972)
2. R. I. PERSONOV, E. I. AL'SHITS, L. A. BYKOVSKAYA, B. M. KHARLAMOV, Zh. Eksper. i Teor. Fiz., 65, 1825 (1973)
3. R. I. PERSONOV, in Spectroscopy and Excitation Dynamics of Condensed Molecular Systems, edited by V. M. Agranovich and R. M. Hochstrasser (North-Holland, Amsterdam, 1983), Chap.10
4. R. I. PERSONOV, Spectrochimica Acta, 38B, 1533 (1983)
5. B. M. KHARLAMOV, R. I. PERSONOV, L. A. BYKOVSKAYA, Optics Commun., 12, 191(1974)
6. A. A. GOROKHOVSKII, R. K. KAARLI, L. A. REBANE, Pis'ma v Zh. Eksper. i Teor. Fiz., 20, 474 (1974)

7. L. A. REBANE, A. A. GOROKHOVSKII, J. V. KIKAS,
   Appl. Phys., B29, 235 (1982)
8. G. SMALL, in Spectroscopy and Excitation Dynamics of Condensed Molecular Systems, edited by V. M. Agranovich and R. M. Hochstrasser (North-Holland, Amsterdam, 1983), Chap. 9.
9. H. FRIEDRICH, D. HAARER, Angevandte Chemie, 23, 113 (1984)
10. B. M. KHARLAMOV, L. A. BYKOVSKAY, R. I. PERSONOV,
    Chem. Phys. Lett., 50, 407 (1977)
11. H. DE VRIES, D. A. WIERSMA, Chem. Phys. Lett., 51, 565 (1977)
12. F. DRIESSLER, F. GRAF, D. HAARER, J. Chem. Phys., 72, 4996 (1980)
13. V. G. MASLOV, A. S. CHUNAEV, Mol. Biol., 16, 604 (1982)
14. K. N. SOLOV'EV, N. E. ZALESSKII, V. N. KOTLO, S. F. SHKIRMAN, Pis'ma v Zh. Eksper. i Teor. Fiz., 17, 463 (1973)
15. S. VÖLKER, J. H. VAN DER WAALS, Mol. Phys., 32, 1703 (1976)
16. B. M. KHARLAMOV, R. I. PERSONOV, L. A. BYKOVSKAYA,
    Opt. Spectrosk., 39, 240 (1975).
17. B. M. KHARLAMOV, E. I. AL'SHITS, R. I. PERSONOV,
    Izv. Acad. Nauk SSSR, Fiz. ser., 48, 1313 (1984)
18. A. A. GOROKHOVSKII, J. V. KIKAS,
    Zh. Priklad. Spektrosk., 28, 832 (1978)
19. A. U. JALMUKHAMBETOV, I. S. OSAD'KO, Chem. Phys., 77, 247 (1983)
20. A. A. GOROKHOVSKII, L. A. REBANE, Izv. Acad. Nauk SSSR, Fiz. Ser., 44, 859 (1980)
21. H. DE VRIES, D. WIERSMA, Chem. Phys. Lett., 51, 565 (1977)
22. A. I. M. DICKER, J. DOBKOWSKI, S. VÖLKER,
    Chem. Phys. Lett., 84, 415 (1981)
23. H. P. H. THIJSSEN, A. I. M. DICKER, S. VÖLKER,
    Chem. Phys. Lett., 92, 7 (1982)
24. I. S. OSAD'KO, S. A. ZHDANOV, Optics Commun., 42, 185 (1982)
25. S. K. LYO, Organic Molecular Aggregates in Sol. State Sciens., 49, (1983)
26. I. S. OSAD'KO, Pis'ma v Zh. Eksper. i Teor. Fiz., 39, 354 (1984)
27. Amorphous Solids, edited by W. A. Phillips (Springer-Verlag, Berlin, Heidelberg, New York, 1981)
28. A. P. MARCHETTI, M. SCOZZAFAWA, R. H. YOUNG,
    Chem. Phys. Lett., 51, 424 (1977)
29. V. D. SAMOILENKO, N. V. RASUMOVA, R. I. PERSONOV,
    Opt. Spektrosk., 52, 580 (1982)
30. F. A. BURKHALTER, G. W. SUTER, U. P. WILD, V. D. SAMOILENKO, N. V. RASUMOVA, R. I. PERSONOV, Chem. Phys. Lett., 94, 483 (1983)
31. U. BOGNER, P. SCHÄTZ, R. SEEL, M. MAIER, Chem. Phys. Lett., 102, 267 (1983)

LASER STUDY OF CONDENSED MATTER 279

32. V. I. IVANOV, R. I. PERSONOV, N. V. RASUMOVA,
Opt. Spectrosk., 58, 6 (1985)
33. O. N. KOROTAEV, N. M. SURIN, A. I. YURCHENKO, V. I. GLYAD-
KOVSKY, E. I. DONSKOI, Chem. Phys. Lett., 110, 533 (1984)
34. A. I. M. DICKER, L. W. JOHNSON, M. NOORT, J. H. VAN DER WAALS,
Chem. Phys. Lett., 94, 14 (1983)
35. A. I. M. DICKER, M. NOORT, S. VOLKER, J. H. VAN DER WAALS,
Chem. Phys. Lett., 73, 1 (1980)
36. A. I. M. DICKER, M. NOORT, H. P. H. THIJSSEN, S. VÖLKER,
J. H. VAN DER WAALS, Chem. Phys. Lett., 78, 212 (1981)
37. N. I. ULITSKII, B. M. KHARLAMOV, A. M. PYNDYK, R. I. PERSO-
NOV, Opt. Spektrosk., 59, (1985)
38. A. I. M. DICKER, S. VÖLKER, Chem. Phys. Lett., 87, 481 (1982)
39. A. A. GOROKHOVSKII, L. A. REBANE, Izv. Acad. Nauk, Fiz. ser.,
44, 859 (1980)
40. J. FRIEDRICH, H. SCHEER, B. ZICKENDRACHT-WENDELSTADT,
D. HAARER, J. Chem. Phys., 74, 2260 (1981)
41. G. CASTRO, D. HAARER, R. M. MACFARLANE, H. P. TROMMSDORFF,
United States Patent, No. 4.101.976 (1978)
42. A. REBANE, R. KAARLI, Chem. Phys. Lett., 101, 317 (1983)
43. R. KAARLI, J. KIKAS, A. REBANE, P. SAARI, Lectures in this
Edition.

3.6 Spectroscopy of Nonlinear Optical Activity in Crystals

N. ZHELUDEV

Laboratory of Nonlinear Optics, Physics Department, Moscow State University, 119899 Moscow, USSR

1. TO THE HISTORY OF THE PROBLEM

Refraction and absorption of a nonlinear medium can depend on the intensity of light. In a similar way, the difference between refractive indexes for circularly polarized waves in a gyrotropic medium should vary in a field of a strong electromagnetic wave. This phenomenon was predicted by S. Akhmanov and V. Zharikov (1967) and independently by S. Kielich (1968) and P. W. Atkins and L. D. Barron (1968).

The first NOA experiments were carried out in strongly absorbing crystals where medium gyrotropy changed under the thermal action of a laser beam. Of course, the main interest was attracted to electronic and not thermal NOA mechanism.

First reliable experiments on nonthermal nonlinear rotation of the polarization plane led to two different mechanisms of this phenomenon - NOA due to spatial dispersion of medium nonlinearity predicted in early works, and NOA due to anisotropy of nonlinear absorption.

At present it is clear that NOA is a fine, but often observed effect of the polarization self-action of lingt. Any chiral media, especially liquid crystals, biological macromolecules, and crystals of most crystallographic classes are nonlinear optically active

281

objects. The progress achieved recently in the technique of polarization (especially pulse polarization) measurements makes NOA a new method of spectroscopy which provides unique information on symmetry, band structure, nonlocal response, and latent crystal anisotropy.

2. ADDITIVE NOA MECHANISMS IN "WEAK" FIELDS

Nonlinear optical activity and the effect of nonlinear rotation of the polarization ellipse are the only "weak" effects of polarization nonlinear optics. As usual, their description is based on the common solution of the wave and material equations[1]

$$[\nabla [\nabla \vec{E}]] - \frac{\omega^2}{c^2} \vec{D} = 0$$
$$\vec{D} = \vec{D}^l(\vec{E}) + \vec{D}^{nl}(\vec{E})$$
(1)

While considering the amplitude effects of light self-action, the material equation is used, as a rule, in the form of an expansion of the electrical induction in power series of $\vec{E}$ up to the third-order terms. There is no necessity of considering spatial dispersion since in most cases the relative contribution of the amplitude effects related to the spatial dispersion of the medium (a/λ) is insignificant ("a" being the characteristic size in the medium-molecule diameter or the unit cell parameter of the crystal). Yet for the description of polarization phenomena spatial dispersion should necessarily be taken into account:

$$D_i = \chi_{ij}^{(1)} E_j + \chi_{ijk}^{(2)} E_j E_k + \chi_{ijkl}^{(3)} E_j E_k E_l + \ldots$$
$$+ \gamma_{ijk}^{(1)} \nabla_k E_j + \gamma_{iljk}^{(2)} E_l \nabla_k E_j + \gamma_{imljk}^{(3)} E_m E_l \nabla_k E_j + \ldots$$
(2)

There is no necessity to take into consideration the terms with the magnetic field of the light wave. The magnetic field is related to the electric one by the Maxwell equations and can be taken into account by the terms with spatial derivatives.

LASER STUDY OF CONDENSED MATTER   283

The first mechanism (NOA-1) of nonlinear rotation of the polarization plane is associated with spatial dispersion of medium nonlinearity described by tensor $\gamma^{(3)}$. As a rule, NOA should be distinguished from the background of natural gyrotropy, yet the high rank of tensor $\gamma^{(3)}$ results in the fact that the number of crystallographic classes having nonzero components of $\gamma^{(3)}$ is larger than for tensor $\gamma^{(1)}$ responisble for linear gyrotropy. Thus, NOA-1 can be observed in nongyrotropic media, i.e., can manifest itself as self-induced gyrotropy.

The second mechanism (NOA-2) of nonlinear rotation of the polarization plane is associated with anisotropy of nonlinear absorption and cannot be observed in isotropic media.

A strong light wave changes the medium refractive index and, consequently, the eigenvalue of the differential operator in the term with $\gamma^{(1)}$ in expansion (4). This mechanism, NOA-3, occurs only in gyrotropic crystals. Natural gyrotropy cannot exist separately from natural circular dichroism. They are described by the real and imaginary parts of tensor $\gamma^{(1)}$. In a similar way, nonlinear optical activity manifests itself as a nonlinear rotation of the polarization plane and as self-induced ellipticity.

3. NOA SPECTROSCOPY IN GALLIUM ARSENIDE

A gallium arsenide crystal, important from the standpoint of practical applications, is a very interesting object for the study by NOA methods[2]. It belongs to the $\bar{4}3$ m class having no natural gyrotropy, $\gamma^{(1)}_{ijk} = 0$, and linear birefrigence ($\chi_{ij} = \delta_{ij}\hat{\chi}$). Yet the symmetry of the crystal admits self-induced nonlinear optical activity of both dissipative (NOA-2) and reactive (NOA-1) types. The contributions of NOA-1 and NOA-2 can readily be distinguished since the corresponding angles of nonlinear rotation differently depend on the angle between the direction of the polarization plane of the excited radiation and symmetry axis of the crystal. Thus, NOA-2 is associated with the anisotropic component of tensor

$\Delta\chi^{(3)}$ while NOA-1 with $\gamma^{(3)}$ ($\bar{k}\parallel[001]$)

$$\binom{\beta}{B} = \binom{Re}{Im}\gamma^{(3)}_{11123}\cos2\beta\frac{12\omega^2\tilde{n} I}{c^3 n} + \binom{Im}{Re}\Delta\chi^{(3)}\sin4\beta\frac{3\omega\tilde{n} I}{c^2 n}$$

$$\Delta\chi^{(3)} = \chi^{(3)}_{1111} - 3\chi^{(3)}_{1122}$$

(3)

Angle β is measured between the direction of the major axis of the polarization ellipse and axis [100] and B is the ellipticity of light.

Such a dependence of the angle of nonlinear rotation on β (0) is confirmed experimentally. The angle of nonlinear rotation of the polarization plane for gallium arsenide single srystal in the frequency range corresponding to the two-photon interband absorption edge $0.9\ E_g < 2\hbar\omega < 1.15\ E_g$ reaches several degrees and can readily be measured for picosecond excitation. Gallium arsenide is a strongly nonlinear crystal - the constant of two-photon absorption k_2 in the indicated region reaches 0.04 cm/MW. Thus for the excitation intensities of an order of 10^8 W/cm^2 nonlinear energy dissipation in a crystal is rather strong (in this frequency range linear one photon absorption can be neglected since $E_g > \hbar\omega$). For the calculation of nonlinear susceptibilities one should take into account the change of the spatial and time beam profile during light propagation in a crystal.

NOA measurements permit us to obtain the valuable information on anisotropy of nonlinear absorption in this crystal (it should be remined that linear-optical properties of this crystal are isotropic). The combination of the components of cubic nonlinear susceptibility tensor $\chi^{(3)}_{1111} - 2\chi^{(3)}_{1212} - \chi^{(3)}_{1122}$ becomes identically equal to zero in an isotropic medium. This anisotropy for a cubic crystal can also be obtained by measurements of the polarization dependence of nonlinear absorption. Then we must separate the anisotropic part from the background of isotropic one, which is

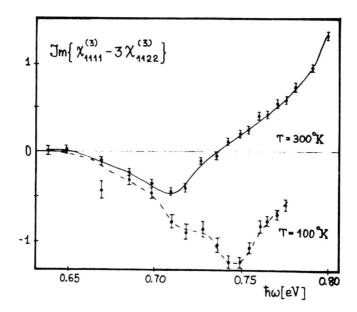

FIGURE 1   Anisotropy of nonlinear absorption in GaAs associated with NOA-2 (arb. units)

polarization independent. Yet the NOA method permits one to directly obtain these data with a better signal/noise ratio. Moreover, independent measurements of the angle of nonlinear rotation and the self-induced ellipticity provide the possibility of calculating both real and imaginary parts of the above anisotropy component $\Delta \chi^{(3)}$. (See Figure 1.) What causes this anisotropy? The energy bands in the Kane model are spherically symmetric and anisotropy of the nonlinear absorption component equals zero. NOA-2 is sensitive to the violation of sphericity. Therefore a corrugated second conductivity band and to a lesser degree, a corrugated band of heavy holes seem to be responsible for the corresponding susceptibility.

NOA spectroscopy can provide the unique information on the spatial dispersion of crystal nonlinearity. The measurements of susceptibility responsible for spatial dispersion of nonlinearity is of special interest for the information on free excitons under the conditions when the application of other methods is difficult, e.g., for massive samples where the exciton line is inhomogeneous broadened and suppressed owing to the presence of local defects, stresses or a random field of impurity bands, while the study of linear absorption spectra is hindered by the competitive process of interband absoprtion. Exciton lines should manifest themselves more distinctly in the susceptibility spectra responsible for spatial dispersion of nonlinearity than in the spectra of conventional nonlinearities, which is caused by essential exciton nonlocality. The order of magnitude of this gain is estimated as (a_{ex}/a) (a_{ex} is the Bohr exciton radius and a is the lattice parameter) and for gallium arsenide is ~ 30. Indeed, no exciton lines were observed in the direct studies of nonlinear absorption in massive gallium arsenide crystals, but these lines were clearly seen on the frequency dependence of $\chi^{(3)}$ in two-photon resonance (Figure 2) despite the significant inhomogeneous and collision broadening at high exciton concentrations. At high excitation levels the exciton line disappears due to the screening of exciton states by the free carrier plasma. The critical density of free carriers, calculated from the condition that the radius of the Debye screening and the Bohr radius of exciton are equal permits us to obtain the estimate for the critical intensity at which the screening arises

$$I = e\, m_{eff} h^{-2} (kT\, 2\hbar\omega / 4\pi\varepsilon k_2 \tau_p)^{1/2} \qquad (4)$$

where m_{eff} is the effective mass of the electron and ε is the dielectric constant. For gallium arsenide the value I is 250 MW/cm^2.

Intensity dependent polarization plane rotation increases of course in the vicinity of one photon absorption interband resonance. A dramatic increase of NOA constant was observed with the increase of light frequency near the band-edge (see Figure 3). We calculated from experimental data for $\hbar\omega$ = 1.40 eV NOA constant to be equal

LASER STUDY OF CONDENSED MATTER 287

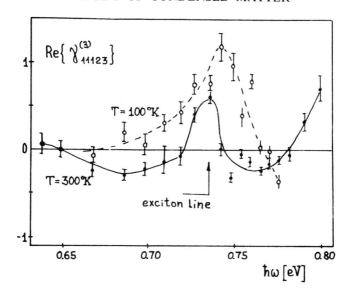

FIGURE 2 Susceptibility connected with spatial dispersion of GaAs nonlinearity (NOA-1, arb. units). One can see line associated with the two-photon exciton resonance.

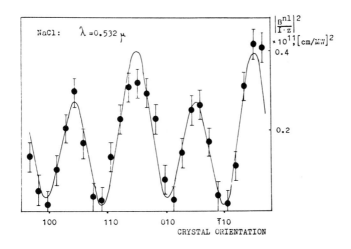

FIGURE 3 Dependence of self-indused elliptisity constant on crystal orientation in NaCl

7.0 deg·cm/MW. This value is 175 times greater corresponding value for two-photon interband excitation ($\hbar\omega = 0.70$ eV, $c^{NOA} \sim 0.04$ deg·cm/MW).

4. POLARIZATION NONLINEAR SPECTROSCOPY OF ALKALI-HALIDE CRYSTALS

The methods of nonlinear optical activity can be quite efficient in the studies of weak nonlinear crystals such as alkali halides[3]. The polarization method provides the study of latent crystal anisotropy which is a touchstone for different models for the potential of ion interaction. We measured anisotropy of the real part of the $\chi^{(3)}$ tensor in alkali halide crystals (NaCl, KBr, KCl, and LiF) at the wavelength $\lambda = 530$ nm by the NOA method. At frequencies of the "two-photon transparency" range of the crystal, $2\hbar\omega \ll E_g$ (which is fulfilled for the indicated wavelength in all the above crystals), the imaginary part of $\chi^{(3)}$ is zero. Thus, in the polarization self-action only nonlinear light depolarization connected with NOA-2 is observed which depends on the orientation of the polarization plane of radiation with respect to the symmetry axes of the lattice.

Alkali halide crystals with the symmetry m3m have both the center of inversion and a symmetry plane, they possess neither natural nor induced gyrotropy related to spatial dispersion. Figure 3 shows the dependence of the nonlienar ellipticity constant for NaCl on crystal orientation ($\bar{k} \parallel [001]$). The obtained experimental data require the reconsideration of the models suggested for the potential of ion interaction for alkali halide crystals, for example, the model of Coulomb interaction, successfully used for the calculation of the components of the tensor of cubic nonlinear nonresonance electronic susceptibility. The values $\chi^{(3)}$ calculated by this model differ from the experimental data only by a factor of two-three, but they give the wrong sign for the component $\chi^{(3)}_{1122}$. In our opinion, this fact directly indicates the essential

LASER STUDY OF CONDENSED MATTER   289

contribution of anharmonicity of short-range forces to $\chi^{(3)}$.
These forces can be taken into account if we proceed from the explicit form of interaction energy

$$\Phi^{ik} = \underbrace{a^i a^k / |\vec{R}^{ik}|}_{\text{Coulomb contribution}} - \underbrace{(\alpha^i + \alpha^k) a^i a^k / 2|\vec{R}^{ik}|^4}_{\text{Charge-dipole interaction}}$$

$$- \underbrace{2\alpha^i \alpha^k a^i a^k / |\vec{R}^{ik}|^7}_{\text{Dipole-dipole interaction}} + \underbrace{b^{ik} a^i a^k / |\vec{R}^{ik}|^9}_{\text{Shell overlapping}}$$

Here α^i and α^k are the electronic polarizabilities of ions, a^i and a^k are the charges of electron shells, and $\vec{R}^{ik}$ is the radius-vector of the i-th ion with respect to the k-th one. This potential allow as to get a correct sign for $\chi^{(3)}_{1122}$ and ($3\chi^{(3)}_{1122} - \chi^{(3)}_{1111}$) and reach a comformity with our experiment.

REFERENCES

1. A. D. PETRENKO, N. I. ZHELUDEV, Optica Acta, 31, 1177 (1984)
2. M. G. DUBENSKAYA, R. S. ZADOYAN, N. I. ZHELUDEV, JOSA-B Optical Physics, 7, (1985) (in press).
3. R. S. ZADOYAN, N. I. ZHELUDEV, L. B. MEYSNER, Solid State Communication, (1985) (in press).

3.7 Nonlinear Spectroscopy of Highly-Excited Molecules and Condensed Media

N. I. KOROTEEV

Moscow State University, 119899, Moscow, USSR

INTRODUCTION

Optical excitation of highly-nonequilibrium states in atoms, molecules and condensed media has undoubtedly become one of the most interesting trends in modern laser physics.

Nonequilibrium states of isolated molecules selectively excited due to linear and, particularly, nonlinear laser absorption are certainly one of the most intersting subjects of modern laser chemistry.

A high optical excitation of a semiconductor surface layer results in a rapid amorphization or epitaxial regrowth of crystals.

Generally the processes mentioned above occur in subnanosecond and even in femtosecond time-scale.

A detailed understanding of their physics calls for the data of the population of levels, of the crystal structure, etc obtained in real time.

The subject of the present paper is to consider the results showing that nonlinear optical spectroscopy is an important and sometimes unique method of studying these short-lived states of substance and dynamic processes.

1. COHERENT ACTIVE (ANTI-STOKES) RAMAN SPECTROSCOPY (CARS) IN STUDYING SELECTIVELY EXCITED MOLECULAR ENSEMBLES

1.1. Kinetics of Population Changes of Two-Photon Raman Excited Vibrational States

The subject of the research by means of CARS[1] with nanosecond time resolution as described in the present section is the ensembles of multiatomic molecules, for example, carbon dioxide (CO_2) and sulfur hexafluoride (SF_6), selectively excited into certain vibrational states. These problems are among the most important ones in vibrational laser photophysics and photochemistry as well as in physical kinetics[2,3].

Nonlinear optics gives an apportunity not only to examine the population exchange between the excited vibrational states using CARS[4-10], but also an opportunity to provide a high selective excitation of certain vibrational modes by means of so called two-photon Raman excitation technique (TRE) in an intensive biharmonic light with specially selected frequencies ω_1, ω_2:

$$\omega_1 - \omega_2 \approx \Omega \qquad (1)$$

where Ω is the frequency of excited Raman active transition.

In CARS a biharmonic light field with frequencies satisfying (1) only couples the phases of molecular oscillators without breaking the equilibrium population distribution of vibrational states. In contrast, a biharmonic field in TRE, being much more intensive, redistributes greatly the populations of vibrational levels thus a considerable excitation of Raman active modes occurs. The condition on the intensity of biharmonic pumping components is as follows:

$$I_1 I_2 \geqslant I_{sat}^2 = \omega_1^3 \omega_2 [16 \pi^2 c^2 (d\sigma/do) T_1 T_2]^{-1} \qquad (2)$$

where I_{sat} is intensity of two-photon Raman saturation, T_1, T_2 are times of "longitudinal" (energy) and "transverse" (phase) relaxation of the vibrational mode under investigation, $d\sigma/do$ is Raman cross section. For typical molecules (N_2, H_2, O_2, ...) at standard pressure, I_{sat} has the order of 0.1 - 1 GW/cm^2. In our experiments we succeeded in effective inducing vibrational and rotational excitation in molecules SF_6, CO_2, H_2 by means of TRE^8. Up to 40% of the molecules could be excited. The nonlinear-optical method of a selective deposition of energy into a certain vibrational mode due to TRE is an alternative to another well-known method of excitation of molecular vibrations due to resonante one-photon absorption of infra-red radiation which frequency coinsides with frequency of an excited transition. Selection rules are responsible for the difference between these methods: in the first case dipole inactive (bur Raman allowed) modes are excited, in the second case, vice versa, IR-active transition (and, as a rule, forbidden in Raman scattering) are excited.

We used the combined TRE - CARS technique to study a vibrational exchange and channels of "thermalization" of energy selectively deposited into totally symmetric excited states $10^0 0$ and $02^0 0$ in CO_2 molecule. (They can't be selectively populated from the ground state by resonant IR absorption because of the "alternative ban" mentioned above.) Although the kinetics of the "collisional" population and deactivation of the upper laser level CO_2 ($00^0 1$) is well established the corresponding data for either of the two low laser levels, $CO_2(10^0 0)$ and $CO_2(02^0 0)$ has been known by the begining of our study in 1982 with a considerably less accuracy just because of the difficulties connected with their selective population.

Using TRE - CARS we discovered the absence of "kinetic manifestation" of Fermi resonance between states $CO_2(10^0 0)$ and $CO_2(02^0 0)^{8,9}$. In striking contrast to the generally accepted but not experimentally confirmed views, the population exchange between these states appeared to occur very slowly, with the rate constant being less than $7 \cdot 10^4$ $s^{-1} Torr^{-1}$. Figure 1 shows the most effective

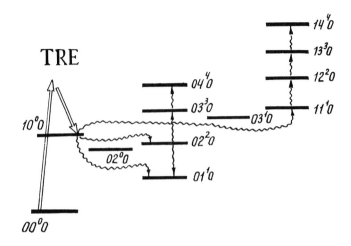

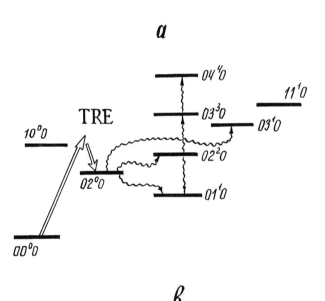

FIGURE 1 The most effective channels (wavy lines, revealed by CARS) of energy (and population) transfer from the states $CO_2(10^00)$ (a) and $CO_2(02^00)$ (b), selectively populated from the ground state $CO_2(00^00)$ by TRE technique.

LASER STUDY OF CONDENSED MATTER

channels of energy relaxation from states (10^00) and (02^00) into other modes of this molecule, determined by TRE-CARS.

1.2. Local CARS-Thermometry of Vibrationally-Excited Molecular Gases

CARS spectroscopy brings much new to the study of energy processes in molecular ensembles subjected to selective multiple-photon excitation (MPE) through IR-active modes by a strong IR resonant radiation. The most complete data currently available concern gas SF_6 resonantly excited by CO_2 laser radiation. Relaxation diagnostics is carried on by means of CARS technique. A variable time delay can be easily introduced between the pulse inducing MPE and the probing pulses. Thus, it is possible to obtain the necessary data at various stages of excitation and relaxation of high vibrational states of the studied molecules. For the first time this technique (MPE-CARS) was reported by us in paper[4], then it was successfully used in the experiments of Lebedev Institute group[5].

The recent papers on this subject were reviewed by S. S. Alimpiev at our seminar. Thus, I do not intend to divel on them in detail, I'd like to stress the fact that equilibrium Boltzman distribution of vibrational level populations has proved to be very fast in collisional regime, namely in the order of 1 μsec at perssure 1 Torr, or even faster in the vibrational subsystem of SF_6 molecule. It undoubtedly is identical to the distribution occurring in the molecule as a result of an ordinary thermal heating when it receives the same energy as in the process of excitation by MPE technique.

When SF_6 gas is heated, the transformation of its CARS spectrum can be calculated by the direct summation of individual contributions of all possible Raman-active transitions "starting" from the populated ro-vibrational levels, into a coherent signal, taking into account the Boltzman distribution[7,10]. In Figure 2

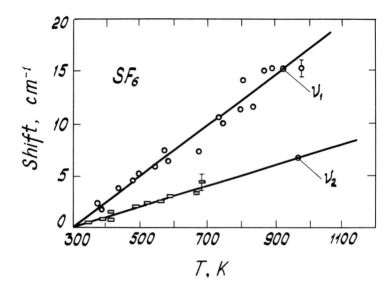

FIGURE 2 Calculated (solid lines) and experimentally determined (dots) dependences of the frequency shift of the Q-bands of totally symmetric (ν_1) and doubly degenerated (ν_2) modes of SF_6 molecules on the temperature of the gas.[11]

solid lines indicate the calculated temperature dependences of peak locations of CARS spectra of totally symmetric ν_1 and doubly degenerated ν_2 normal modes of SF_6 molecular gas.

The dots shows the experimental results with the thermally heated gas. A good correlation is obvious, though the ro-vibrational spectra and the corresoponding calculations are exclusively complicated.

CARS also allows to carry on the local thermometry of gas samples, transiently heated, for example, when the shock wave travels in the gas. Figure 3 shows the time evolution of the CARS signal from nitrogen molecules in air when a strong shock wave caused by the breakdown of gas near the metal target by the 3 J pulse of TEA CO_2 laser propogated through the volume of probing (that is 0.25 mm away from the target as shown in Figure 3). It is not only possible to detect the presence of shock wave but also to

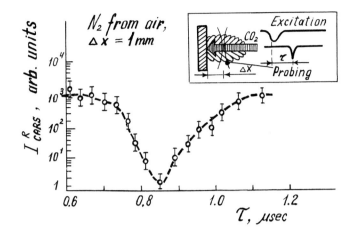

FIGURE 3 The influence of the shock wave propagation in air on the CARS signal from nitrogen molecules. One can easily see the variation of the CARS intensity with the time delay, τ, between the pulse, causing air breakdown, and probing pulse, as shock wave travels through the probing volume (see insert for details of the experimental conditions).

measure its velocity and other parameters[12].

2. OPTICAL NONLINEARITY OF LASER PLASMA

Until recently optical nonlinearities of homogeneous plasma were thought to be small since they are due to the relatively small Lorentz and hydrodynamic nonlinearities. In view of this fact nonlinear-optical methods of plasma diagnostics have not been considered to be promising, and there have been practically no experiments on the study of nonlinear optical plasma response (except[14]). At the same time there are many theoretical papers on this subject (see[15] and the references given there).

We managed to find experimentally that the optical nonlinearity of plasma created by laser breakdown of gases is, in contrast, relatively large and posesses a number of puzzling properties[19].

In the experiment, plasma was created by the laser breakdown

298 LASERS IN ATOMIC, MOLECULAR AND NUCLEAR PHYSICS

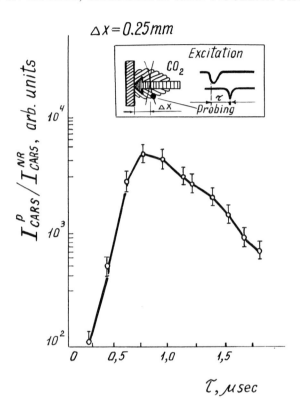

FIGURE 4 The enhancement of "non-resonant" CARS signal from air in the presence of the target-assisted laser breakdown as a function of time delay between CO_2 laser pulse (initiating the gas breakdown) and a pair of pulses, used for CARS probing.

of the gas in the vicinity of a metal target. The breakdown was produced either by a Q-modulated Nd:Yag laser with pulse energy of up to 0.5 J or by a TEA CO_2 laser pulse with energy up to 3 J. The CARS signal ($\omega_a = 2\omega_1 - \omega_2$) was studied when $\omega_1 - \omega_2$ did not coinside with any frequency of neutral molecular resonances, i.e. when the anti-Stokes signal from the neutral gas was generated by "nonresonant" electron gas susceptibility, $\chi^{(3)NR}$. As plasma was created, the intensity of the "nonresonant" signal I_{CARS}^p sharply increased, see Figure 4.

LASER STUDY OF CONDENSED MATTER   299

The enhancement of the CARS signal in the presence of a laser breakdown was found to have the universal character: it was observed both in molecular (SF_6, O_2, N_2, CO_2) and atomic (Ar) gases at the atmospheric or at higher pressure; in the self-breakdown of a gas (i.e. in case of initiating a breakdown by one of the beams used in CARS scheme) as well as in a spark caused by an independent laser source (irrespective of this source wavelength, $\lambda = 1.06\,\mu m$ or $\lambda = 10.6\,\mu m$); in a "free spark" occuring directly in the gas, or in a breakdown near metal or dielectric targets and even in case of a breakdown in the vapours of a metal target placed into a evacuated volume.

Nonlinear-optical plasma response proved to posess a slight frequency dependence, decreasing a little both at $|\omega_1 - \omega_2| \to 0$ and at $(\omega_1 - \omega_2)/2\pi c \gtrsim 3000$ cm^{-1}, but still easily detectable at frequency detuning as large as $(\omega_1 - \omega_2)/2\pi c \approx 8000$ cm^{-1}. Even the enhancement of third harmonic generation in the presence of laser breakdown of a gas has been experimentally observed, although the net enhancement was not so greatly pronounced as in the case of CARS-type mixing process.

We are leaning now towards describing the observed effect in terms of self-focusing of the probing beams due to ponderomotive forces [15], which become especially high when the difference frequency, $\omega_1 - \omega_2$, is tuned to the frequency of Langmuir's oscillations of a plasma, Ω_p:

$$\Omega_p = (4\pi Ne^2/m)^{1/2} \qquad (3)$$

where N is the electron density, e and m are electron charge and mass, respectively.

At a complete single-fold ionization of a gas at atmospheric pressure, $\Omega_p/2\pi c = 1650$ cm^{-1}. The absence of well-pronounced frequency dependence of the effect in the vicinity of Ω_p can be ascribed to the high degree of inhomogeneity and nonstationarity

300 LASERS IN ATOMIC, MOLECULAR AND NUCLEAR PHYSICS

of real laser-produced plasma. A small enhancement (up to 10 times) of the CARS intensity have been experimentally observed also in a plasma created by electric discharge with electron density not more a few times 10^{17} cm^{-3}, $30\,\mu sec$ after the leading front of the discharge.

It is also possible to study the nonlinear optical response of ionic component of a plasma.

3. NONLINEAR-OPTICAL DIAGNOSTICS OF A STATE AND OF FAST LASER-INDUCED PHASE TRANSFORMATIONS OF A SEMICONDUCTOR SURFACE

3.1. Harmonic Generation in Reflection

Methods of optical spectroscopy have been playing an important role in the study in "real time" of processes, occuring in a pulsed laser annealing (PLA) of semiconductor surfaces and other laser-induced phase transformations of a solid surface (see, for example,[16]). Nonlinear-optical methods have gained recently even greater significance. The latter are extremely informative, local, fast and, which is the most important, they supply inaccessible by other optical methods structural information of the processes occurring on the surface[16,17].

In the recent papers[17,18] carried out in our laboratory the phenomenon of second harmonic generation (SHG) at the reflection from a noncenter-symmetrical GaAs crystal was applied to study the dynamics of pulsed laser annealing (PLA) in nanosecond time-scale.

The essence of the effect used consists in the following: the surface layer of a crystal as a result of melting under PLA (or as a result of amorphization in the condition of ion bombarding used for ion implantation) becomes center-symmetrical and thus makes no dipole contribution to the reflected SH. Crystal lattice regrowth in the course of PLA causes the SH reflected wave of probing incident on the studied surface with a small variable time delay with respect to the annealing pulse. Measuring the intensity of this SH as a function of the time delay allows to study the dynamics of melting and the following fast recrystalization of the sample surface during PLA. Analysing polarization characteristics of SH allows to obtain information of the regrown crystal lattice quality after PLA as well as of the degree and character of the surface amorphization before laser annealing.

Soon after publishing our first results on SHG in GaAs[17] there appeared a number of papers[20] considering SHG in reflection from silicon crystal posessing the center of inversion due to the presence of quadrupolar second-order nonlinear polarization (see below) in the conditions close to PLA. The effect was (apparently by mistake) interpreted in the terms of varying local symmetry of the subsurface layers of a crystal (before its melting) and of generating Frenkel's excitations. Shank and others[21] using the data obtained by Tom and others[22] correctly interpreted SHG in Si as a consequence of the quadrupole nonlinearity. They used this effect for the diagnostics of the Si surface structural changes at the initial PLA stage-melting the surface layer-in femtosecond time scale.

3.2. Nonlinear Surface Sourse of Reflected Optical Harmonics

To describe the phenomena of second (SHG) and third (THG) harmonic generation and that of sum (SFG) and (or) difference (DFG) frequencies under reflection from the surface of a solid we shall introduce, as is conventional in nonlinear optics, the nonlinear polarization of a surface part of a medium induced by incident fields in the form of power series taking into account dipole and quadrupole contributions:

$$P_i^{NL} = P_i^{(2)} + P_i^{(2)S} + P_i^{(3)} + P_i^{(3)S} + \ldots \qquad (4)$$

where

$$P_i^{(2)} = P_i^{(2)D} + P_i^{(2)Q} = \chi_{ijk}^{(2)D} E_j E_k + \chi_{ijkl}^{(2)Q} E_j \nabla_k E_l \qquad (5)$$

- bulk second-order and

$$P_i^{(3)} = P_i^{(3)D} + P_i^{(3)Q} = \chi_{ijkl}^{(3)D} E_j E_k E_l + \chi_{ijklm}^{(3)Q} E_j E_k \nabla_l E_m \qquad (6)$$

- third-order nonlinear susceptibilities of dipole (D) and quadrupole (Q) types, and

$$P_i^{(2)S} = \chi_{ijk}^{(2)S} E_j E_k \qquad (7)$$

$$P_i^{(3)S} = \chi_{ijkl}^{(3)S} E_j E_k E_l \qquad (8)$$

- analogous surface polarization terms taking into account dipole contributions only (see, for instance,[23]). Identical indexes in eqs. (5)-(8) imply summation from 1 to 3. The symmetry and structure of the tensors of nonlinear susceptibilities $\chi_{ijk}^{(2)D}$, $\chi_{ijkl}^{(2)Q}$, $\chi_{ijkl}^{(3)D}$, $\chi_{ijklm}^{(3)Q}$, ... are determined by the point symmetry group of the crystal in the bulk, and those of the surface susceptibilities $\chi_{ijk}^{(2)S}$ and $\chi_{ijkl}^{(3)S}$ - by the symmetry of corresponding surface subgroup of the crystal point group.

In cubic crystals with the center of inversion of class m3m (which Si and Ge crystals belong to) and isotropic media expression (5) can take the following vector form:

$$P_i^{(2)Q} = \beta E_i (\vec{\nabla}\vec{E}) + \gamma \nabla_i (\vec{E}\vec{E}) + (\delta - \beta - 2\gamma)(\vec{E}\vec{\nabla}) E_i + \\ + \eta_Q E_i \nabla_i E_i ; \qquad (9)$$

$$P_i^{(2)D} = 0 ;$$

where

$$\beta = \chi_{1122}^{(2)Q}, \quad \gamma = \frac{1}{2} \chi_{1212}^{(2)Q}, \quad \delta - \beta - 2\gamma = \chi_{1221}^{(2)Q} \quad \text{and}$$

$$\eta_Q = \chi_{1111}^{(2)Q} - (\chi_{1122}^{(2)Q} + \chi_{1212}^{(2)Q} + \chi_{1221}^{(2)Q})$$

- anisotropy parameter of quadrupole nonlinearity (in linear optics these crystals are isotorpic); in isotropic media $\chi_Q = 0$.

In cubic crystals of class $\bar{4}3m$ lacking the center of inversion (semiconductors GaAs, GaP, and others belong to them) there is a dipole contribution to second-order polarization, as there is a nonvanishing component $\chi_{123}^{(2)D}$ of the tensor of the second order nonlinear susceptibility (other components of $\chi_{ijk}^{(2)D} \neq 0$ can be obtained from $\chi_{123}^{(2)D}$ by permutation of indexes).

3.3. Studying Structural Changes of the Surface by SHG and SFG

The crystalline anisotropy of class m3m and $\bar{4}3m$ crystals which are isotropic from the point of view of linear optics, manifests itself in the dependence of the intensity of the reflected SH and other nonlinear signals upon the crystal orientation with respect to the probe radiation plane of incidence i.e. upon angle ψ.

In this case the symmetry of $I_{SH}(\psi)$ dependence reflects the local symmetry of the crystal surface, from which reflection occurs.

In particular, melting the surface layer of noncenter-symmetrical crystal GaAs subjected to PLA results in forming center-symmetrical liquids giving no dipolar SH at all.

The structural change of GaAs surface subjected to ion-implantation by fast ions results in a partial or even complete loss of the far order in the crystal lattice, i.e. in the amorphization of the surface layer. This is easily observed by SHG: SH efficiency decreases in the course of reflecting from the amorphous section[18]. Nonlinear-optical technique of the surface state probing is proved in this case to be much more sensitive to small implantation doses than the traditionally used for these purposes methods[18].

The structure of partially amorphous GaAs surface, as our polarizational measurements by means of SHG and SFG have shown, retains the main symmetry elements of the initial crystal lattice,

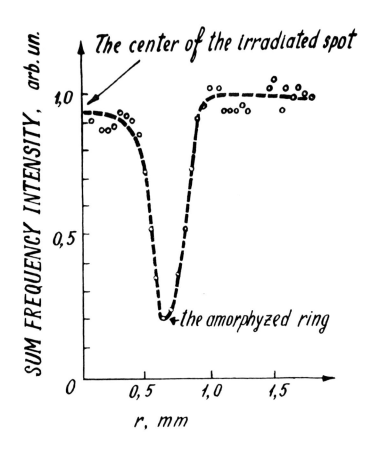

FIGURE 5 Sum-freequency generation (SFG) in reflection from (100) GaAs surface as a technique for the charcterization of the presence of a thin amorphyzed layer caused by a picosecond UV pulsed irradiation. The SFG intensity is shown as a function of the sample displacement across the focused probe laser beam.

i.e. amorphyzation under ion implantation occurs in "spots" surrounding the trace of a fast particle, whereas the rest of the surface remains unchanged.

GaAs surface amorphization can also be carried out by means of laser radiation. For example the action of strong enough pulse of the fourth harmonic of picosecond Nd:YAG laser on the (100) GaAs surface results in forming a thin amorphous film (thickness d = 15 nm) either in the center or in the periphery of the irradiated spot. This film was discovered in our SFG experiments: $\omega_{SF} = 2\omega + \omega$, where ω, 2ω are the fundamental and second harmonic frequencies of Nd:YAG probing laser (see Figure 5).

"Quadrupole" SH generation in reflecting from the plane (111) of a center-symmetrical silicon crystal in the course of PLA was used by Shank and others to register directly the surface melting with time resolution of 100 fsec.[21]

REFERENCES

1. S. A. AKHMANOV, N. I. KOROTEEV, Methods of Nonlinear Optics in Light Scattering Spectroscopy (Moscow, Nauka Publ. House, 1981) (in Russian).
2. V. S. LETOKHOV, Nonlinear Selective Photoprocesses in Atoms and Molecules (Moscow, Nauka Publ. House, 1983) (in Russian).
3. B. F. GORDIETZ, A. I. OSIPOV, L. A. SHELEPIN, Kinetic Processes in Gases and Molecular Lasers (Moscow, Nauka Publ. House, 1980) (in Russian).
4. R. V. AMBARTZUMIAN, S. A. AKHMANOV, A. M. BRODNIKOVSDII, N. I. KOROTEEV, et al., JETP Letters, 35, 170 (1982); Izv. Akad. Nauk SSSR, Ser. Fiz., 47, 1931 (1983).
5. S. S. ALIMPIEV, S. I. VALYANSKII, S. M. NIKIFOROV, et al., JETP Letters, 35, 291 (1982; ibid., 38 (1983).
6. A. OWYOUNG, P. ESHERICK, Opt. Lett., 5, 421 (1980).
7. S. A. AKHMANOV, V. N. ZADKOV, S. M. GLADKOV, et al., IEEE J. Quant. Electron., QE-20, 424 (1984).
8. A. M. BRODNIKOVSKII, S. M. GLADKOV, M. G. KARIMOV, N. I. KOROTEEV, JETP, 84, 1664 (1983).
9. S. M. GLADKOV, M. G. KARIMOV, N. I. KOROTEEV, Opt. Lett., 9, 298 (1983).

10. A. A. PURETZKII, V. N. ZADKOV, Appl. Phys. B, 31, 89 (1983).
11. S. A. AKHMANOV, S. M. GLADKOV, V. N. ZADKOV, et al.,
    Paper GG-5, Pres. at XIII-th IQEC, Annaheim, CA, June 1984.
12. S. M. GLADKOV, N. I. KOROTEEV, B. A. CHUPRYNA, et al.,
    Preprint, Phys. Dept., Moscow State University, N°20 (1984).
13. N. BLOEMBERGEN, Y. R. SHEN, Phys. Rev., 141, 298 (1966).
14. B. L. STANSFIELD, R. A. NODWELL, J. MEYER,
    Phys. Rev. Lett., 26, 1219 (1971).
    L. A. GODFREY, R. A. NODWELL, F. L. CURZON,
    Phys. Rev. A, 20, 567 (1979).
15. Y. R. SHEN, Principles of Nonlinear Optics (J. Wiley & Sons, N. Y., 1984), Ch. 28.
16. S. A. AKHMANOV, V. I. EMELJANOV, N. I. KOROTEEV,
    V. N. SEMINOGOV, Usp. Fiz. Nauk, December (1985).
17. S. A. AKHMANOV, M. F. GALJAUTDINOV, N. I. KOROTEEV, et al.,
    Opt. Commun., 46, 214 (1983); Sov. J. Quant. Electron., 10, 1077 (1983).
18. S. A. AKHMANOV, N. I. KOROTEEV, G. A. PAITIAN, et al.,
    JOSA B (Opt. Phys.), 2, 350 (1985);
    Izv. Akad. Nauk SSSR, Ser. Fiz., 50, 1000 (1985).
19. A. M. BRODNIKOVSKII, S. M. GLADKOV, V. N. ZADKOV, et al.,
    Sov. Phys. Tech. Phys. Letters, 8, 497 (1982).
20. D. GUIDOTTI, T. A. DRISCOLL, H. J. GERRITSEN,
    Solid State Comm., 46, 337 (1983).
21. C. V. SHANK, R. YEN, C. HIRLIMANN,
    Phys. Rev. Lett., 51, 900 (1983).
22. H. W. K. TOM, T. F. HEINZ, Y. R. SHEN,
    Phys. Rev. Lett., 51, 1983 (1983).
23. T. F. HEINZ, M. M. T. LOY, W. A. THOMPSON,
    Phys. Rev. Lett., 54, 63 (1985).

4. LASER STUDY OF PHOTOSYNTHESIS

4.1 Use of Lasers in Photophysical Research of Photosynthesis

G. LACZKÓ, P. MARÓTI, L. SZALAY

Department of Biophysics, József Attila University, Szeged, Hungary

During the past decade, extensive research has been carried out to utilize the unique characteristics of laser light. Mainly the high intensity, short pulse and small divergence of the laser beam have been made use of, and some of the results are briefly summarized here. The effect of the coherence of actinic light on the primary photochemical charge separation is discussed in some detail.

INTRODUCTION

A brief survey of the most recent reviews, monographs and congress reports on the rich and complex subject of the photophysics of photosynthesis reveals that the use of lasers has opened up new and highly exciting avenues in this fiels[1-5]. Many of the topics of current interest owe their existence to the application of laser techniques providing characteristics of actinic and measuring light which are not available with conventional light sources. Because of the vast number of relevant papers, we shall confine ourselves to selected topics and to a reasonable amount of the available information.

1. ULTRAFAST AND MULTIPHOTONIC PROCESSES

The successful production of shorter light pulses and higher powers, especially after 1975, encouraged several laboratories to carry out research into hitherto unexplored regions relating to primary events. With very short single exciting pulses, it is possible to study time-resolved kinetics of fluorescence rise time and decay (the dynamics of exciton transfer) and the dynamics of the early steps of charge separation. At very high exciting light intensities, a number of different non-linear phenomena appear, which depend among others on the pulse duration; an example is bimolecular exciton-exciton annihilation.

Let us consider a few examples. Fluorescence spectroscopic studies were surveyed quite recently[5]. Energy transport characteristics have been determined from fluorescence life time and quantum yield[6], and from delayed luminescence[7]. The use of short flashes led to successful study of the primary photochemical charge separation in photosystem-2, and to the establishment of the role of pheophytin-a as primary acceptor in the reaction center[8]. Photosystem-2 reactions have been studied by laser flash-induced 150 ns luminescence[9]. Since the processes in the reaction centers are very fast, 5-50 ps pulses should be used in order to prevent (or at least reduce) the change of the photochemical state of the centers during the flash excitation[10]. With very intense exciting pulses, a "photosynthetic reaction center parameter" has been suggested for discrimination between "photosynthetic" and "non-linear" regions[11]. The non-linear chlorophyll-a absorption has been found to depend on the organization of the chlorophyll-a[12]. The yields of both fluorescence and photochemical reaction products should depend on the excitation intensity[13]. It must be borne in mind that the effects of ultrafast and extremely intense pulses are not directly related to in vivo processes which occur with continuous excitation and moderate light intensity under physiological conditions.

LASER STUDY OF PHOTOSYNTHESIS 309

2. RE-REDUCTION OF THE PRIMARY DONOR IN PHOTOSYSTEM-2 IN THE NANOSECOND TIME RANGE FROM ABSORPTION CHANGE

The very small divergence and effective focusing of the laser beam opened up the way for the study of very fast absorption changes. A brief description of the problem may be given. In recent years, sound evidence has accumulated in favor of the role of pheophytin-a (Pheo) as the electron acceptor between P_{680} (a special chlorophyll-a) and Q (a plastoquinone molecule)^8. According to the new concept, the primary charge separation is $P_{680}\text{Pheo} \xrightarrow{h\nu} P_{680}^+ \text{Pheo}^-$, followed by the re-reduction of P_{680}^+ to P_{680}. The main phase of this process at physiological temperatures is in the ns time range. Until a few years ago there was only indirect evidence of the existence of the ns phase, based upon the hypothesis that chlorophyll-a fluorescence is quenched by P_{680}^+. It has long been known that the absorption spectrum of P_{680} depends on its redox state (its name is derived from the location of the absorption peak exhibiting maximum change on oxidation-reduction). Direct evidence of the ns re-reduction of P_{680}^+ requires determination of the absorption kinetics at 680 nm in the ns time range, in contrast to the indirect evidence from fluorescence measurements. The absorption changes cannot be measured without the use of special lasers, for the following reasons. In order to obtain well-defined initial conditions, saturating actinic flashes of high intensity should be used, which lead to intense prompt fluorescence with maximum intensity at around 680 nm (and are also highly scattered by the sample), a region where the absorption changes are to be measured. No optical filtering is possible, and therefore the absorption changes cannot be measured in the ns time range. Even modulation of the measuring light allows measurements only in the μs range^{13}. There is another maximum change in P_{680} absorption at around 820 nm. However, even in this spectral region ns time resolution cannot be attained with a conventional measuring beam. Mathis and van Best[14] suggested the use of a CW-operated gallium aluminium arsenide injection laser as measuring light source. The very thin beam of this laser allows decrease of the solid angle of light collection by the photodetector, so that only a very small amount of fluorescence and scattered light reaches the detector. The (10 mW) measuring light does not disturb the dark adaptation of the sample, but its intensity is high enough to ensure the low photon noise level of the beam. The avalanche photodiode detector meets the requirements of the measurement: it has a high quantum efficiency (85% at 820 nm), wide-band amplification, low noise and a fast rise time, without a post-impulse tail. A block diagram of the apparatus (after[14]) is shown in Figure 1. The main phase of the re-reduction of P_{680}^+ (with 30 ns half time) has been observed with this apparatus. A similar set-up was later used with repetitive actinic flashes and the results led to the assumption that the rate of reduction of P_{680}^+ is different in the different S states of the O_2-evolving complex[15]. This assumption was corroborated by

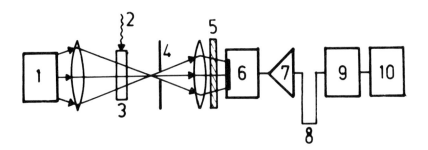

FIGURE 1 Apparatus for the measurement of absorption changes after laser flash excitation in the 20 ns - 10 μs time range (from[14]). 1 - laser diode (820 nm), 2 - exciting laser flash, 3 - cell, 4 - diaphragm, 5 - interference filter, 6 - avalanche photodiode, 7 - 30 MHz amplifier, 8 - delay line, 9 - digitizer, 10 - averager

LASER STUDY OF PHOTOSYNTHESIS 311

the results of direct experiments[16]. For measurement of the dependence of the reduction kinetics on the number of actinic flashes, the signal to noise ratio and other characteristics were improved. The kinetics in the S_0 and S_1 states is monophasic (half time 20 ns), and in S_2 and S_3 states biphasic (half times 50 and 300ns). All these new measurements contributed to a better understanding of the function of the O_2-evolving system.

3. COHERENT EXCITATION AND THE UTILIZATION OF LIGHT ENERGY IN THE PHOTOSYNTHETIC PRIMARY PROCESSES

The most significant property of laser light is the coherence, since high intensity, a short pulse, a small devergence and a high monochromaticity can in principle be produced with conventional light sources (though with much difficulty in practice), whereas coherent light can never be. Any specific biological effect of laser light can therefore be attributed to the coherence. We decided to study the effect of the coherence of light on the utilization of light in photosynthesis[17]. We do not know of any other attempt to investigate the biological effects of coherence, and we therefore report our studies in some detail. We presume that the ability of a photosynthetic organism to differentiate betweer coherent and incoherent light is based upon the coherence-dependent photon distribution of actinic light. Since the pthotosynthetic units (PSU-s) (fairly well separated parts of the pigment system as concerns the uptake of light energy) cannot use more than one photon during their turnover time (about 1 ms), independently of the actual number of incident photons (i.e. the function of a PSU is similar to that of a photon counter), it can be assumed that the utilization of light depends on the photon statistics. For differences in photon statistics to be attained within the space--time volume determined by the size and the turnover time of the PSU (or the time of illumination, if this is shorter), we need

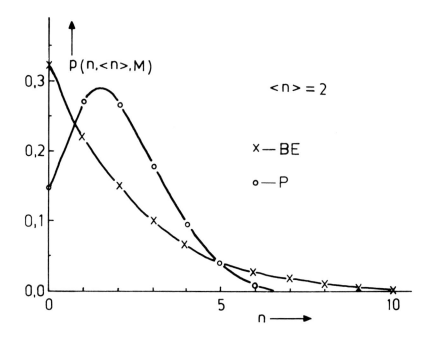

FIGURE 2 Probability of detection of n photons for M = 1 (BE distribution) and M = ∞ (P distribution). Average intensities are equal ($\langle n \rangle$ = 2)

light with a phase cell (coherence volume) including this space-time volume. This condition is easily met for the spatial coherence, as the PSU is several orders of magnitude smaller than the volume relating to the spatial coherence. However, the temporal coherence can be ensured with lasers only.

a) <u>Photon distribution and coherence</u>. For a closer consideration of the conditions of finding the expected effect, we summarize the properties of photon distribution in relation to the coherence.

Mandel[18] states that the probability of finding n bosons (photons) over a number of phase cells M (M ≫ 1) is

$$p(n, \langle n \rangle, M) = \frac{\Gamma(n + M)}{n! \, \Gamma(M)} (1 + \frac{M}{\langle n \rangle})^{-n} (1 + \frac{\langle n \rangle}{M})^{-M} \qquad (1)$$

LASER STUDY OF PHOTOSYNTHESIS 313

where $\langle n \rangle$ is the expectation value (average number) of the number of photons. This equation gives Bose-Einstein (BE) distribution for $M = 1$, and Poisson (P) distribution for $M = \infty$. These distributions are shown for $\langle n \rangle = 2$ (for the same average intensity) in Figure 2. It is worth remembering that both for few and for many (e.g. $n = 0.1$ and 7) photons the probability is higher when $M = 1$. This fact is important in the photon counter model of the PSU.

A more practical form of equation (1) is obtained if time parameters are introduced. Bedard[19] gives

$$M = \frac{1}{2}(\frac{2T}{\tau_c})^2 / [\frac{2T}{\tau_c} - 1 + \exp(-\frac{2T}{\tau_c})] \qquad (2)$$

i.e. M depends on the exposure time T, and the coherence time τ_c. If this expression is substituted into equation (1), we obtain the relation between photon distributions and coherence.

To find the range of T/τ_c values in which the BE distribution goes over to P, the standard deviation of (1)

$$\sigma^2 = \langle n \rangle (1 + \frac{\langle n \rangle}{M})$$

should be plotted versus T/τ_c. If $T/\tau_c \ll 1$, virtually BE distribution is observed whereas when $T/\tau_c \gg 1$ the distribution is virtually P. Figure 3 shows that the transition of the distribution from BE to P is practically complete when T/τ_c is varied within 1-2 orders of magnitude on either side of 1. In other words, a photodetector having a surface area smaller than the coherence area will "feel" BE or P photon distribution for coherent or incoherent light, respectively.

b) <u>A photon counter model of the photosynthetic unit</u>. For the sake of simplicity, let us suppose that there is <u>no transfer of electronic excitation energy</u> between the PSU-s. As mentioned earlier, a PSU cannot utilize more than one photon until its turnover time is completed. If it absorbs n photons, n-1 photons will be reemitted as fluorescence; if the probability of this event is p(n),

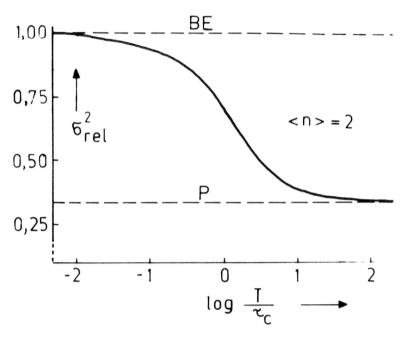

FIGURE 3 Dependence of the relative standard deviation of the distribution function (eq. (1)) on the ratio of the exposure time (T) and the coherence time (τ_c)

the total number of photons used in fluorescence is $F = \sum_{n=2}^{\infty} (n-1)p(n)$, or, after minor transformations, $F = p(0) + \langle n \rangle - 1$. The relative fluorescence yield is $\Delta\Phi = dF/d\langle n \rangle$. For a rectangular exciting light pulse, $\langle n \rangle = It$ (I is the rate of excitation in number of photons per unit time, while t denotes time). From equations (1) and (2), the time-dependence of the fluorescence yield (induction) can be obtained for BE and P statistics:

$$\Delta\Phi_P = 1 - \exp(-It) \quad \text{and} \quad \Delta\Phi_{BE} = 1 - 1/(1 + It)^2 \quad (3a, 3b)$$

As illustrated in Figure 4, the half rise time is about 60% greater and the initial slope is smaller for P statistics than for BE. A comparison of the complementary areas shows that in the range It < 3 the photosynthetic energy conservation is more effective in the case of P distribution.

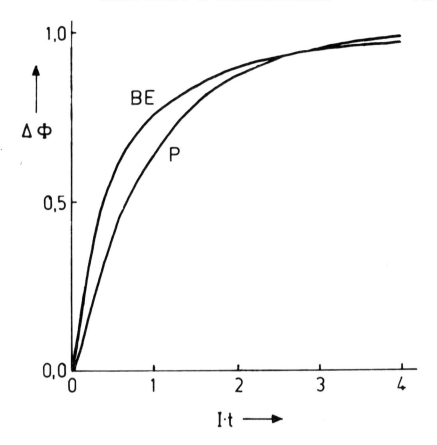

FIGURE 4 Fluorescence induction curves of the photosynthetic unit for Bose-Einstein and Poisson distributions of the exciting light. I is the rate of excitation; t denotes time.

If energy transfer occurs (and in fact it does), the excitons originating from the excess number of photons $F = p(0) + \langle n \rangle - 1$ can migrate to neighboring PSU-s with probability η. A PSU having c neighbors receives no excitons from them with a probability $(1 - \eta)^{c[p(0) + \langle n \rangle - 1]}$, since the individual misses are independent events. Receiving at least one exciton via energy transfer results in a charge separation only if the reaction center of the receiver PSU is open, and therefore the probability of a charge

316 LASERS IN ATOMIC, MOLECULAR AND NUCLEAR PHYSICS

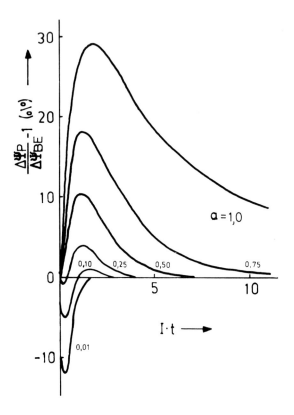

FIGURE 5 Relative photon utilization of the photosynthetic units for Bose-Einstein and Poisson distributions of the exciting light on different couplings of the units (a = 1: no coupling, no energy transfer; a < 1: coupling of neighboring units)

separation due to the energy transfer is $p(0)[1-(1-\eta)^{c(p(0)+\langle n\rangle-1)}]$
A PSU absorbs at least one photon with a probability $1 - p(0)$, and thus the overall photon utilization is

$$\Delta\Psi = 1 - p(0)(1 - \eta)^{c[p(0) + It - 1]} \qquad (4)$$

Figure 5 shows the relative values of this function for BE and P distributions at several $a = (1 - \eta)^c$ values as a function of It. For isolated PSU-s (when $a = 1$ and there is no energy transfer) at It =2, the difference is 30% for the two distributions; in coupled units the difference decreases and becomes less than 10% (independently of It) when $a < 0.5$.

c) <u>Experimental</u>. In order to check the above predictions, a light source is needed having variable coherence time (in the us-ms range) with constant intensity and geometry of illumination. These requirements are met if the light of a He-Ne laser operating in a single TEM-00 mode is passed through a rotating ground glass disc[20]. The frequency spectrum of the scattered laser light is Gaussian[21] and characteristic of the thermal light, but its coherence time is much longer. The coherence time of this pseudo-thermal light source can be varied by changing the linear velocity (v) of the illuminated area on the rotating disc[17]:

$$\tau_c = \frac{R\lambda}{a}\frac{1}{v}$$

Where R is the distance between the sample and the disc, λ is the wavelength of the light and <u>a</u> is the average diameter of the scattering grains of the ground glass. By measuring the average half-times of the intensity fluctuations, we found that the coherence time of our light source could be varied in the range $10^{-5} - 10^{-1}$ s. Using a phosphoroscope[17], we measured the chlorophyll fluorescence induction and delayed fluorescence of green plant leaves, green algae and spinach chloroplasts in the μs and ms time ranges. With a given illumination time, the coherence time of the pseudo-thermal light source was varied between the two limiting photon distributions and the results were compared. The illumination time was decreased down to 50 μs, the lower limit of our instrument.

The experiments gave negative results: no significant differences could be observed. This may be ascribed to the coupling of the PSU-s, which leads to a decrease of the predicted effects of the different statistics. The experiments gave positive results too: from equation (4) we can estimate the value of η, the probability of transfer between the PSU-s. The statistical error of our experiments was about 5%; with this error, taking four neighbors into account (c = 4), the transfer probability is at least η = 0.3. Though this is indirect evidence of transfer, it may be of interest, for it is independent of the other (also indirect) evidence.

REFERENCES

1. Light Reaction Path of Photosynthesis, edited by F. K. Fong (Springer Verlag, Berlin, Heidelberg, New York,1982)
2. G.RENGER, Photosynthesis, Biophysics,edited by W. Hoppe, W. Lohmann, H. Markl, H. Ziegler(Springer Verlag, Berlin, Heidelberg, New York, Tokyo, 1983) p. 515
3. Photosynthesis, edited by Govindjee (Academic Press, New York, 1982)
4. Advances in Photosynthesis Research, edited by C. Sybesma (Martinus Nijhoff/Dr W. Junk Publ., The Hague, 1984)
5. I. MOYA, Time-resolved Fluorescence in Photosynthesis, Time-resolved Fluorescence Spectroscopy in Biochemistry and biology, NATO ASI Series A: Life sciences, Vol. 69, edited by R. B. Cundall, R. E. Dale (Plenum Press,New York, London, 1983) p. 755
6. L. B. RUBIN, B. N. KOWATOVSKY, O. V. BRAGINSKAJA, V. Z. PASCHENKO, H. PAERSHKE, V. B. TUSOV, Mol. Biol., (Moscow) 14, 575 (1980)
7. V. I. GODIK, A. YU. BORISOV, Biochim. Biophys. Acta, 590, 182 (1980); L. N. M. DUYSENC, Prompt and Delayed Fluorescence from Photosystem-2, Oxygen-evolving System of Photosynthesis (Acad. Press Japan Inc., 1983) p. 3
8. V. V. KLIMOV, A. A. KRASNOVSKY, Photosynthetica, 15, 592 (1981); Biofizika (Moscow) 27, 179 (1982); V. V. KLIMOV, S. I. ALLAKHVERDIEV, V. A. SHUVALOV, A. A. KRASNOVSKY,

Dokl. Akad. Nauk SSSR, 263, 1001 (1982)
9. L. N. M. DUYSENS, A. SONNEVELD, Photoelectrochemical and Photobiological Processes, Proc. EC Contractors' Meeting, Brussels, 1982, Vol. 2, Series D Solar Energy R and D in the European Community, edited by D. O. Hall, W. Palz and D. Pirrwitz (D. Reidel Publ. Co., Dordrecht, 1983) p. 188
10. J. DEPREZ, A. DOBEK, N. E. GEACINTOV, G. PAILLOTIN, J. BRETON, Biochim. Biophys. Acta, 725, 444 (1983)
11. S. A. ACHMANOV, A. YU. BORISOV, R. V. DANIELUS, R. A. GADONAS, V. S. KOZLOWSKI, A. S. PISKARKAS, A. P. RAZJVIN, Studia Biophys., 77, 1 (1979)
12. B. VOIGT, D. LEUPOLD, B. HICKE, P. HOFFMANN, Studia Biophys., 75, 93 (1979); B. HIEKE, P. HOFFMANN, D. LEUPOLD, S. MORY, J. SCHOTTE, Photosynthetica, 13, 37 (1979); B. HIEKE; D. S. MATORI, B. VOIGT, D. LEUPOLD, Biol. Resch., 17, 194 (1979); D. MAUZERALL, Photochem. Photobiol., 29, 169 (1979)
13. G. RENGER, H.-J. ECKERT, H. E. BUCHWALD, FEBS Lett., 90, 10 (1978)
14. J. A. VAN BEST, P. MATHIS, Biochim. Biophys. Acta, 503, 178 (1978)
15. K. BRETTEL, H. T. WITT, Photobiochem. Photobiophys., 6, 253 (1983)
16. K. BRETTEL, E. SCHLODDER, H. T. WITT, Advances in Photosynthesis Research, edited by C. Sybesma (Martinus Nijhoff/ Dr. W. Junk Publishers, The Hague, 1984)
17. P. MARÓTI, A. RINGLER, L. VIZE, L. SZALAY, Acta Phys. Chem. Szeged, 23, 155 (1977); P. MARÓTI, Thesis, Szeged (1981)
18. L. MANDEL, Proc. Phys. Soc., 72, 1037 (1958)
19. G. BEDARD, J. CHANG, L. MANDEL, Phys Rev., 160, 1496 (1967)
20. W. MARTINSSEN, E. SPILLER, Am. J. Phys., 32, 919 (1964)
21. E. ESTES, M. NORDUCCI, A. TUFT, J. Opt. Soc. Am., 61, 1301 (1971)

4.2 Primary Processes of Photosynthesis Studied by Fluorescence Spectroscopy Methods

A. FREIBERG

Institute of Physics, Estonian SSR Acad. Sci., Tartu 202400, USSR

1. INTRODUCTION

Photosynthesis is one of the superior accomplishments in the evolution of Nature. It is thanks to photosynthesis that the human race ixists and has reached its present stage of development. The amount of energy annually utilized by photosynthesizing organisms ($\sim 3 \cdot 10^{21}$ J) exceeds by about an order the amount of energy consumed by mankind[1] ($\sim 4 \cdot 10^{20}$ J in 1980). These global as well as some purely cognitive factors urge the elucidation of photosynthesis mecanisms

The utter complexity of the problem has necessitated an incoporation in the photosynthesis science a number of other sciences, such as physics, chemistry, biology, and some interdisciplinary sciences as well. Exact physical methods have proved their actuality in a number of cases. Among them a more informative one is the fluorescence spectroscopy method. Its use is based on a circumstance that after absorbing the light by photsynthesizing organisms, because of the interaction with the electromagnetic field, a part of the absorbed energy is indispensably emitted in the form of photons (in the particular case of singlet-singlet transition it is fluorescence). The fluorescence intensity in each moment of time is proportional to the concentration of excited species. The light-excited molecule in a photosynthetic unit (PSU) (the definition of PSU, see part 3 of this paper) can lose energy via various channels like light quantum emission, radiationless dissipation at the expense of nonadiabaticity of electronic states or in the process of transferring energy to other molecules. The "aim" of the so-called primary processes of photosynthesis is to bring the absorbed energy with possibly smaller losses (i.e. as quickly as possible) to a special molecular complex, a photochemical reaction centre, where it is stabilized for a sufficiently long time (for a fraction of second). Fluorescence decay after a short pulse excitation gives, in the real time scale information on the balance between different relaxation channels and allows the efficiency of the so-called photochemical relaxation channel, which is the most important among the others, to be measured directly.

The duration time of primary processes, characteristically less than 10^{-8} s, asserts high demands to the experimental technique, particularly excitation sources, spectral devices, recording systems and data processing. To describe thoroughly the situation by means of fluorescence a full set of its parameters, such as decay time τ, yield ϕ, polarization, excitation and emission spectra, etc., to be measured.

This paper does not aim at a comprehensive survey of the fluorescence studies of primary processes because of its limited content. The rather that the relevant surveys can be found in literature[3-10]. Instead we shall focus our attention on some actual problems related with the study of energy transfer processes by fluorescence kinetics of basic photosynthetic pigments chlorophyll (Chl) and bacteriochlorophyll (BChl) in vivo. Among these a special consideration is given to methodological problems of picosecond measurements, as a number of inconsistent (or seemingly inconsistent) experimental data have appeared recently. To introduce the subject we should immediately point at an important methodological evidence. The τ of the emission of chloroplasts and chromatophores (and also their fragments) is known to be a function of several parameters: $\tau = \tau(A,B,C,D,..?)$, where A is the sample and its state (the following aspects are of special importance here: method of preparation, dilution medium, temperature, state of the photochemical apparatus, etc.); B are the excitation conditions (intensity, duration, spectral composition, excitation homogeneity, etc.); C are the recording conditions (spectral composition of the radiation recorded, presence of spurious radiation, etc.); D is the method of recording (meaning that systematic errors and uncertainty of results peculiar to various methods, vary)[2]. Unfortunately, not always the whole set of factors listed is duely taken into account. Then some experimental results obtained in the Institute of Physics of the Estonian Academy of Sciences (in collaboration with Moscow State Univerity) on the primary processes of photosynthesis in bacteria are discussed. Our approach proceeds from the desire to study most important natural phenomena in the conditions possibly closer to the natural ones. A spontaneous (in contrast to a stimulated) spectral and time response of the sample to a weak (average intensity $\leq$ the average day-time solar radiation intensity on the Earth's surface) picosecond excitation at the ambient medium temperature is recorded.

2. EXPERIMENTAL METHODS

Almost all methods used in pico-nanosecond nonstationary fluorescense spectroscopy can be distinguished by their high (10^4-10^8 Hz) and low (0.1-10 Hz) repetition rates of operation cycles. The former case allows the use of efficient signal averaging and noise suppression methods, in the latter case, the insufficient signal--to-noise ratio is, as a rule, compensated by the growth of excitation power (proceeding from a rough calculation that the excitation energy average in macroscopic time lapse be preserved).

In high repetition rate mode operate phase fluorometers (the present-time limit resolution 10-50 ps), single-photon counting systems (50-100 ps), synchroscan streak camera systems (1-5 ps), and also some systems based on up-conversion technique and fluorescence correlation function measurements (in these methods time resolution is determined by the excitation pulse duration). For high-frequency excitation source serve the electrical spark discharge, various continuous-light sources whose light flux is chopped by an external modulator (mainly in phase fluorometers) and, since 1969[11], continuous-wave (CW) mode-locked picosecond lasers (HeNe, Ar^+, Kr^+, and dye lasers), which by the present moment are in the most widely extended use. (Regardless the sufficiently high time resolution of phase fluorometers and their monentuous processes it should be noted that their results are reliable and unambiguous only in case of single exponential decay. Therefore, more direct measurement techniques are to be preferred.)

With low repetition rates work fluorometers based on solid state mode-locked picosecond lasers and optical parametric oscillators, and also on nitrogen- or excimer-laser-pumped dye lasers. As emission registers high-speed photodetectors together with oscillographs, the Kerr cells, and traditional single-shot streak cameras are used. Their common drawback is a relatively low dynamic range, also difficulties in synchronizing the moment of detection with excitation. Nevertheless, the introduction of streak camera systems in 1975[12-15] denotes a big onward leap in the development of the experimental technique for fluorescence studies of the primary processes in photosynthesis. Contemporary streak cameras allow the whole fluorescence decay curve to be recorded within a single laser shot, they have high sensitivity and excellent

time resolution (~0.5 ps).
Experiment shows[16] that the combination of picosecond time resolution with high spectral resolution (≲1 nm) meats great difficulties, which, as shown in[17], to a certain limit $\Delta\omega\Delta t \approx 1$ ($\Delta\omega$ and Δt resp. the spectral and time resolution) is reduced to the technical perfectness of separate units. In real spectral devices the area of the resolution element $\Delta\omega\Delta t$ exceeds the theoretical (diffraction) limit by several orders because of the presence of inescapable aberrations and other faults. This is immaterial when low-spectral-resolution devices (prism and also interference filters, etc.) are used, as then the time resolution is limited by the detector response. In case a better spectral resolution is needed, i.e. when grating spectrometers are used, the excess broadening of their time response (up to 10^3 ps) is removable, as shown in[17], by using double monochromatization in subtractive dispersion scheme. In conformity with[18], if the product of experimental $\Delta\omega$ and Δt tends to the theoretical limit, a spectrochronogram is measured. It is possible to demonstrate that the information content of the spectrochronogram is maximal.

An approach based on picosecond spectrochronography was used in investigating the fluorescence kinetics of pigments <u>in vivo</u> in 1981[2]. In a standard version our picosecond spectrochronograph consists of three basic parts (Figure 1).

(1) a picosecond CW (pulse repetition rate 82 MHz) synchronously--pumped dye laser (tuning range 685-850 nm operating on three dyes: oxazine 1, oxazine 750, and styrile 9; maximum average power up to 300 mW; pulse duration ≲ 3 ps). To obtain excitation in the near UV region of 345-420 nm the second harmonic generation in $LiIO_3$ crystal is used with the conversion efficiency up to 1%;
(2) subtractive dispersion double monochromator mounted from two

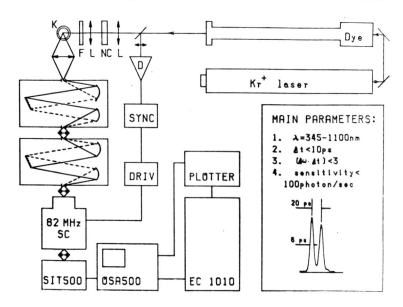

FIGURE 1 Picosecond spectrochronograph for investigating nonstationary spectra[19]. In the insert two 3 ps laser pulses mutually shifted by 20 ps, measured by means of a spectrochronograph, are depicted.

high-luminosity (1:2.5) single-grating monochromators (reverse dispersion 2 nm/mm with the grating 600 grooves/mm);
(3) a recording system based on a streak camera in synchroscan operation mode, which records the fluorescence excited by each laser pulse. The accumulation of successive traces on the streak-camera screens enables the investigation of extra weak signals with high dynamic range (~3 orders of magnitude). A single count level of the SIT vidicon videosignal from the streak camera screen is provided by a flux of 100-1000 incident photons on the streak camera photocathode depending on the type of the cathode used. High precision of synchronization by laser light pulses[19] ensures the overall time resolution of the spectrochronograph 1-5 ps. For collecting and operative processing of a huge mass of data the spectrochronograph is connested to a central computer complex.

LASER STUDY OF PHOTOSYNTHESIS   327

It is to the purpose here to compare, though briefly, the fluorescence method with another equally widespread absorption method. We would like to underline three essential aspects:

(1) in a differential absorption method by means of a weak probing beam the change in the optical density of the sample as a result of interaction with a comparatively powerful pumping pulse is measured

$$\Delta D(t) = \lg(I_o^c/I_o) - \lg(I^c(t)/I(t)) = \lg(I(t)I_o^c/I^c(t)I_o)$$

where I_o and I_o^c are the probing beam intensities in the channels of measurement and comparison, respectively, before the arrival of the pumping pulse; $I(t)$ and $I^c(t)$ are the same after pumping. The measurement error depends on the relative errors of measuring all the four values and it cannot be made infinitely small. The mean square error of $\Delta D(t)$ is

$$\Delta[\Delta D(t)] = [(\Delta I^c(t)/I^c(t))^2 + (\Delta I(t)/I(t))^2 + (\Delta I_o^c/I_o^c)^2 + (\Delta I_o/I_o)^2]^{1/2}.$$

It is, as a rule, larger than that on measuring an essentially background-free fluorescence singal, the rather that the relative light-induced absorption changes in photosynthetic systems are $\geqslant 100$ times less than relative fluorescent changes;

(2) the relaxation investigation method based on absorption measurement requires, unlike the fluorescence method, an application of complementary probing actions, which in principle can distort the course of the processes studied. In this sense one can refer even to a forced character of the system response;

(3) due to the population of the excited states by the pumping pulse, the time-dependent absorption from these always exists. It is difficult to decide, however, at what relaxation stage and between which states the absorption occurs. These difficulties become explicit if, instead of the common scheme of energy levels, a more realistic picture of potential energy curves is used. In contrast,

328 LASERS IN ATOMIC, MOLECULAR AND NUCLEAR PHYSICS

the interpretation of the spontaneous fluorescence signal, as a rule, is unambiguous, including the interpretation of weak transient spectral components emitted before the thermal equilibrium has been established[20].

The statements on literature[21,22] as if fluorescence does not render adequately the processes occurring in PSU or as if its data are interpreted incorrectly, must be considered an exaggeration. On the other hand, the absorption method has its indisputable advantages over the fluorescence one, as it is correctly referred to in[21,22]. The only constructive approach here lies joining the strong sides of both methods to solve such difficult problem like elucidating the details of the primary processes in photosynthesis.

3. ENERGY TRANSFER IN PHOTOSYNTHESIS

At present the following construction and functioning model of PSU is generally accepted. The unit contains at least two types of Chl-protein (or BChl-protein) complexes. The first-type complexes are the so-called light-harvesting antenna complexes (LA). These serve to collect light and transfer excitations to the second-type complexes - reaction centers (RC) where a multistage rediationless energy transformation occurs in the form of charge separation in RC. In this scheme the RC appears as a trapping center for excitations in LA. The typical ratio of the number of Chl molecules in LA and the number of RC in bacterial photosynthesis is 50 and 100-400, in the photosynthesis of higher plants. It is shown that the Chl molecules in RC and LA are identical and therefore their ability to perform various functions should rather be explained by different arrangement of Chl molecules in pigment-protein complexes and by the peculiarities of interaction between the complexes. This explains also the absence of fluorescence self-quenching due to high concentration of Chl <u>in vivo</u> molecules ($\sim 0.1 M \simeq 6 \cdot 10^{20}$ cm^{-3})[4]. Within the given simplifications, which are more justified on case of bacterial photosynthesis, the energy transfer problem divides into two: energy transfer along LA to RC and its trapping on RC. In more realistic models the heterogeneity and anisotropy of the antenna are to be taken into account, also, the matter that in reality various PSU are connected at the level of energy transfer, forming the so-called domains (see, e.g. [23-26] and below), and other complicating factors [27].

The common viewpoint is also that the absorbed energy in LA is localized on separate pigment molecules (or small molecular aggregates) and is transferred in the form of singlet excitation via inductive resonance [23-28]. Indeed, one can hardly expect that in a disordered or only partially arranged conglomerate of molecular complexes, yet more at room temperature, there could exist non-localized excitations and, consequently, an essentially coherent energy transfer [29,30]. In conformity with [30,31] the excitation migration along LA can be considered coherent only during the phase memory time $1/\Gamma$, where Γ is the width of the pigment molecule spectrum (in the case of BChl it is $\sim 10^{-14}$ s at room temperature).

In incoherent approximation and at low level of excitation intensity $I(t)$ the probability of the i-th molecule excitation ρ_i is determined by the solution of a system of kinetic equations (which inexplicitly contain also equations for the trapping centre):

$$\frac{\partial \rho_i}{\partial t} = - \frac{\rho_i}{\tau_{oi}} - \sum_{j=0}^{N} (F_{ji}\rho_i - F_{ij}\rho_j) + I_i(t) \qquad (1)$$

where $1/\tau_{oi}$ is the monomolecular decay rate of i molecule, that takes into account fluorescence and intramolecular conversion; F_{ji} is the rate of energy transfer from i molecule to j molecule (usually only the transfer between nearest neighbours is taken into account); N is the number of molecules (compexes) in PSU. In the approximation of weak dipole-dipole coupling

$$F_{ij} = (1/\tau_{oi})(\bar{R}_o/R_{ij})^6 \qquad (2)$$

where $\bar{R}_o$ is the distance between the energy donor and acceptor, where the transfer rate equals $1/\tau_{oi}$. The critical distance $\bar{R}_o$ is related with the overlapping of the donor fluorescence spectrum $F(\omega)$ and the acceptor absorption spectrum with the cross-section $\sigma(\omega)$ as follows[32]:

$$\bar{R}_o = [3\phi\alpha c^4/(4\pi\omega^4 n_o) \int F(\omega)\sigma(\omega)d\omega]^{1/6}$$

α is the multiplier allowing for the anisotropy of dipolar interaction, n_o is the refration index of the solvent, and c the light velocity. In more exact version of the theory possibility of energy transfer from the vibrationally non-equilibrium excited electronic state is also taken into account[33] and then, instead of $F(\omega)$, the formula of $\bar{R}_o$ should include the whole spectrum of the donor resonance secondary emission which, besides the ordinary luminescence, contains also hot luminescence and resonant Raman scattering.

System (1) is not solvable in its general form. It depends on the arrangement of LA as well as on initial excitation conditions. Therefore one is usually satisfied with computer calculations of various definite models[27,34,35]. Yet, in a particular case of a regular lattice an analytical solution of system (1) exists, that helps one to a deeper understanding of the processes under investigation. The soulution is given in the terms of the effective lifetime of the excitation T and of the full probability P(t) that excitation is still in LA (and not trapped by RC or decayed by other reasons)[36,37].

$$T = \int_0^\infty P(t)dt \qquad (3)$$

$$P(t) = \sum_{m}^{N-1} C_m \exp[-(\gamma_m + 1/\tau_{om})t] \qquad (4)$$

γ_m and C_m depend on F_{ij}, C_m depends also on initial excitation conditions. Qualitatively, the time T is the sum of two times: the

LASER STUDY OF PHOTOSYNTHESIS   331

average time t_a for excitation to reach at RC, if it was generated somewhere in the antenna, and the time t_t, if RC was excited directly[36-38].

$$T \simeq \langle n \rangle / \langle F \rangle + [1 + (F_D/F_T(N-1)]/k_p \qquad (5)$$

where $\langle n \rangle$ is the average number of jumps in LA (in the case of two-dimentional quadratic lattices, it is $NlnN/\pi + 0.2N$ [39]); $\langle F \rangle$ is the average over LA transfer rate constant, F_D/F_T is the ratio of some average rate constants of excitation detrapping and trapping, respectively, at the RC, k_p is the charge separation rate constant in RC. It is taken here that in the conditions of photosynthesis $\tau_o \gg T$, that is usually well fulfilled. In the case of $t_a \gg t_t$ one can speak about the transfer where its rate is limited by the excitation diffusion in LA, in the opposite case, about the transfer that is limited by excitation trapping on RC. As maintains the author of[37], the fluorescence decay law in the case of a regular matrix arrangement is practically always exponential and the detection of nonexponentiality indicates deviations from regularity. The opposite statement is certainly not valid.

The above said immediately suggests an experimental scheme to elucidate the limiting stages of the primary energy transfer rate in photosynthesis. For that a successive selective excitation of LA and RC is needed and the obtained lifetimes should be compared [26,36,40].

So far nothing has been said about the influence of the RC state upon fluorescence. Until recently for bacteria the dependence on the RC state was quantitatively established only for the fluorescence yield[23]. The ϕ and τ dependences on the RC state can be explained by the following simplified scheme of processes in RC

$$DPQ \xrightarrow{h\nu} DP^*Q \rightarrow DP^+Q^- \rightarrow D^+PQ^-$$

where P are the photochemically active RC molecules (considered to be $(BChl)_2$ in bacteria); P* is the photoexcited state of P; D and Q are respectively the donor and acceptor of the electron. In the so-called closed RC state, P^+Q^-, the fluorescence yield exceeds several times (approaching ~5%) that in the open or photochemically active state, PQ, whereas, as shows experiment, the dependence in the share of closed centres P^+/P_o in the amount of the culture studied is smooth. The definite form of this dependence, however, is determined by the structure and build-up of the pigment system as well as by the details of energy transfer process. The influence of the pigment system build-up can be elucidated by considering two simple physical models: (1) different PSU in chromatophores or chloroplasts do not exchange mutually energy; (2) several PSU constitute a unit system at the level of energy transfer (a multicentre or lake model). In the first case, the fluorescence yield is a linearly growing function of P^+/P_o, $\phi = \phi_o + aP^+/P_o$, where ϕ_o corresponds to the fluorescence yield in case all RC are in the state PQ and a is a constant. In the case of heterogeneous PSU with various number of LA molecules on RC, the dependence is superlinear[41]. In the simplest version the fluorescence decay is a two-component one, where the short time corresponds to the PSU emission with RC in the PQ state and the long one, in the P^+Q^- state. In the second case, a hyperbolic equation $\phi = a/(1 - pP^+/P_o)$[23,41] is observed, where p is a constant which depends on k_p. The fluorescence decay in a trivial version is single-exponential and the course of τ is proportional to that of ϕ. In formula (5) the change of RC redox state shows itself through the change of N. (In a multicenter case N is redetermined as a number of LA molecules per RC in PQ state.)

To conclude this chapter it is to be noted that PSU represents a model system with essentially nonlinear properties, which appear already at a rather delicate level of illumination ($\sim 10^{-2}$-10^{-1} W/m^2 instead of usual MW/m^2 in physics).

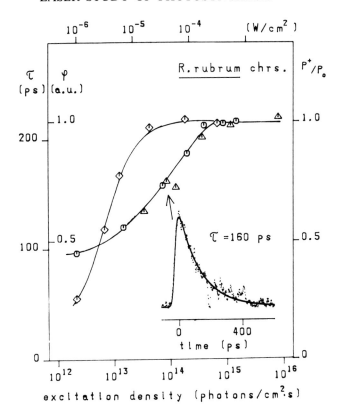

FIGURE 2 R. rubrum chromatophotes fluorescence lifetime τ (△-△-△) and yield ϕ (o-o-o) as a function of excitation density. (◇-◇-◇) - the normalized fraction of photooxidized RC, P^+/P_o. In the insert: example of fluorescence kinetics at excitation density of ~10^{14} phot/cm^2·s [42].

4. INVESTIGATION OF BACTERIAL PHOTOSYNTHESIS

In the light of what is said above the dependence of picosecond fluorescence kinetics on the RC state of cells and chromatophores of the purple bacteria Rhodo spirillum rubrum and Rhodopseudomonas sphaeroides was investigated[26,40,42]. The measurements, performed by the spectrochronograph described in part 2 (Figure 1), were carried out with the average excitation density from 10^{13} phot/cm^{2}s to 10^{17} phot/cm^{2}s. The maximum energy of each pulse was 10^{-14} -

334 LASERS IN ATOMIC, MOLECULAR AND NUCLEAR PHYSICS

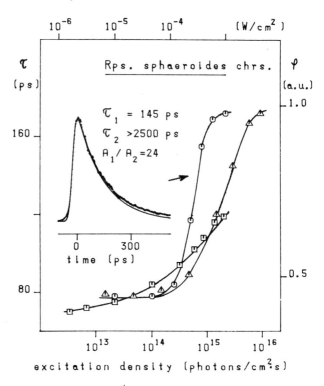

FIGURE 3 The same (excluding P^+/P_0 curve) as in Figure 2 for Rps. sphaeroides chromatophores with a $5 \cdot 10^{-5}$ M TMPD addition[42]. The curve (□-□-□) represents ϕ for chromatophores without additions.

10^{-10} J/cm^2, that fully excluded undesirable nonlinear excitation effects. In this intensity interval τ grows together with excitation intensity approximately from 60 ps to 210 ps (Figures 2 and 3). The fluorescence yield for all the bacteria studied, measured on continuous as well as on picosecond excitation, changed analogously. Parallel measurements of the ratio of photooxidized RC show that the increase of fluorescence is related with the RC passing into a closed, photooxidized state. The fluorescence decay in chemically pure chromatophores without any chemical additions was well approximated with a single exponential component at low as well as at saturating intensities. In the emission of intact cells and chromatophores with reductive substanc-

LASER STUDY OF PHOTOSYNTHESIS 335

es, a nanosesond component appeared when a fraction of RC passed into the state PQ^-. At low excitation level the amplitude of that component amounted to several % of that of the picosecond component and grew by 2-3 times with the growth of exciting light density, while its lifetime probably remained unchanged. Thus, an explicit dependence of τ on the light intensity and its parallel behaviour with ϕ have been detected, that, together with the characteristic dependence of τ and ϕ on RC state, suggest the multicentricity of the build-up of the photosynthetic bacterial system.

These data allow numerical estimations of a number of important physical parameters. Thus, an estimation of the maximum charge separation quantum yield in RC $\phi_p^{max} = (\tau_a - \tau^{min})/\tau_a$ (τ_a is the excitation lifetime in isolated antenna complexes) gives 0.95±0.02 [42]. By assuming that RC is an absolute trap for excitations ($k_p \to \infty$), from formula (5) we obtain an estimation of the excitation stay-time on a BChl molecule of LA in a two-dimensional case, $\langle F \rangle^{-1} \gtrsim$ 1 ps and, respectively from formula (2), an estimation of the average intermolecular distance R, 16-20 Å. Further, it is possible to calculate the coefficient of isotropic excitation diffusion $D = R^2 \langle F \rangle / 4 \simeq 8 \cdot 10^{-3}$ cm^2/s and the diffusion path-length $L = (D\tau)^{1/2} \simeq 70$ Å [42]. The latter value agrees poorly with the above conclusion about the multicentre arrangement of the bacterial PSU. On these grounds the validity of the assumption of RC being an absolute trap is to be revised [40]. If the observed lifetime is limited by photoconversion rate in RC (the second term prevails in formula (5)) and not by diffusion along LA, all the constants obtain quite reasonable values $(k_p)^{-1} \simeq 3$ ps, $\langle F \rangle^{-1} \simeq 0.1$-0.5 ps, $R \simeq 12$-15 Å, $D \simeq 4 \cdot 10^{-2}$ cm^2/s, $L \simeq 160$ Å. Numerical values of L and R allow one to infer that in the domain 10-13 PSU may be connected on the level of energy transfer. Naturally, these estimations are too crude to make any far-reaching conclusions. Nevertheless, proceeding

336 LASERS IN ATOMIC, MOLECULAR AND NUCLEAR PHYSICS

from another type of experiments on R. rubrum, the authors of [43] reached qualitatively the same conclusion (the domain includes 14-17 PSU). It is interesting to note that analogous results have been obtained for chloroplasts by investigating excitation annihilation[44,45].

Some remarks on the differences between the single-pulse(SP) and CW picosecond excitation methods in the case of a multicentre PSU organization should be made[40,43]. In a SP mode, as a rule, it is difficult to work with the densities of $<10^{13}$-10^{14} phot/cm^2 per pulse. Considering the minimum antenna lifetime of 60 ps, all these quanta are virtually simultaneously absorbed, and, for instance, with the Chl (BChl) absorption cross-section of $\sim 10^{-16}$ cm^2, one molecule out of 1000 or 100 molecules, respectively, is excited. This means that even with minimum SP excitation densities there is at least one excitation per domain and on further density increase an appearance of nonlinear excitation interaction effects can be expected. It also becomes clear from here that the commonly used weak excitation criterion - one quantum per RC - does not suffice and, as a minimum, one must stick to a more strict condition - one quantum per domain. In our CW version, the average density of 10^{13}-10^{14} phot/cm^2·s makes only $\sim 10^5$-10^6 phot/cm^2 per pulse, that constitutes quite negligible excitation concentration in the domain and, consequently, an inappreciable probability of their meeting. However, as the pulse sequence period ($\sim$12 ns) is much less than the time of restoring RC into a state able to trap the next excitation, starting from a definite intensity, in full analogy with stationary excitation, there occurs an accumulation of RC in a closed state. This results in an observable growth and the subsequent saturation of τ and ϕ intensity dependences. The above-said is illustrated in Figure 4, where the expected dependences of τ

LASER STUDY OF PHOTOSYNTHESIS

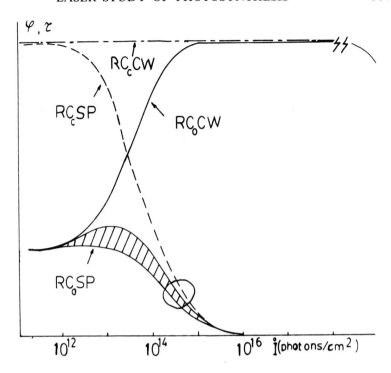

FIGURE 4 Model dependences of τ and ϕ on the incident light intensity. The initial state of RC (O - open, C - closed) and the picosecond excitation mode (SP - cingle pulse, CW - continuous-wave excitation) are denoted. The course of the curve in the striped area depends on the competition between the trapping on RC and annihilation. The latter inevitably prevails in the case of sufficiently high excitation densities. The course of ϕ under stationary excitation and in case of RC_O initial conditions corresponds to the curve RC_OCW. The approximate intensity scale is in phot/cm^2 per pulse for SP case and in phot /cm^2·s for CW case.

and ϕ on the excitation nature and density are presented. Note that the measurements in the encircled area are rather insensitive to the RC state, that may explain the corresponding results of[22]. Due to the rather high density of SP excitation used there the influence of excited state absorption is also possible, causing the observed deformation of the absorption spectrum.

At high excitation frequency there is a danger of the triplete state accumulation and the following singlet-triplet annihilation (see, e.g.[8] and refernces therein). This should be highly important in the case of closed RC. No systematic investigations of this problem have been carried out yet. However, the fact that no decrease in fluorescence τ has been observed in R. rubrum chromatophotes up to average density of $\sim 5 \cdot 10^{18}$ phot/cm$^2 \cdot$s[40] suggests that it is negligible.

5. CONCLUSION

Our data as well as those of other authors are in qualitative agreement with the following simple model. Photons are absorbed by LA molecules and the excitations created reach RC in a very short time, $\sim$1-10 ps. The irreversible trapping on RC, however, proceeds in 50-70 ps, since, figuratively speaking the excitation hits and jumps out of RC for several times. One may wonder what is the biophysical sense of such manifold detrapping? Calculations[35,46] show that the negative effect of detrapping on the photochemical yield is negligible and the only idea is that the Nature has probably failed to "design" a 100%-efficiency trap of the first hit. The rise in excitation intensity results in an ever-growing number of closed RC and, consequently, in an increase of excitation lifetime in the domain, where on the level of energy transfer about ten PSU have been connected. The present level of understanding of the problem is essentially due to the progress in experimental methods and tecnique.

The author is indebted to K. Rebane for his kind support and valuable advice, to A. Yu. Borisov and V. I. Godik for many useful discussions, to R. Avarmaa for reading and criticizing the manuscript, to V. I. Godik, K. Timpmann, and R. Tamkivi for cooperation in experiments and to T. Altukhova and E. Vaik for drawing up the manuscript.

REFERENCES

1. Bol'shaya sovetskaya entsiklopediya v. 27 (Soviet Encyclopedia, Moscow, 1977), p. 592.
2. A. FREIBERG, K. TIMPMANN, R. TAMKIVI, R. AVERMAA, Eesti NSV TA Toimetised Füüs. Matem., 31, 200 (1982)
3. A. Yu. BORISOV, V. I. GODIK, Biochim. Biophys. Acta, 301, 227 (1973)
4. J. LAVOREL, A.-L. ETIENNE, in Primary Processes in Photosynthesis, edited by J. Barber (Elsevier, Amsterdam, 1977), p. 203
5. A. J. CAMPILLO, S. L. SHAPIRO, in Ultrashort Light Pulses, edited by S. L. Shapiro (Springer-Verlag, Berlin, Heidelberg, 1977), p. 317.
6. A. B. RUBIN, Photochem. Photobiol., 28, 1021 (1978)
7. J. AMESZ, in Photosynthetic Bacteria, edited by R. K. Clayton and W. R. Sistrom (Plenum Press, New York, 1978), p. 333.
8. J. BRETON, N. E. GEACINTOV, Biochim. Biophys. Acta, 594, 1, (1980)
9. F. PELLEGRINO, R. R. ALFANO, in Biological Events Probed by Ultrafast Laser Spectroscopy, edited by R. R. Alfano (Academic Press, New York, 1982), p. 27.
10. G. S. BEDDARD, R. J. COGDELL, in Light Reaction Path of Photosynthesis, edited by F. K. Fong (Springer-Verlag, Berlin, 1982), p. 46.
11. H. MERKELO, S. R. HARTMAN, T. MAR, G. S. SINGHAL, GOVINDJEE, Science, 164, 301 (1969)
12. V. Z. PASCHENKO, S. P. PROTASOV, A. B. RUBIN, K. N. TIMOFEEV, L. M. ZAMZOVA, L. B. RUBIN, Biophys. Acta, 408, 143 (1976)
13. S. L. SHAPIRO, V. H. KOLLMAN, A. J. CAMPILLO, FEBS Lett., 54, 358 (1975)
14. J. BRETON, E. ROUX, in Lasers in Physical Chemistry and Biophysics, edited by J. Joussot-Dubien (Elsevier, Amsterdam, 1975), p. 379.
15. G. S. BEDDARD, G. PORTER, C. J. TREDWELL, J. BARBER, Nature, 258, 166 (1975)
16. N. H. SCHILLER, R. R. ALFANO, Opt. Commun., 35, 451 (1980)
17. P. SAARI, J. AAVIKSOO, A. FREIBERG, K. TIMPMANN, Opt. Commun., 39, 94 (1981)
18. A. FREIBERG, P. SAARI, IEEE J. Quant. Electron., QE-19, 622 (1983)
19. A. O. ANIJALG, K. E. TIMPMANN, A. M. FREIBERG, Pis'ma Zh. Tehn. Fiz., 8, 753 (1982)
20. A. O. ANIJALG, P. M. SAARI, T. B. TAMM, K. E. TIMPMANN, A. M. FREIBERG, Kvantovaya Elektronika, 9, 2449 (1982)
21. R. DANIELIUS, A. PISKARSKAS, V. SIRUTKAITIS, Kvantovaya Elektronika, 9, 2491 (1982)

22. A. Yu. BORISOV, R. V. DANIELIUS, A. S. PISKARSKAS, A. P. RAZ-
    JIVIN, R. I. ROTOMSKIS, Kvantovaya Elektronika, 10, 1531 (1983)
23. W. J. VREDENBERG, L. N. DUYSENS, Natute, 197, 355 (1963)
24. A. JOLIOT, P. JOLIOT, C.R. Acad. Sci. Paris, 258, 4622 (1964)
25. L. A. TUMERMAN, E. M. SOROKIN, Molek. Biol., 1, 628 (1967)
26. A. FREIBERG, V. GODIK, K. TIMPMANN, in Advances in Photosynthetic Research, vol. 1, edited by C. Sybesma (Martinus Nijhoff/Dr. W. Junk Publishers, Hague, 1984), p. 45.
27. Z. G. FERISOVA, M. V. FOK, A. Yu. BORISOV, Molek. Biol., 17, 437 (1983)
28. V. M. AGRANOVICH, M. D. GALANIN, Electronic Excitation Energy Transfer in Condensed Matter (North-Holland Publ. Comp., Amsterdam, 1982)
30. V. M. KENKRE, in Exciton Dynamics in Molecular Crystals and Aggregates, edited by G. Hohler (Springer-Verlag, Berlin, 1982), p. 1.
31. R. S. KNOX, Ref. 4, p. 56.
32. Th. FÖRSTER, Ann. Physik, 2, 55 (1948)
33. I. J. TEHVER, V. V. HIZHNYAKOV, Zh. Eksp. Teor. Fiz., 69, 599 (1975)
34. G. W. ROBINSON, Brookhaven Symp. Biol., 19, 16 (1967)
35. L. SHIPMAN, Photochem. Photobiol., 31, 157 (1980)
36. R. M. PEARLSTEIN, Photochem. Photobiol., 35, 835 (1982)
37. R. M. PEARLSTEIN, in Photosynthesis: Energy Conversion by Plants and Bacteria, edited by Govindjee (Academic Press, New York, 1982), p. 293.
38. G. PAILLOTIN, Theor. Biol., 58, 219 (1976)
39. E. W. MONTROLL, J. Math. Physics, 10, 753 (1969)
40. V. GODIK, A. FREIBERG, K. TIMPMANN, Proc. Acad. Sci. Estonian SSR, Phys. Math., 33, 211 (1984)
41. L. N. M. DUYSENS, in Chlorophyll Organization and Energy Transfer in Photosynthesis, Ciba Foundation Symposium 61 (new series) (Excerpta Medica, Amsterdam, 1979), p. 323.
42. A. Yu. BORISOV, A. M. FREIBERG, V. I. GODIK, K. K. REBANE, K. E. TIMPMANN, Biochim. Biophys. Acta, (1985), submitted for the publication.
43. J. G. C. BAKKER, R. VAN GRONDELLE, W. T. E. DEN HOLLANDER, Biochim. Biophys. Acta, 725, 508 (1983)
44. G. PAILLOTIN, J. Theor. Biol., 58, 237 (1976)
45. N. E. GEACINTOV, J. BRETON, C. E. SWENBERG, G. PAILLOTIN, Photochem. Photbiol., 26, 629 (1977)
46. D. MAUZERALL, Ref. 9, p. 215.

4.3 Influence of Structural Heterogeneity on Energy Migration in Photosynthesis

L. VALKUNAS

Institute of Physics, Academy of Sciences of the Lithuanian SSR, Vilnius, USSR

The picosecond laser spectroscopy was applied in biology first of all in the studies of the primary stages of photosynthesis. With the help of multiple investigations which were carried out during the last 15 years a number of kinetic and spectroscopical parametres of the photosynthetic objects were defined. Partly, the mean times of the excitation energy transfer in the light harvesting antenna (LHA) as well as the charge separation in the reaction centre (RC) are known. The spectral forms of the chlorophyll molecules and their sequence of a participation in the evolution of the absorbed solar energy are also known. However, in spite of apparent evidence of the role played by each spectral form, the cause as well as the mechanicm of such effective light-harvesting are not completely clear. The measurement of the quantum yield of the charge separation in the RC, for instance, gets $\mu > 90\%$. The very primitive valuation of this quantity while modeling the LHA as homogeneous matrix of chlorophyll molecules and the RC being the trap for an excitation, shows that such high magnitude of the quantum yield of photosynthesis could be reached while supposing only the extremely short excitation trapping mean time.

It can be demonstrated in the following way. Let us assume that the excitation migration rate in the LHA is very fast and the excitation lifetime (T) in the system is limited by the excitation trapping rate (τ_{RC}^{-1}) on the RC, i.e. $T^{-1} = \tau_o^{-1} + \tau_{RC}^{-1}/N$ (τ_o being the excitation lifetime on the isolated pigment molecule, N number of molecules which take part in the energy transfer). Then the excitation lifetime on the RC states which are in resonance with the excited states of antenna equals to T/N and the charge separation quantum yield can be calculated thus:

$$\mu = \tau_{RC}^{-1}/(N/T) \equiv (1 + \tau_{RC} \tau_o^{-1} N)^{-1} \qquad (1)$$

From formula (1) it follows that in the case of N being large, the high (near 100%) quantum yield is possible only assuming rather strict requirements to the ratio τ_{RC}/τ_o.

The present report is devoted to the discussion of the structural heterogeneity of the energy migration process in the LHA and the excitation traping on the RC. The accepted analysis lets us give unequivocal interpretation to the experimentally defined parametres as well as solve some contradictions while explaining the mechanism of the considered processes.

1. STRUCTURAL DATA

It is known that a photosynthetic apparatus both of bacteria and plants is build up by pigment-protein complexes (PPC) which are situated in the lipid membrane. The above-mentioned is approved by the following data.

The X-ray crystallographic investigations of the green bacteria (Prosthesochloris aestuarii) antenna complexes[1,2] enable us to establish the mean distance between the centres of chlorophyll molecules: within the PPC it equals to 12 Å and between them 24 Å. This fact points to the inhomogeneous location of antenna chlorophyll molecules. The experiments carried out with bacteria Rhodopseudomonas sphaeroides[3], corresponding to 880 nm and 800-850 nm respectively testify weak intercomplex pigment interaction. The change of the PPC relation shows the spectrum variations being additive. Otherwise the experiments on lin-

LASER STUDY OF PHOTOSYNTHESIS   343

near and circular dichroism of individual comlexes[4] point to the strong (exciton-like) interaction between pigments.

It is possible to judge about the interlocation of the PPC in photosynthetic system from the experiments on electron microscopy. The measurements of bacteria[5], for instance, show a fine hexagonal structure. Besides, it must be mentioned, that each RC together with surrounding antenna forms so called photosynthetic unit (PSU).

2. THEORETICAL MODEL

On the grounds of the above-mentioned experimental data it is natural to propose the existance of two energy migration processes which are determined by intercomplex and intracomplex interactions. The assumption that the excitation decay time within the PPC (τ_{ex}) being much less than the intercomplex energy migration mean time (τ_{hop})[6],

$$\tau_{ex} \ll \tau_{hop}, \qquad (2)$$

expresses the strong (exciton-like) pigment interaction within the PPC and the weak intercomplex interaction. In this case τ_{ex} is conditioned by electron-phonon interation and τ_{hop} incoherent excitation transfer between the PPC time (hopping time). The assertion of the intercomplex energy transfer being incoherent, is based on the comparison of the kinetic constants and the overlap of the chlorophyll molecules spectra[7]. Naturally, it turns out, that the excitation energy transfer to the molecule of the distance of 25 Å° and even more, proceeds the Foerster-like mechanism. Now, when inequality (2) is satisfied, each of the PPC can be characterized approximately by a single level. In this case the evolution of the excitation can be described by the following balance-type equation set:

$$\frac{d}{dt} a_n^i(t) = \sum_j \sum_m^{N_c} H_{nm}^{ij} a_m^j(t) \qquad (3)$$

where a_n^i is the probability of n-th PPC to be in the i-th excited state, N_c number of complexes in the PSU, H_{nm}^{ij} matrix describing the relaxation constants as well as the excitation transfer probabilities. It is enough to characterize the LHA complexes by a single level (i=1) and the RC - by three (i=0,1,2). If we take into consideration the symmetry of the PSU structure[5] the kinetic problem becomes and can be expressed as an energy transfer problem in one-dimensional structure where nodes are RC and its corresponding surrounding spheres[8]. It must be mentioned that the excitation transfer within a certain sphere does not play any role.

The essential moment of such consideration is the fact that N_c is not a large parametre ($N_c \simeq 10$). Therefore effects of the local anisotropy of kinetic constants in the vicinity of the RC can influence on the excitation lifetime in the LHA. On the contrary, the homogeneous antenna model where a number of nodes taking part in the energy transfer process is fairly large ($N \gtrsim 100$), in case of 2- and 3-dimensional structures is characterized practically by one-exponential decay kinetics the same time being insensitive to the local anisotropy effect[9]. Direct solution of set (3) in the case of the PSU with one or two surrounding spheres[8] indicates to a two-exponential excitation decay kinetics. Besides, at any relation of the kinetic constants, these exponents practically are connected separately with the excitation transfer processes to the RC and with the trapping. Consequently, for such structures with a small number of nodes N_c, where the excitation proceeds the random walk, its decay kinetics can not be described by a single mean time T, to formula (1) is not applicable.

In fact the quantum yield of photosynthesis is determined by the occupation of state 2 (see Figure 1) at extremely large time (t), when t is much more than characteristic time of passing processes in the PSU:

LASER STUDY OF PHOTOSYNTHESIS 345

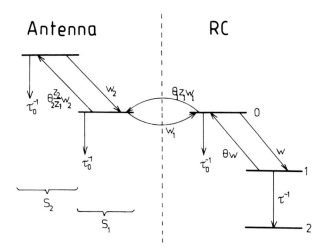

FIGURE 1 Scheme of energy levels of the PSU containing two
surrounding spheres.
RC being characterized by three levels (0,1,2), S_1
near sphere to the RC, S_2 extrenal sphere. Rows
point out the energy transfer and charge separation
processes. W_i is the rate of intercomplex energy
transfer from S_i sphere to the RC direction, z_i the
PPC number in S_i sphere, $\mathcal{T}_o$ the excited PPC lifetime, W the rate of primary charge separation in
the RC, $\mathcal{T}$ the separate charge fixation time, θ_i
and θ parametres characrerizing migration anisotropy

$$\mu = \int_0^\infty a_o^1(t)/\mathcal{T} \, dt \qquad (4)$$

$\mathcal{T}$ being the separate charge fixation time in the RC.

Let us restrict ourselves for the case of two surrounding
spheres. Naturally, such a model expresses qualitatively the process considered. As it can be seen directly from Figure 1, we will
have 4 equations for the mathematical descrition, 2 of them describing the excitation evolution in the LHA and two others - on
the RC.

While determining the quantum yield of photosynthesis, there
is no need to solve the set of equations. So it is possible to express μ via coefficients at various power of the characteristic

equation of the eigenvalue:

$$\mu = \frac{1 + Q/(N_c \tau_o)}{1 + (\tau_{RC} N_{ef} + F)/\tau_o} \quad (5)$$

where τ_o is the excitation decay time in the PPC, $\tau_{RC} = \theta\tau + W^{-1}$ the state of separate charge stabilization time in the RC, $N_{ef} = 1 + \theta_1 z_1 + \theta_1 \theta_2 z_2$, z_i the PPC number in the S_i-th sphere, $\theta_i = \exp(-\Omega_i/kT_o)$ parametres describing anisotropy of the kinetics in the direction to the RC and back, Ω_i the energy distance between corresponding levels, kT_o the temperature (in energy units), Q is determined by the initial conditions:

$$Q = \begin{cases} 0 & \text{excited only sphere } S_2 \\ z_1/W_2, & \text{excited both } S_1 \text{ and } S_2 \\ z_1/W_2 + F, & \text{uniform initial condition} \end{cases} \quad (6)$$

$$F = W_2^{-1} + (1 + \theta_2 z_2/z_1)/W_1$$

In case of more than two surrounding spheres the value μ is exprssed by the analogical formula as (5).

If we assume the excitation migration rate being extremely high, i.e. $W_1, W_2 \to \infty$, the value μ can be expressed thus:

$$\mu = (1 + \tau_{RC} \tau_o^{-1} N_{ef})^{-1} \quad (7)$$

In the case of $N = N_{ef}$ formula (7) coincides with expression (1).

If we assume now $\tau_o = 1$ ns, $\tau = 200$ ps, $W^{-1} = 7$ ps, $\Omega = 0.1$ eV, $\tau_{RC} = 10$ ps, $\mu = 0.9$ (parametres of the bacterial photosynthesis) so from formula (7) we get $N_{ef} = 11$. Since $\theta_1 \approx 1$ and if we assume that all the spheres are spectrally homogeneous ($\theta_2 = 1$), so in this case the system should contain the RC and only a single surrounding sphere.

Let us analyze a more complicated situation of two surrounding spheres, assuming that $\theta_2 < 1$ which is inherent in real photosynthesis. Then, while changing the parametres we get $\mu = 0.82$ to 0.99 (see Figure 2).

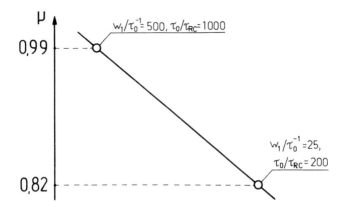

FIGURE 2   Schematical representation of change limits of the quantum yield of photosynthesis

It must be mentioned that, while taking into account the structural heterogeneity of the PSU, the high quantum yield of photosynthesis could be reached without any limitation for the kinetic parametres. The infinite rate of the excitation migration is not the propitions case for the increase of effectivity. The quantum yield of process essentally increases if the "focussing" effect, which manifests itself both in the process of rapid relaxation on the lowest excited state within the PPC parametres as well as in the anisotropy of the kinetics of intercomplex transfer takes place.

3. ANALYSIS OF KINETIC PARAMETRES

The presented model of the PSU gives a good explanation to the excitation transfer mechanism in the LHA. The comparison of the calculated fluorescence decay time[6] with the experimental measurements lets us determine a single intercomplex hopping time (τ_{hop}). From the experiment with bacteria Rhodospirillum rubrum it follows that T_{of} = 50 ps [10]. Then the single hopping time is of the order of 10 ps. Such a result completely coincides with the initial assumption of incoherent process of the excitation migration between the PPC. Contrary, in the case of the homogeneous model of the PSU, the analogical estimation of the single hopping gives value less than 0.5 ps. Such a result testifies to impossibility of the energy migration incoherent-ixciton-like[11]. Naturally, it could be assumed that the energy migration is determined by coherent exciton[12]. However, this idea is not without contradictions. Partly, in this case we have to speak about a very close-packed structure of chlorophyll molecules in the PSU with a mean distance of order 10 A° which contradicts to the above-mintioned structural data.

The essential point of the given model containing one or two surrounding spheres is the fact that fluorescence decay kinetics should be two-exponential like. However, the majority experiments were carried out up to this time with exciting pulses of 20 ps duration or more. Therefore the kinetic curves difference from a single exponential function is difficult to determine. The recent experiments with 5 ps pulse duration[10] point out the single exponential kinetics could be explained by: i) a number of surrounding spheres being more than two, ii) the meantime of one of exponents being less than 5 ps. The very fact that the excitation decay time does not depend upon priority the LHA or the RC to excite[10], could testify to a very small parametre, of one of exponents ($<$ 5 ps). Besides, there is no necessity to assume that the excitation migration in tha LHA should be very fast. That is because the number of the PPC of the first surrounding sphere z_1 and the ratio z_2/z_1 are the factors which exceed the excitation departure from the RC to the LHA. In the case of hexagonal structure (z_1 = 6), for instance, while assuming the single hopping time to be 10 ps, the time of the excitation transfer from the RC to the first surrounding sphere, would be equal to 1.6 ps. Consequently, it is possible that the fluorescence decay kinetics satisfies the two exponential law at any initial conditions that influence the relative contribution of each of them.

The fluorescence quantum yield measurements with the increase of the excitation intensity witness the manifestation of annihilation effects. During the LHA excitation, the occupation of the first singlet state (S_1) of the PPC is created. The process of the singlet-singlet annihilation makes excitation pass from one excited PPC to the other excited one, letting the latter get to a higher excited state (e.g. in bacterial photosynthesis these states correspond to the Soret band of the absorption spectrum). Besides, such a process takes part not in an isolated PSU but in the domain being a set of the PSU on which free hopping of the excitation occurs. From that high energy state a very rapid intramolecular relaxation of excitation to the S_2 state of the LHA and subsequent relaxation from the S_2 to the S_1 state occurs, the lifetime of the former apparantly being relatively considerable[13,14]. Particularly, as it is evident from the analysis of bacterial photosynthesis (Rhodospirillum rubrum)[15] the dimension parameter r equals 1.1 (r=$2\gamma_1/\gamma_2$, γ_1 and γ_2 being the rates of the linear decay of S_1 excitation and annihilation of a pair of excitations correspondingly in the domain). Then we get $\gamma_2^{-1} \simeq 200$ ps. Assuming the fact that in the whole thickness of the sample the excited molecules are distributed homogeneously, one can estimate the mean time of the annihilation process by the formula:

$$\tau_a^{-1} = \frac{1}{2}\gamma_2 \lambda n_{PSU} \qquad (8)$$

n_{PSU} being the average number of absorbed photons per PSU in the whole thickness of the sample, λ the number of PSU in the domain. While inserting the value $\lambda = 16$ [15] and $n_{PSU} = 1$ we get $\tau_a = 25$ ps. However, straight from the measurements of the value γ_2 (as well as τ_a) it is difficult to judge about the parametres of the migration process though some authors have undertaken such attempts. Naturally, the annihilation time of a pair of excitation as it is seen from the analysis[15,16] is of order or less than 5 ps. Therefore the supposition of the diffusion limited approximation to the annihilation process in this case is not applicable.

Recent investigations of the spectral and kinetic properties of the PSU by the difference absorption spectroscopy have appeared [17,18]. They show certain deviations from the fluorescence data:

1. In the difference spectrum of chromatophores absorption the minor bacteriochlorophyll spectral form having a longer wavelength than the absorption band of the RC photodonor is observed, and the excitation energy transfer occuring through the minor form to the RC is shown.

2. The mean time of the excitation decay in the LHA is essentially less than the corresponding time obtained from the fluorescence measurements. Besides, the kinetics of the excitation decay does not depend on the state of the RC.

In connection with minor spectral form of the LHA chlorophyll, the hypothesis of the pericentral complex which is responsible for such a form was presented. Such an assumption casts out the doubt on the generally accepted conception of the multicentricity in the frame of the globular model of the LHA. That's why it is necessary to discuss these results in a more detailed way.

The measurements of the difference absorbance of the chromatophores[17,18] testify different temporal courses as well as light dependencies of the optical density of the sample at low ($I_o \simeq$ $\simeq 10^{14}$ hν/cm^2) and high ($I_o \gtrsim 10^{16}$ hν/cm^2) excitation intensities (normalization of the intensity per single pulse of the duration $\tau_o \simeq 25$ ps is used everywhere in text). The main argument, that the kinetics of the spectral changes in the case of $I_o \simeq$ $\simeq 10^{14}$ hν/cm^2 reflects the picture of the natural photosynthesis, lies in the fact that the quantum yield of the charge separation is $\mu > 0.5$. In the case of used experimental parametres (the samples thickness ~ 1 mm, the optical density of about a unit on

LASER STUDY OF PHOTOSYNTHESIS   351

the wavelength ≈ 880 nm and the number of bacteriochlorophyll per
RC ≈ 50) such an intensity corresponds to n_{PSU} ≈ 1. However, as it
follows from the above-presented discussion and formula (8) such
intensities get τ_a ≃ 25 ps. So the kinetics of the difference absorption can be explained in another way.

As it was mentioned above, the fluorescence decay time in the
case of the open RC approximately equals to τ^o = 50 ps [10]. Therefore, while all the RCs are opened, the initial excitation decay
kinetics is mainly determined only by the singlet-singlet annihilation process because $\tau_a < \tau^o$. In course of time the contribution of annihilation reduces and the quenching of excitation by
the open RC prevails. At still longer times the closed RCs start
to influence the excitation decay process (the mean time of fluorescence quenching by the closed RCs τ^c = 180 ps [10]). Actually,
it is not difficult to show numerically that three exponential
kinetics with indexes mentioned above (τ_a = 25 ps, τ^o = 50 ps and
τ^c = 180 ps) easily approximates the experimental kinetic curves
[17,18]. Besides, it must be mentioned that the pre-exponential factor of the "rapid" exponent occurs to be essential (≥0.5) while
the calculation of the quantum yield of charge separation for
n_{PSU} = 1 giving the value μ = 0.5.

The presented three-exponential kinetics of the excitation
decay in the LHA enables us easily to understand its independence
of the redox state of the RC. The excitation decay kinetics with
the increase of the fraction of the closed RCs changes by the interplay of relative weights of excitation quenching by the open
(τ^o) and the closed (τ^c) RCs. In the limit of all RCs closed the
"rapid" (annihilation) and the "slow" (excitation quenching by the
closed RC) parts of the kinetics remain only. Consequently, within
the limits of the time resolution restricted by the duration of
the probing pulse (τ_o ≃ 25 ps) the kinetics for the open or the

closed RC must differ only in the relative contribution of the "slow" exponent the fact being noticed in the experiment.

The origin of the minor spectral form could be also understood while analizing the difference absorption spectrum which could be derived thus[19]:

$$\Delta A(\lambda_{pr}) = \alpha [n_0 \sigma_0(\lambda_{pr}) + n_1 \sigma_1(\lambda_{pr}) + n_2 \sigma_2(\lambda_{pr}) - \sigma_0(\lambda_{pr})] \quad (9)$$

α being the numerical coefficient, n_i the occupation of the ground (i=0) the first (i=1) and the second (i=2) excited singlet states, $\sigma_i(\lambda_{pr})$ is the cross-section of the light absorption (and stimulated emission in case of I ≠ 0) of the i state at the wavelength λ_{pr} of the probbing pulse. The temporal dependence of ΔA is determined by the state occupations n_i, the exact values of which can be obtained by solving the corresponding kinetic equations.

In case of low intensities of the exciting light pulse $n_2 \ll 1$, hence, $n_1 + n_0 \simeq 1$. Then from (9) it follows that:

$$\Delta A(\lambda_{pr}) = \alpha n_1(t) [\sigma_1(\lambda_{pr}) - \sigma_0(\lambda_{pr})] \quad (10)$$

σ_1 dependence on λ_{pr} is unknown. However, it should be pointed out that the transition from the first singlet state to the Soret band (in case of bacterial photosynthesis) is slightly blue-shifted in comparison with the maximum of the LHA absorption from the ground state σ_0. Bearing in mind that σ_1 in order of magnitude is equal to σ_0[13], and the above said, it follows directly that the minor spectral form is caused by the difference of two absorption spectra: $\sigma_1(\lambda_{pr}) - \sigma_0(\lambda_{pr})$. It must be pointed out, that while taking into account the heterogeneous structure of the LHA the additional phenomenon should be displayed in the difference absorption spectrum. Particularly, for one excitation in the dimer owing to the resonance interaction the optical transition to the

LASER STUDY OF PHOTOSYNTHESIS 353

lowest singlet state could be red-shifted in comparison with that in the monomer. However, for the absorbed second light quantum the resonance interaction is absent, therefore on the dimer (as well as in the oligomers) at low excitation intensities the difference absorption spectrum will always possess the minor component in the longwavelength side of the spectrum. In such a case the light curves (e.g. Figure 3 in[17]) also become comprehensible. With the growth of the excitation intensity the process of singlet-singlet annihilation becomes more extensive, which leads to the nonlinear n_1 dependence on I_o. Besides, at very high intensities, when $n_{PSU} \gg 10$, it becomes possible for a new non-linear mechanism to come - the absorption of the second light quantum with the transition from the first singlet state to a higher one as well as to the ground state during the action of the exciting pulse. The nonlinear processes tend to diminish the difference between the occupations of the ground and the first singlet states, but at very high intensities owing to the relatively long lifetime of excitation in the second singlet state (see, e.g.[13,14]) a considerable occupation of the state S_2 occurs. Thus the change of difference absorption of the main band in the case of a very high excitation intensity could be explained. The lifetime of the state S_2 determines then the delay of the appearance of the minor component in the difference absorption spectrum.

An analytical solution of the kinetic equations for the occupations of the singlet states of the LHA while taking into account the pulse duration of the right-angled form with the width of 25 ps in case of excitation intensities $I_o = 10^{14}$ to 10^{16} hν/cm^2 gives the dependencies $n_1^{max} \sim I_o^{1/2}$ and $n_2^{max} \sim I_o$. Then from formula (9) it follows directly that with the increase of excitation intensity the occupation number n_2 starts to grow as soon as augmentation of n_1 (the signal of the minor spectral component) slackens. In the discussed region of the spectrum (880 nm) the absorption cross-section σ_2 apparently does not possess resonances and therefore the signal of the main component of the difference absorption spectrum increases, thus qualitatively reproducing the experimental light curves.

We must point out two moments which can decisively affect while reaching the quantitative agreement between the model calculations and the experimental data. Firstly, the duration and the shape of the exciting and probbing pulses must be included. The use of the δ- like exciting pulse in the consideration mentioned above gives for $n_1(t)$ the dependence on I_o in a state of saturation, where t is the time passed after excitation. Secondly, as the nonlinear quenching should be taken into account, the inhomogeneous distribution of excitation in the PPC in the sample as well as the fluctuations in the domain can influence the result.

Thus the observable discrepancies of the results obtained by fluorescence and difference absorption spectroscopy in fact are apparent and qualitatively could be solved while taking into account the absorption of the probing pulse from the excited states of the LHA as well as the singlet-singlet annihilation in the description of the excitation evolution. Besides, the description did not contain any fitting parametres - all kinetic constants are taken from the fluorescence data.

In conclusion we will mention that the characteristic changes of the transitions $S_o - S_2$, $S_1 - S_2$, S_2 - the Soret band which must appear in the fluorescence and difference absorption spectra (of the regions 600 nm, 2000 nm, and 1240 nm respectively) can serve as confirmation of the explanation presented above. The investigation of their kinetics at the difference excitation intensities might give an adequate answer.

REFERENCES

1. R. E. FENNA, B. W. MATTHEWS, Nature, 258, 573 (1975)
2. B. W. MATTHEWS, R. E. FENNA, M. C. BOLOGNESI, M. F. SCHMID, J. M. OLSON, J. Mol. Biol., 131, 259 (1979)

3. R. M. BROGLIE, C. N. HUNTER, P. DELEPELAIRE, R. A. NIEDERMAN, N.-H. CHUA, R. K. CLAYTON, Proc. Nat. Acad. Sci. USA, 70, 87 (1980)
4. K. D. PHILIPSON, K. SAUER, Biochemistry, 11, 1880 (1972)
5. K. R. MILLER, Nature, 300, 53 (1982)
6. S. KUDZMAUSKAS, L. VALKUNAS, A. Y. BORISOV, J. Theor. Biol., 105, 13 (1983)
7. L. L. SHIPMAN, D. L. HAUSMAN, Photochem. Photobiol., 29, 1163 (1979)
8. L. VALKUNAS, S. KUDZMAUSKAS, V. LIUOLIA, Advance in Photosynthesis Research (Martinus Nijhoff/Dr W. Junk, The Hague, 1984) v.1, p. I.1.41.
9. R. M. PEALSTEIN, in Excitons (North-Holland Publishing Co, 1982) p. 735.
10. V. GODIK, A. FREIBERG, K. TIMPMANN, Proc. Sci. Ectonian SSR, 33, 211 (1984)
11. A. Y. BORISOV, in Photosynthetic Bacteria (Plenum Press, New York, 1978), p. 323.
12. G. W. ROBINSON, Brookhaven Symp. Biol., 19, 16 (1967)
13. J. F. SHEPANSKI, W. R. ANDERSON, J. Chem. Phys. Lett., 78, 165 (1981)
14. R. P. TAMKIVI, R. A. AVARMAA, Izv. Akad. Nauk SSSR Fiz., 42, 268 (1978)
15. J. G. C. BAKKER, R. VAN GRONDELLE, Den Holander W.T.F. BBA, 725, 508 (1983)
16. A. I. ONIPKO, Phys. Stat. Sol. (b), 73, 699 (1976)
17. A. P. RAZJIVIN, R. V. DANIELIUS, R. A. GADONAS, A. Y. BORISOV, A. S. PISKARSKAS, FEBS Lett., 143, 40 (1982)
18. A. Y. BORISOV, R. A. GADONAS, R. V. DANIELIUS, A. S. PISKARSKAS, A. P. RAZJIVIN, R. I. ROTOMSKIS, Biofizika, 20, 398 (1984)
19. S. KUDZMAUSKAS, L. VALKUNAS, Studia Biophysica, 96, 213 (1983)

4.4 Picosecond Spectroscopy of Photoreceptor Molecules

F. R. AUSSENEGG, M. E. LIPPITSCH and M. RIEGLER

Institut für Experimentalphysik, Karl-Franzens-Universität, Universitätsplatz 5, A-8010 Graz, Austria

1. INTRODUCTION

Light is an indispensible prerequisite for life. In the early days of prebiotic evolution, complex organic molecules were formed under the influence of solar UV radiation. Some billions of years ago the invention of photosynthesis by cyanobacteria and green algae laid the basis for the evolution of aerobic organisms. Sunlight is still the only energy source for all kinds of living systems. Eventually, the possibility to form living structures with their high degree of complexity is derived only from the increase in entropy provided by solar radiation while dropping in radiation temperature from the value of the hot surface of the sun to that of the relatively cool one of our planet when reradiated from earth into space. In accordance with the unrivaled importance of light for all kinds of life, sophisticated mechanisms have been developed in the course of evolution to collect and perceive this unique form of high-valued energy.

Light is utilized by living beings mainly for two purposes: to consume energy, or to obtain information. Accordingly, we find two differnt strategies in photoreception. Energy harvesting is most efficiently done by using a large mumber of antenna pigments to funnel excitation into a reaction center, where the conversion

358 LASERS IN ATOMIC, MOLECULAR AND NUCLEAR PHYSICS

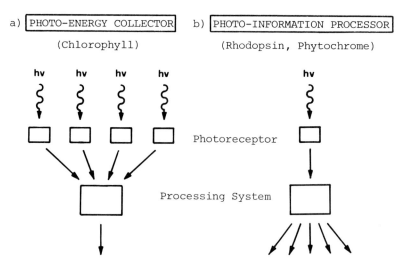

FIGURE 1 Different strategies in light utilization by living organisms a) energy collection, b) information processing

to chemical energy is performed (Figure 1a). This strategy is followed by all photosynthetic organisms. Most surprisingly, apparently only one kind of molecule, chlorophyll, has been invented by evolution to serve as a reaction center. The function of antenna pigments is accomplished by different pigments, e.g. again chlorophyll (in higher plants), or biliproteins (in cyanobacteria and algae). This use of light remained restricted to plants, while using light as a medium for information transport is most conspicious in animals. The sense of vision has evolved on different groups of animals and has reached a high degree of perfection in invertebrates (squids, insects) as well as vertebrates, including man. Nevertheless, also plants are able to perceive light and thereby collect information. Photomorphogenesis, the regulation of morphological development in higher plants by light, is the most important example. Mechanisms for acquiring information work in a way different from that of energy collectors. Usually a single--photon event is sufficient, which triggers a cascade of dark

LASER STUDY OF PHOTOSYNTHESIS 359

reactions, forming some kind of signal amplification (FIGURE 1b).

Photoreception was the first biological problem to which picosecond spectroscopy had been applied. Busch et al. showed[1] that the primary photochemical event in vision proceeds in less than 6 ps. In the meantime picosecond measurements have been proven useful for studying various photoreceptive systems, e.g. rhodopsin[2], chlorophyll[2], phytochrome[3], and phycobiliproteins[4].

There exist three different approaches to investigate primary events in photoreception. The first one is to start from the intact photoreceptor pigment, in vitro or even in vivo. This of course gives the most complete picture of the whole process, but usually it is difficult to analyse the highly complex results from measurements of this kind. A method more remote from the intact system, but easier in understanding, is to study only isolated chromophores. In some cases, even the chromophore is too complex or too difficult to handle, so that spectroscopic studies have to be conducted on the level of model compounds and their partial structures. The most complete knowledge can be obtained applying all three methods to the same case.

In the following a limited account will be given of picosecond work done in our laboratory on all three levels of complexity. The photoreceptive systems conserned are the visual system of animals and the photomorphogenetic system of higher plants. It will become apparent that it is most important in picosecond measurements not to rely on a single spectroscopic method but to gain information from different kinds of spectrocopy on different levels of complexity of the system under examination.

2. EXPERIMENTAL

To produce light pulses of picosecond duration, mainly two kinds of lasers are currently being used: synchronously pumped dye lasers, and passively mode-locked solid-state lasers. Because of

360 LASERS IN ATOMIC, MOLECULAR AND NUCLEAR PHYSICS

the low pulse energies the synchronously pumped systems are only moderately capable of generating new frequencies by nonlinear processes. The spectral region in which these systems are useful is thus more or less limited to the tuning range of common laser dyes like rhodamine. Generation of broadband pulses, as is necessary for single-shot absorption measurements, is hardly possible and can only be achieved by sophisticated amplification techniques. However the high repetition rate of synchronously pumped lasers allows data accumulation and averaging. Thus these systems can be used particularly for picosecond fluorescence studies in photochemically and thermally stable molecules. Passively mode-locked solidstate lasers, on the other hand, due to their high power allow effectice frequency transformations to discrete frequencies (by parametric processes as well as by stimulated scattering) as well as into a broadband continuum (by self-phase modulation and parametric four-photon processes). This opens the possibility in a single-shot mode to measure time-dependent fluorescence, absorption, Raman, and infrared spectra. Thus the versatility of this type of lasers is higher than that of synchronously pumped lasers, even if the pulse-to-pulse reproducibility is inferior. The use of optical multichannel analysers in connection with laboratory computers has opened up the possibility to take spectra over a range of ceveral hundreds of nanometers with a single laser shot. Since the event of reliable streak cameras also the whole time scale from a few picoseconds to many nanoseconds (and longer) has become accesible to single-shot measurements. Careful set-up and calibration of the measuring system now gives the opportunity to obtain with a single laser shot the time-evolution of a complete near UV/visible spectrum (absorption or fluorescence) up to some ten nanoseconds.

The experimental equipment used in our work on photoreceptor

LASER STUDY OF PHOTOSYNTHESIS

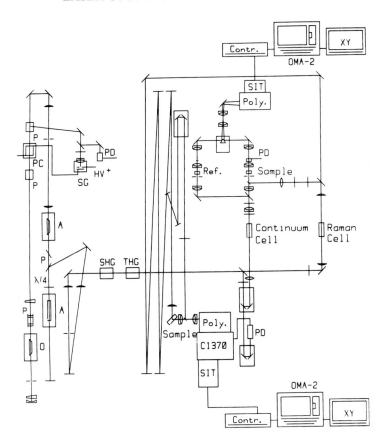

FIGURE 2 Experimental setup for measuring picosecond transient absorption and fluorescence spectra. O laser oscillator, A laser amplifiers, P polarizers, PC Pockels cell, SG spark gap, HV high voltage, λ/4 quarter-wave plate, SHG frequency doubler, THG frequency tripler, PD photodiode. Poly Polychromator, C 1370 streak camera, SIT Silicon intensified target vidicon detector, OMA optical multi-channel analyser, XY plotter

molecules is shown schematically in Figure 2. A passively mode--locked Nd-phosphate glass laser (Hoya LHG 8) delivers a train of pulses with duration of ∼6 ps each. An electrooptical switch selects a single pulse from the leading edge of the pulse train, having a typical energy of ∼100 µJ. Despite the fact that by an intracavity diaphragm the laser is forced to oscillate in the

TEM_{00} mode, spatial filtering is performed after the single-pulse selector to provide optimum beam quality. The pulse is amplified to ~1 mJ in a first amplifier stage. In a second stage the pulse is doubly amplified and coupled out by a combination of a quarter--wave plate and polarizer. After passing another spatial filter, the pulse has a final energy of up to 100 mJ. This energy is high enough to allow all necessary nonlinear frequency conversions with very good efficiency. Frequency doubling and tripling is achieved in nonlinear crystals (KD*P), and further shifting of the frequency can be performed by stimulated Raman scattering in various liquids. The different frequencies are used to excite the sample and thus to initiate the photophysical and photochemical processes under examination. The remaining IR pulse, after appropriate delay, is either used to trigger a 2 ps streak camera (Hamamatsu C1370) when fluorescence is to be measures, or to generate a picosecond continuum (useful range 350 - 800 nm) in the case of absorption measurements.

In fluorescence experiments the exciting pulse is sent along an optical path of ~20 ns to allow for the delay of the steak camera. A small part of the pulse is split off before reaching the sample to act as a time marker (prepulse) later on. The fluorescence emitted from the excited sample is collected to the slit of an ISA HR 320 polychromator, additionally corrected for astigmatism (low dispersion grating, 150 lines/mm). The spectrum is imaged to the cathode of the streak camera. The streak camera output is viewed using a Silicon intensified target vidicon in connection with an optical multichannel analyser (OMA 2, Princeton Applied Research). The digitized timeresolved spectrum then is transmitted to a small laboratory computer (Corvus Concept) or to the central computing facility (Univac 1100/81), respectively, for further processing.

LASER STUDY OF PHOTOSYNTHESIS 363

For measurements of transient absorption, the continuum pulse is split into two pulses moving along distinct, but optically similar ways. One pulse transverses the excited sample, while the other one passes an unexcited one, acting as a reference. Both pulses are focussed to the slit of an ISA UFS 200 polychromator. The spectra are registrated and processed by a similar equipment as for the fluorescence measurements. The time resolution in this case is accomplished by delaying the continuum with respect to the exciting pulse step by step, so that in this case a larger number of shots is necessary to get the time evolution of the spectra than in the case of fluorescence measurements. The change in optical density is calculated from the spectra of sample and reference following a procedure given by Greene et al.[5].

The result is a plot of fluorescence intensity or transient absorption vs. wavelength vs. time after excitation. A section parallel to the time axis gives the time evolution for a given wavelength. This can be analysed by a leastsquare routine, fitting an arbitrary function (most frequently a single exponential or a sum of exponentials, but also non-exponentials, if physically reasonable) to the experimental points. A section along the wavelength axis gives the spectrum at a given moment, which also can be fitted using appropriate functions (Gaussian, Lorentzian, and sums hereof), yielding position and halfwidth of transient spectral features. This evaluation process has to be done very carefully, always starting from a reasonable physical model of the kinetics. Since the fitting procedures not always yield unequivocal results, it is not advisable to blindly trust the computer output.

3. THE PLANT PHOTORECEPTOR MOLECULE PHYTOCHROME

Phytochrome is a plant pigment governing photomorphogenesis in higher plants. Among the processes depending on this pigment are germination, phototropism, and photoperiodism, all of practical interest in crop growing. Phytochrome acts as a photoreceptor due to the fact that it can exist in two modifications (P_r and P_{fr}) interconvertable by light. One modification, P_r, absorbs around 660 nm and is physiologically inactive. Absorption of a red photon transforms it into P_{fr}, which has an absorption band at 730 nm and is the compound triggering the photomorphogenetic actions. It can be converted back to P_r by a dark reaction or upon absorption of a dark-red photon.

Phytochrome is a chromoprotein, containing a chromophore consisting of four pyrrol rings connected by methine bridges. The protein structure is largely unknown. The P_r chromophore structure has recently been elucidated[6,7], while the structure of the P_{fr} chromophore is still a matter of intense investigations. On the primary photoprocess in phytochrome a number of hypotheses have been proposed. A definite decision has been impediated by the lack of time-resolved spectroscopic information. To fill this gap has been the reason of our work on this photoreceptive system.

Since the chromophore structure is not yet fully understood,

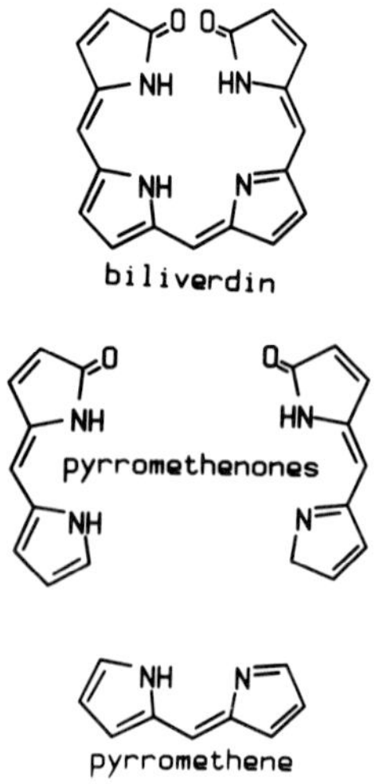

FIGURE 3 Backbone of biliverdin molecule, and partial structures

LASER STUDY OF PHOTOSYNTHESIS

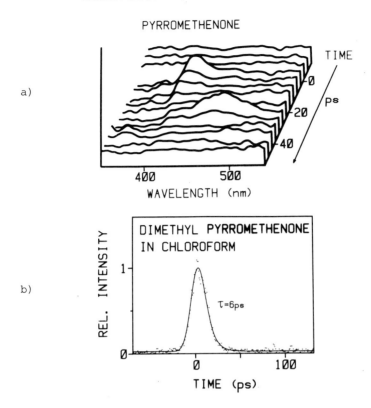

FIGURE 4 Picosecond results for dimethyl pyrromethenone: a) Absorption change vs. wavelength vs. time after excitation. b) fluorescence intensity vs. time after excitation. Fluorescence maximum 420 nm, excitation wavelength 354 nm

the only possibility to really understand time-resolved spectra is by comparison with spectra of well-defined model compounds. Thus, prior to starting with native phytochrome, we decided to conduct a broad picosecond study on pyrrol pigments, four-ringed species as well as their two-ringed partial structures. A short account of our results on this group of compounds shall be given in the following.

The chromophore in the phytochrome molecule is known to have biliverdin structure. This structure can be regarded to be composed of a pyrromethen and two pyrromethenone partial structures, as schematically shown in Figure 3. So we started our investigations studying primary photoprocesses in these two-ringed molecules. Figure 4a shows a plot of the transient absorption of dimethyl pyrromethenone in chloroform after excitation with a pulse of 354 nm wavelength. Two spectrally overlapping bands can be observed. The first one is centered at 410 nm and shows, after a buildup too fast too resolve (~ 2 ps), a decay time of ~ 6 ps. The second band is rather broad and has its center around 460 nm. It appears with a delay of ~ 30 ps and decays within 4 ps. Bleaching of the ground state is observed to last for 40 ps, which is approximately the sum of delay and decay times obseved in the transient absorptions. Figure 4b shows, for comparison, the decay of the fluorscence at 420 nm. The decay time was the same for any part of the fluorescence band, hence no spectrally resolved curve is shown. The observed time constant of 6 ps matches very well the life time of the first transient absorption. Since the fluorescence usually gives the life time of the first excited state S_1, the corresponding absorption is attributed to a $S_1 - S_n$ transition. To understand the delayed absorption, we have to take into account the fact, that pyrromethenone is able to photoisomerise around the exocyclic double bound. A thermal barrier has to be overcome to convert the molecule into a 90° configuration. From Fokker-Planck calculations the measured time of 6 ps seems to be reasonable for this process. Obviously this twisted configuration does not show up within the spectral region covered by our experiment. The broad second transient absorption in our opinion is due to $S_o - S_n$ transitions starting from highly excited vibrational levels, which are populated by internal conversion in the 90° configuration. The delay between the decay of the originally excited A_1 state and the appearance of the new absorption is interpreted as the life time of the twisted excited state. The short time of 4 ps observed for the decay of the second transient absorption is interpreted as the vibrational relaxation in the ground state, leading to the original Z or the isomerised E configuration. Thus the whole time-course of the isomerisation can be followed by picosecond spectroscopy.

The behaviour of pyrromethene, despite its structural resemblance to pyrromethenone, is quite different. It is known from steady-state experiments that pyrromethene does not isomerise. The fluorescence decay for this partial structure is below 2 ps, and no transient absorption could be observed within the time and wavelength region observed. So we have to conclude that the relaxation from the excited state proceeds very fast, namely in less than 2 ps. We think that this fast relaxation is due to the mobility of the molecule around the methine single bond. This assumption is corroborated by a comparison with a pyrromethene BF_2 complex. By chelation the molecule is now fixed in a flat conformation, for which we would expect a long life time. This expectation is verified experimentally. Ground-state bleaching as well as fluorescence last for 5 ns now (Figure 5).

As an example of an integral pigment suitable as a model chromophore, etiobiliverdin has been investigated. The transient absortion spectrum is given in Figure 6. To allow better interpretation, also spectra of N-methylated and protonated etiobiliverdin have been measured (Figures 7,8). In all three molecules the original ground-state absorption after bleaching recovers within 30 ps. The fluorescence life times are, at least for the protonated and N-methylated molecules, singificantly shorter (27, 11, and 9 ps). Transient and delayed transient absorptions can be observed for all three molecules, none of which have a time behaviour corresponding to the observed fluorescence decay. From this fact it is concluded, in agreement with prior suggestions[8], that the fluorescence observed is due only to a minority of molecules, the majority being essentially non-

368 LASERS IN ATOMIC, MOLECULAR AND NUCLEAR PHYSICS

a)

b)

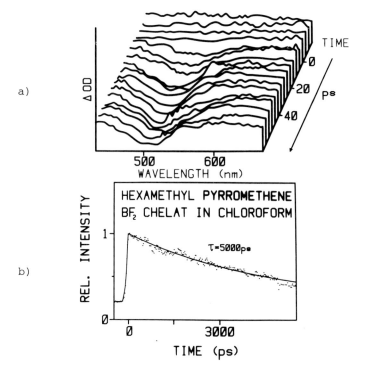

FIGURE 5 Picosecond results for pyrromethene BF_2: a) absorption change vs. wavelength vs. time after excitation, b) fluorescence intensity vs. time after excitation. Fluorescence maximum 540 nm, excitation wavelength 527 nm.

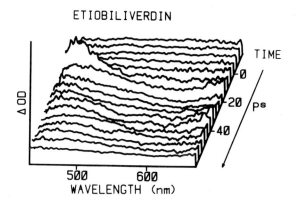

FIGURE 6 Absorption change vs. wavelength vs. time after excitation for etiobiliverdin. Excitation wavelength 621 nm.

N-METH. ETIOBILIVERDIN

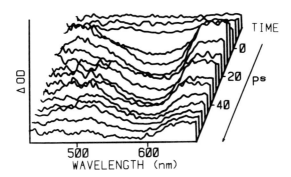

FIGURE 7 Absorption change vs. wavelength vs. time after excitation for N-methyl etiobiliverdin. Excitation wavelength 527 nm.

PROT. ETIOBILIVERDIN

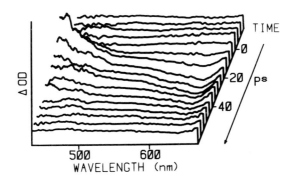

FIGURE 8 Absorption change vs. wavelength vs. time after excitation for protonated etiobiliverdin. Excitation wavelength 621 nm.

-fluorescent. This is a good example proving the application of a single spectroscopic method to be insufficient for analysing the ultrashort time behaviour of complex molecules.

From the results obtained for partial structures, it seems reasonable to assume isomerisation and single-bond rotation to be important also in integral pigments. From steady-state experiments it is known that etiobiliverdin makes no photoisomerisation around the methine single bond in the pyrromethene fragment of the molecule. Obviously the excited state is non-fluorescent due to vanishing Franck-Condon factors. Transient absorptions with life times of 7, 11, and 15 ps in the three molecules, respectively, are attributed to S_1-S_n absorptions. The S_1 state relaxes, most probably by internal conversion, to a twisted ground-state, to which the delayed transient absorptions are attributed. From force-field calculations[9] it can be assumed that a small barrier (~ 2400 cm^{-1}) has to be overcome for return to the original ground-state conformation. The observed life times of the delayed transient absorptions (~ 25 ps for all three molecules) are reasonable as concluded from Fokker-Planck calculations. A computer simulation based on PPP and force-field calculations[9,10] and assuming a single-bond rotation gives good agreement between measured and calculated spectra, so that single-bond rotation seems to be verified as the relaxation mechanism. An intramolecular proton transfer, as discussed in the literature[11], can be definitely excluded as a relevant contribution, since it is impossible in N-methyl etiobiliverdin.

Thus, using well-defined phytochrome chromophore model compounds and their partial structures, we were able to show that isomerisation and single-bond rotation are the prevailing relaxation mechanisms in pyrrol pigments.

On the basis of these results on model compounds and their partial structures, it seemed possible to understand picosecond data of native phytochrome itself. The measurements were performed using large phytochrome isolated from rye[12]. Upon excitation with picosecond pulses with a wavelength of 621 nm, a time-resolved fluorescence was observed as shown in Figure 9. The decay was found to be not single-exponential. Following a recent paper[3] we uses a trial function of three exponentials to fit the experimental

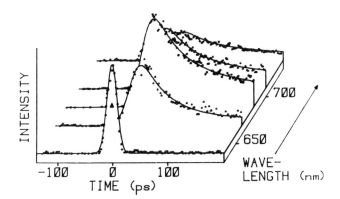

FIGURE 9 Fluorescence intensity vs. wavelength vs. time after excitation for large phytochrome. Excitation wavelength 621 nm. ▲ exciting pulse.

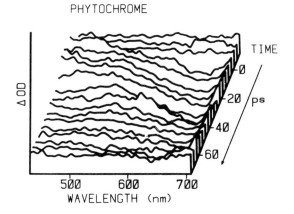

FIGURE 10 Absorption change vs. wavelength vs. time after excitation for large phytochrome. Excitation wavelength 621nm.

values, yielding time constants of ~ 40, 350, and 1250 ps, the shortest decay having the largest amplitude. An interpretation in terms of molecular kinetics based only on these results is impossible, however. Picosecond absorption measurements (Figure 10) showed an immediate bleaching of the 660 nm absorption after excitation, accompanied by a broad transient absorption centered at about 500nm. This transient absorption has a life time of ~ 35 ps.

The bleaching disappears nearly completely within ~ 90 ps. It recovers, however, with a time constant exceeding the full time scale of our experiment (1 ns). It should be pointed out, that we cannot follow the relatively slow formation of intermediates described by other authors[13] on our time scale. Anyway, finally the well-known 730 nm absorption of P_{fr} is observed also in our experiment. The disappearance and reappearance of bleaching can only be understood assuming an intermediate with an absorption very nearly similar to that of P_r (absorption maximum ~ 690 nm), forming within ~ 100 ps and decaying with a time constant $\gg 1$ ns. The absorption centered at 500 nm, from its spectral similarity with equivalent absorptions in the model compounds as well as from its decay time, equalling approximately that of the short-lived fluorescence, can be identified as P_r S_1-S_n absorption. The long-lived fluorescence components might be attributed to impurity emissions, as already argued in ref.[3]. The transient red absorption evolving within ~ 100 ps seems to represent a very early electronically relaxed intermediate. From the resemblance of its absorption to that of the original P_r and the lack of spectral features known to be characteristic for single-bond rotation from our previous experiments, it follows, that in this intermediate conformational changes can be ruled out as relaxation mechanisms. For possible mechanisms leading to this intermediate only proton transfer or Z-E isomerisation remain as reasonable explanation. Proton transfer, as shown above, plays no major role as a relaxation mechanism in pyrrol pigments, however. Furthermore the time of ~ 100 ps is much larger than expected for a proton transfer. So we conclude that the observed intermediate is an isomere, most probably formed by Z-E isomerisation at the C-15 double bond[14]. The process resembles, with a somewhat slower time course, very much the one described above for pyrromethenone, where also a S_1-S_n absorption and a delayed (intemediate) ground-state absorption had been observed. The slowing down has to be attributed to the stabilizing action of the protein. Thus, the comparison of these phytochrome results with previous results obtained on model compounds gives strong evidence that in phytochrome the S_1 state lives about 40 ps and a Z-E isomerisation at C-15 is accomplished within ~100 ps. So isomerisation seems to be the primary photochemical event, as proposed recently by Rüdiger et al.[14].

4. THE VISUAL CHROMOPHORE RETINAL

Most visual systems investigated up to now use retinal as a chromophore responsible for light reception. Usually it is bound to a protein, forming a chromoprotein called rhodopsin. That the primary photochemical event in rhodopsin goes on in a few picoseconds has been shown a decade ago[1] and repeatedly verified in the meantime[2]. Yet, despite a great number of investigations, also with different picosecond techniques, the process is not fully understood due to the complexity of the problem. It could be assumed that extensive knowledge of the properties of the chromophore itself should be indispensable as a sound basis for an interpretation of the complex results in the chromoprotein. Most surprisingly, however, only very few picosecond investigations have been conducted on retinal[15,16]. Also from those results no clear picture of the primary photoreaction in retinal has emerged. Thus new spectroscopic data with a time-resolution of picoseconds are urgently needed to help elucidating this interesting problem. Especially needed are methods providing new structural information, not obtained by the usual picosecond fluorescence and absorption measurements.

As a new method we have introduced linear dichroism spectroscopy to the picosecond regime. In this method, molecules are incorporated into a polyethylene film and oriented by stretching the film. From previous work[17] the orientation of retinal in the film

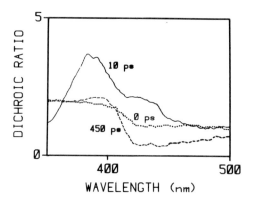

FIGURE 11   Dichroic ratio of all-trans retinal after excitation with 354 nm

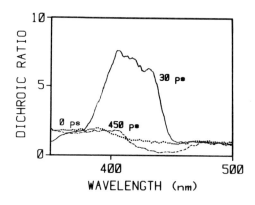

FIGURE 12   Dichroic ratio of 11-cis retinal after excitation with 354 nm

is known. So by measuring transient absorption spectra polarized parallel and perpendicular to the streching direction, the dichroic ratio $D = OD_\parallel / OD_\perp$ gives information on the direction of the transition-dipole moment and hence on the molecular geometry in the excited and intermediate states. Examples are given in Figures 11 and 12. It is found that a short-lived ($\sim$35 ps) S_1-S_n absorption has a transition-dipole moment oriented well parallel to the stretching direction, indicating that retinal is more planar in

the excited than in the ground state. In the film as well as in
n-hexane solution, intersystem crossing is the most dominant relaxation mechanism. The triplet absorption appears within ~ 30 ps.
The triplet-triplet absorption observed at 400 nm has a low dichroic ratio (<1), indicating its transition dipole being directed more or less perpendicular to the stretching direction. The
spectral features and relaxation times we observed are in good
agreement with values given in ref.[15], but the information on the
orientation of the dipole moments could not be obtained previously
by other methods. From these results we have to conclude that in
discussing the primary event in vision not only geometrical
changes at the single and double bonds in the retinal polyene
chain should be considered, but also the orientation of the chain
with respect to the ring and hence the planarity of the chromophore. This fact has been largely neglected in the theoretical
papers on intramolecular motion in retinal, for recent reviews see
e.g.[2]. An extension of the picosecond linear dichroism technique
to protonated and unprotonated Schiff bases of retinal should give
further valuable information.

5 CONCLUSION

Photoreceptor molecules are of high biological interest. Their
primary photoreactions proceed on a picosecond time scale. Single-
-pulse picosecond measurements are especially suited to investigate these processes. In this paper it has been shown, that a reliable understanding of primary photoreactions requires the combined use of different techniques of picosecond spectroscopy, applied at different molecular levels of complexity. The example of
the phytochrome molecule gives good evidence, how knowledge gained
on the partial-structure and model-compound level is indispensible
to interpret spectra obtained from the native molecule. By the example of the visual chromophore, retinal, it has been shown that

the extension of new spectroscopic techniques into the picosecond regime makes obtainable information on the chromophore previously unaccessible, which sheds new light on proposed theories on the primary event in vision.

Acknowledgment

The authors thank Prof. H. Falk, Linz, Austria, for providing the pyrrol pigment samples and for valuable discussions; Prof. E. Müller and Dr. G. Hermann, Jena, German Demovratic Republic, for preparation of phytochrome; Prof. Y. Mazur, Rehovot, Israel, for providing the retinal samples; Dr. L. Margulies, Rehovot, Israel, for introducing us to the technique of linear dichroism spectroscopy, preparing the stretched films, and for valuable discussions. We are grateful to acknowledge financial support of this work by the Austrian Fonds zur Förderung der wissenschaftlichen Forschung (grants nos. 4031, 4894) and the Jubiläumsfonds der Österreichischen Nationalbank (grants nos. 1814, 2111).

REFERNCES

1. G. E. BUSCH, M. L.APPLEBURY, A. A. LAMOLA, R. M. RENTZEPIS, Proc. Nat. Acad. Sci. USA, 69, 2802 (1972)
2. Biological Events Probed by Ultrafast Laser Spectroscopy, edited by R. R. Alfano(Academic Press, New York, 1982)
3. J. WENDLER, A. R. HOLZWARTH, S. E. BRASLAVSKY, K. SCHAFFNER, in print, Biochim. Biophys. Acta (1984)
4. A. R. HOLZWARTH, J. WENDLER, W. WEHRMAYER, Biochim. Biophys. Acta, 724, 388 (1984)
5. B. I. GREENE, R. M. HOCHSTRASSER, R. B. WEISMAN, J. Chem. Phys., 70, 1274(1979)
6. W. RÜDIGER, T. BRANDLMEIER, I. BLOSS, A. GOSSAUER, J.-P. WELLER, Z. Naturf., 35 C, 763 (1980)
7. J. C. LAGARIAS, H. RAPOPORT, J. Am. Chem. Soc., 102,4821(1989)
8. A. R. HOLZWARTH, H. LEHNER, S. B BRASLAVSKY, K. SCHAFFNER, Liebigs Ann. Chem., 1978, 2002 (1978)
9. H. FALK, N. MÜLLER, Monatsh. Chemie, 112, 791 (1981)
10. H. FALK, G. HÖLLBACHER, Monatsh. Chem., 109, 1429 (1978)

11. H. FALK, K. GRUBMAYER, F. NEUFINGERL, Monatsh. Chem., 110, 1127(1979)
12. G. HERMANN, E. MÜLLER, Biochem. Physiol. Pflanzen, 175, 755(1980)
13. W. RÜDIGER, Struct. Bonding, 40, 101 (1980)
14. W. RÜDIGER, Phil. Trans. R. Soc. Lond., B303, 377 (1983)
15. R. M. HOCHSTRASSER, D. L. NARVA, A. C. NELSON, Chem. Phys. Letters, 43, 15 (1976)
16. A. G. DOUKAS, M. R. JUNNARKAR, D. CHANDRA, R. R. ALFANI, R. H. CALLENDER, Chem. Phys. Letters, 100, 420 (1983)
17. L. MARGULIES, N. FRIEDMAN, M. SHEVES, Y. MAZUR, M. E. LIPPITSCH, M. RIEGLER, F. R. AUSSENEGG, to be published

4.5 Electrolyte Control of Photosynthetic Electron Transport in Cyanobacteria

GEORGE C. PAPAGEORGIOU

Nuclear Research Center Demokritos, Department of Biology, 153 10 Athens, Greece

INTRODUCTION

The foundations of the modern theory of photosynthesis, namely the concepts of the photosynthetic unit[1], the minimum requirement of 8 photons per evolved O_2 [2], and the existence of two distinct photoreactions[3,4] are based on experiments whose design, execution and interpretation made no reference to a biological membrane. In the late sixties and early seventies, however, two new concepts emerged which profoundly modulated scientific thinking in biological sciences in general, and photosynthesis in particular: the lipid bilayer model of biological membranes[5-7] and the chemiosmotic coupling of electron transport to the generation of energy-rich phosphate bonds[8]. In relatively short time, light harvesting pigments, electron carriers and enzyme activities have found a place within a asymmetric thylakoid membrane. The membrane defines an outer and an inner aqueous phase, a membrane phase, and it is indispensable for the transduction of photonic energy to reduction and electrochemical potential differences.

Like all biological membranes, thylakoid membranes expose the polar moieties of the protein and lipid constituents to the aqueous

Abbreviations: Chl - chlorophyll; $FeCN^{3-}$ - ferricyanide; $FeCN^{4-}$ - ferrocyanide; Hepes - N-2-hydroxyethyl piperazine-N-ethane sulfonic acid; PD_{ox} - ferricyanide-oxidized p-phenylene diamine.

phases, and bury the hydrophobic moieties in the membrane phase. Since ionizable acidic groups (phospho- and sulfolipid headgroups; pKa = 2.0; protein carboxyls, pKa = 3.8-4.8) predominate over basic groups (amino-, guanidino-, imidazolylo-; pKa = 9.2-12.5), biological membranes carry negative fixed charges above approx. pH 4[9]. The surface charge density of higher plant thylakoids is in the range 0.0046-0.036 $C \cdot m^{-2}$ [10], of Rhodopseudomonas spheroides chromatophores 0.022-0.038 $C \cdot m^{-2}$ [11] and of cyanobacterial thylakoid fragments 0.09-0.11 $C \cdot m^{-2}$ (K. Kalosaka and G. C. Papageorgiou, unpublished experiments).

Higher plant and cyanobacterial thylakoids have the same gross composition in ionizable and nonionizable constituents[12]. Proteins make 68-70% of membrane mass, polar lipids 6-8%, and the remainder is nonpolar constituents, galactolipids, carotenoids and chlorophylls. Carboxyls are, then, the main contributors to the thylakoid surface charge, a conclusion supported also by the proximity of the isoionic point of thylakoid particles (pI = 3.8-4.3) to the pKa values of protein carboxyls.

Masamoto et al.[13] have calculated an average surface area of 500-625 (mean 560) $Å^2$ per carboxyl group, which amounts to 0.029 $C \cdot m^{-2}$ of surface charge. An approximate contribution of the polar lipids can be calculated by making the reasonable assumption that if these lipids make 7% of the membrane mass, then they should occupy 7%/5 = 1.4% of the membrane surface. It is also known that a polar lipid headgroup occupies an area of approx. 70 $Å^2$ [9]. From these values, we may calculate a polar lipid contributions of 0.003 $C \cdot m^{-2}$ to the surface charge. The total calculated surface charge density 0.032 $C \cdot m^{-2}$, the sum of the two contributions, is in the range of experimentally determined values. It is equivalent to a hexagonal lattice of point charges, fixed at a distance of 12.6 Å from each other.

The surface electricity generates an electrostatic potential

LASER STUDY OF PHOTOSYNTHESIS   381

whose magnitude at each point depends on the electrolyte structure of the aqueous phase bathing the membrane. Drastic alterations of of membrane electric properties occur upon transition from a state of rest to a state of activity, or vice versa. Upon illumination of a darkened chloroplast sample, the following processes, in temporal sequence, are known to occur: pigment excitation, excitation energy migration and trapping; charge separation and shift of electrons from a domain near the inner thylakoid surface to a domain near the outer surface; diffusion of H^+ from the outer aqueous phase to the inner aqueous phase; cotransport and counter-transport of other ions in order to maintain a state of quasi-neutrality at the bulk aqueous phases.

These dynamic prosesses drastically change the inner face potential (positive shifts) and modify the transmembrane electrochemical potential difference. Assuming equilibrium, the latter is expressed as follows:

$$\Delta \mu = RT \ln \frac{[H^+]_i}{[H^+]_o} + F \Delta \Psi$$

where $\Delta \Psi$ is the electric potential difference between the aqueous phases.

MEMBRANE CATION INTERACTION

There are two basic ways a cation may interact with a negative membrane surface: it may accumulate in the next to the membrane aqueous phase without making any permanent contact with membrane sites, or in addition to this it can also bind to the membrane sites. In the first instance, the cations screen the surface charge from the bulk phase and thus lower the absolute value of the electrostatic potential, but the surface charge density does not change. In the second instance, both the surface charge density and the potential are lowered.

The electrostatic properties of a system consisting of an

382 LASERS IN ATOMIC, MOLECULAR AND NUCLEAR PHYSICS

electrically charged flat dielectric surface (the membrane) in contact with a solution of a symmetric electrolyte (z-z) can be described in terms of the equations shown in Figure 1. The Gouy-Chapman equation relates the surface charge density (σ) and the surface potential (Ψ_o) to the bulk phase concentration of the electrolyte (C_∞). The electrostatic potential along the normal to the membrane surface (Ψ_x) is described by Poisson-Boltzmann equation. The quantity k^{-1} is known as the Debye distance, and represents the distance at which the potential drops to e^{-1} of its initial value. As shown in the Figure 1, k^{-1} becomes shorter with increasing electrolyte concentration and valence. For example, at 0.1 M and 0.2 M KCl, the Debye distance is 9.6 and 6.8 Å respec-

Gouy-Chapman Equation

$$\sigma = 2\left[\frac{RT\varepsilon\varepsilon_o}{2\pi}\right]^{\frac{1}{2}} \sinh\left[\frac{ZF\Psi_o}{2RT}\right] C_\infty^{\frac{1}{2}}$$

$$\sigma = 0.1174 \sinh\left|\frac{Z\Psi_o}{51.7}\right| C_\infty$$

Poisson-Boltzmann Equation

$$\Psi_{(x)} = \frac{2RT}{ZF} \ln\left[\frac{1-\alpha e^{-\kappa x}}{1-\alpha e^{-\kappa x}}\right]$$

$$\alpha = \left(e^{ZF\Psi_o} - 1\right)/\left(e^{ZF\Psi_o} + 1\right)$$

$$\kappa^2 = \frac{2Z^2F^2}{RT\varepsilon\varepsilon_o} C_\infty \qquad \frac{1}{\kappa} = \frac{3}{Z\sqrt{C_\infty}} \text{ Å}$$

Boltzmann Equation

$$C_{(x)} = C_\infty e^{-\frac{Z_i F\Psi_{(x)}}{RT}}$$

FIGURE 1 Equations describing the electrostatic properties of a membrane surface in contact with a solution of a symmetric (z-z) electrolyte. Symbols are defined in the text, or they have their usual meaning. Numerical values are calculated for T = 298 K, σ in C·m^{-2}, and Ψ in mV.

tively, while for 0.1 M and 0.2 M $MgSO_4$ it is 4.8 and 3.4 Å. These distances are shorter than the calculated separation of fixed negative charges on the thylakoid surface, which therefore cannot be considered as homogeneous. Finally, the distribution of cations along the normal to the surface is described by the Boltzmann equation. It predicts that cations should exceed anions near the negative surface. Taking T = 298 K and ψ_o = -60 mV, for example, we may calculate that monovalent cations exceed monovalet anions near the surface by a factor of 100.

The derivation of these equations is based on several unrealistic assumptions (see references 9 and 14 for detailed discussion). It is indeed surprising that they fit experimental data so often. These assumptions include: (i) Membranes are considered to be flat. (ii) Ions are not allowed to penetrate to $x < 0$. (iii) Ions are considered to be point charges. (iv) The surface distribution of charge is considered to be homogeneous. (v) Specific binding of cations is ignored. (vi) The dielectric constant of the solution is considered to be constant up to $x = 0$. (vii) The generation of image charges in the membrane dielectric is ignored. (viii) Only one electrolyte is assumed to interact with the membrane.

According to the equations of Figure 1, symmetric electrolytes exert the same effects on the electrostatic properties of the system membrane-solution when they belong to the same valence group, that is, the particular chemical character of the ions is not expressed. The latter, however, manifests itself when there is specific binding of ions to membrane sites. Eisenberg et al.[14] treated the case of a monovalent cation associating with a membrane site with the intrinsic association constant K (in M^{-1}). The lowering of the surface charge density is described in terms of a Langmuir isotherm,

384 LASERS IN ATOMIC, MOLECULAR AND NUCLEAR PHYSICS

$$\bar{\sigma} = \sigma_{max}/(1 + KC_o)$$

where $C_o = C_\infty \exp(-F \psi_o/RT)$ is the cation concentration at the solution-membrane interface.

INTERACTIONS OF ELECTROLYTES WITH THYLAKOID MEMBRANES

Table I displays a partial list of thylakoid membrane properties that are influenced by the electrolyte structure of the suspension medium. In a very general sense, two types of forces are involved: forces originating from electrostatic interactions and forces originating from hydrophobic interactions[16]. Recently, Barber[10] summarized the electrostatic effects on thylakoid membranes, distinguishing between those due to the negative surface potential and

TABLE I Properties of higher plant thylakoid membranes which are affected by the cations present in the suspension medium.

Property	Effect
Biophysical	
Chl a fluorescence (prompt)	Stimulation
" " (delayed)	"
Δ A515	Suppression
Excitation spillover (PSII $\longrightarrow$ PSI)	Increase
Surface area pH	Decrease
Biochemical	
Coupling of electron transport to phosphorylation	Enhancement
Interactions with anionic/cationic cofactors	"
Interactions with nonionic cofactors	"
Structural and Morphological	
Lateral movements of proteins	Influenced
Thylakoid stacking	Enhancement
Nonosmotic shrinking	"

those due to the space charge density.

The bulk phase concentration of electrolyte is an experimentally controlled magnitude. By increasing it, it is possible to lower the absolute value of the thylakoid surface potential (positive shift) and thus bring more anions close to the solution-membrane interface. This explains why electrolytes stimulate electron exchanges between thylakoid membranes and anionic acceptors (e.g. $FeCN^{3-}$ [17-19]) and anionic donors (e.g. $FeCN^{4-}$ [20]; plastocyanin and cytochrome c-550 [21-24]). The outer surface potential influence also the transmembrane potential, since in the absence of ionophores membranes are impermeable to ions. Finally, it affects also the redox potential of membrane bound electron carriers ($E_m = E^o + \psi_o$ [25]).

Membrane embedded proteins and proteolipids are subject to electrostatic repulsive forces, which depend on the surface charge they carry and on the space charge density at the interface region, and to attractive forces which originate from hydrophobic interactions. Similar forces exist between membrane surfaces. The space charge density at the interface ($F \sum_i z_i C_i$) is a function of the electrolyte concentration (C_∞) at the aqueous bulk phase. Interactions between hydrophobic subunits are enhanced in the presence of "salting-out" anions (e.g. HPO_4^{2-}, SO_4^{2-}) and diminished in the presence of "salting-in" anions (e.g. SCN^-, NO_3^-, Cl^-). The interplay of atractive and repulsive forces influences the structure and the structure-linked properties of thylakoid membranes. Through such forces, thylakoid stacking, nonosmotic thylakoid shrinking, lateral movements of protein subunits and excitation spillover (cf. Table I) are controlled by the electrolytes of suspension medium.

INTERACTIONS OF ELECTROLYTES WITH CYANOBACTERIAL THYLAKOIDS

Nearly all the information we have on the electrostatic properties of thylakoid membranes derives from experiments with thylakoid fragments from higher plants. Results obtained with such preparations reflect, at best, an average of the outer and the inner thylakoid surface properties. In addition, the chaotropic and antichaotropic effect of anions is routinely ignored, although they are often present in significant concentrations in the suspension media.

The cyanobacteria is a class of oxygenic photosynthetic organisms about whose thylakoid electric properties we know very little. Cation effects on electron transport, phosphorylation and spectroscopic properties have been studied by several authors, and more recently by Binder et al.^{17}, Piccioni and Mauzerall^{26}, Brand^{27}, Yu and Brand^{28}, Brand et el.^{29}, and Wavare and Mohanty^{30,31}, but they have never been correlated with the suppression of the negative surface charge and potential.

The cyanobacteria, evolutionary predecessors of algae and higher plants, share with them common mechanisms of electron and proton transport, oxygen evolution and photophosphorylation, yet they differ from them in several key aspects. The cyanobacteria are prokaryotes. Therefore, only one permeability barrier intervenes between the suspension medium and the thylakoid surface, the cell envelope consisting of the Gram negative cell wall and the cell membrane. In contrast, two such barriers, the cell envelope and the chloroplast envelope are present in the eukaryotic higher plants and algae.

In cyanobacteria, the light-harvesting pigment-protein complex of higher plants, an integral component of thylakoid membranes, is missing. Its role, as a source of electronic excitation for the photosystem II centers, is performed by the phycobilisomes, proteinaceous organelles that are external to the membrane (reviewed in 32). Phycobilisomes are assembled from hexameric phycobiliprotein subunits. Allophycocyanin and phycocyanin are present

LASER STUDY OF PHOTOSYNTHESIS   387

in all species, phycoerythrin complements them in some species. These phycobiliproteins are held together noncovalently, mostly by hydrophobic interactions. Noncovalent is also the attachment of phycobilisomes to the outer thylakoid surface, where they form two-dimensional regular patterns. It is on account of these binding forces that phycobilisome structure and attacment to the membrane are very sensitive to the chaotropic effect of low ionic strength ambience.

The interactions of electrolytes with cyanobacterial thylakoids are difficult to research because the bacterial envelope is imperme to ions. In our laboratory, we initiated a series of experiments with which we seek first to permeabilize the cell envelope to ions, with minimal damage to the structure and activities of thylakoids, and then to use these cells in order to study the effects of electrolytes. Relevant results appeared in[12,19,33,34].

PERMEABILIZATION OF ANACYSTIS NIDULANS TO IONS

Cyanobacterial thylakoids can be made accessible to ions either by cell disruption, or by cell envelope permeabilization. The first has been achieved by the following methods: blending with abrasive materials[35-37]; ultrasonic cavitation[38,39]; sudden decompression of pressurized liquid[40] of frozen[33] cell suspensions; and hypoosmotic lysis of spheroplasts[17,40-42]. Outside the cell, however, cyanobacterial thylakoids are very fragile[17,33,40,42]. Cell permeabilization is achieved by the enzymatic hydrolysis of the cell wall peptidoglycan with lysozyme. Some cyanobacteria, however, and among them <u>Anacystis nidulans</u>, the unicellular species we use in our experiments, exhibit unusual resistance to lysozyme. This is attributed to the difficulty with which the enzyme molecules penetrate through the outer membrane of the cell wall in order ot reach their substrate, the peptidoglycan layer. Thus, excessive amounts of lysozyme and extraordinarily long incubations are required, and often the permeabilized cells are photosynthetically inactive.

To overcome such problems, we developped a procedure for rapid (approx. 15 min) enzymatic permeabilization of Anacystis[19]. The progress of permeabilization is monitored in terms of $FeCN^{3-}$-dependent O_2 evolution (Figure 2A). Initially, the evolution rate is low because the anionic acceptor cannot permeate through the cell envelope in order to reach the thylakoids. The rate increases with progressing permeabilization, but soon reaches maximum and then declines.

Figure 2A illustrates that $FeCN^{3-}$ supports photoinduced O_2 evolution only when electrolytes are present (here $CaCl_2$) in order to auppress coulombic repulsions between thylakoids and acceptor

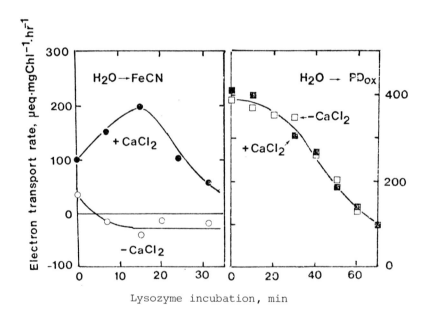

FIGURE 2   The rate of photosynthetic O_2 evolution, or uptake, by Anacystis nidulans in the presence either of $FeCN^{3-}$ (1 mM: Fig. 2A), or of p-phenylene diamine (0.5 mM) and $FeCN^{3-}$ (1.5 mM: Fig. 2B), as a function of the duration of the lysozyme reaction. 20 mM $CaCl_2$ was present where indicated. Assay buffer, sorbitol 0.5 M, Hepes. NaOH 5 mM, pH 7.5.

LASER STUDY OF PHOTOSYNTHESIS 389

molecules. In a low electrolyte medium, $FeCN^{3-}$ does not support O_2 evolution, although the cells are permeabilized. In contrast, when the lysozyme reaction is monitored in terms of the lipophilic oxidant PD_{ox} (Figure 2B), electron transport starts from a high rate value, but after a stationary phase declines. Presence or absence of $CaCl_2$ is immaterial here, since electrons are taken from the thylakoid membrane by the nonionic acceptor p-phenylene diamine. The rate decline shown in Figure 2A and 2B is attributed to the exportation of low molecular weight solutes. The site of inactivation is close to the water-splitting complex of the photosynthetic electron transport chain[19].

In our experiments, we employed permeable cells corresponding to peak activity (Figure 1A). The quality of these cells is characterized in terms of several structural and functional indices (Table II). The dissociation of phycobilisomes from the thylakoid surface and their subcequent dissociation proves that the low ionic strength of the suspension medium has been communicated to the cytoplasm, hence the cells have been permeabilized. The photosynthetic control, defined as the ratio of photosynthetic electron

TABLE II Properties of rapidly permeabilized Anacystis cells

Property	Effects of rapid permeabilization
Cell wall as exoskeleton	Virtually intact
Permeability to small solutes	Permeable
Permeability to macromolecules	Impermeable
Phycobilisome system	Disintergrated
Photosynthetic electron transport	Intact
Photosynthetic control ratio	2-2.4

Permeatilization by means of 15 min incubation at 37°C in a mixture containing: Anacystis 0.15 mg Chl a; EDTA 1 mM; lysozyme 10 mg/ml; and Hepes. NaOH 0.05 M, pH 7.5.

390 LASERS IN ATOMIC, MOLECULAR AND NUCLEAR PHYSICS

transport in the presence and in the absence of photophosphorylation cofactors (ADP, phosphate, Mg^{2+}) proves that thylakoids of permeabilized cells (permeaplasts) are impossible to ions. Hence, in the absence of ionophores, medium electrolytes probe only the outer thylakoid surface, This property, and the survival of photosynthetic electron transport prove the presence of virtually intact thylakoids in permeaplasts. Detachment of phycobilisomes from the thylakoid surface was deduced on the basis of the failure of phycobiliproteins to sensitize Chl a fluorescence. Finally, dissociation of hexameric C-phycocyanin and allophycocyanin to smaller oligomers was deduced on the basis of the 2nd derivative absorption specta, according to the method in[50].

INTERACTIONS OF ELECTROLYTES WITH THE OUTER THYLAKOID SURFACE

Cyanobacteria thylakoids carry more negative electricity than higher plant thylakoids[33]. Hence, their surface electric potential will be more negative, and their surface concentrations of anions will be smaller. Table III lists the suface concentrations of $FeCN^{3-}$ and the stoichiometric ratios $FeCN^{3-}$: Chl for suspensions of Anacytis and spinach thylakoids of equal Chl content (10 $\mu g/ml$). At 0.01 M of a 1-1 electrolyte (e.g. KCl) there is virtually no $FeCN^{3-}$ in contact with the cyanobacterium thylakoids, but there are approx. 2 acceptor radicals per electron transport chain in the spinach thylakoids. This situation adequately explains why electrolytes are so critical for the $FeCN^{3-}$ Hill reaction in cyanobacteria[17,19,51] but not in higher plant chloroplasts.

Anacystis permeaplasts, prepared by the rapid enzymatic permeabilization procedure, is a suitable system to investigate the effects of cations on the interactions with ionic electron transport cofactors. Employing such cells, we made the following experimental observations:

(1) Cations stimulate the rate of $FeCN^{3-}$-supported photosyn-

TABLE III Surface electric potential (Ψ_o), surface concentration (C_o) of $FeCN^{3-}$ and stoichimetric ratio $FeCN^{3-}:P700$ for suspensions of thylakoid fragments from Anacystis nidulans and Spinacea oleracea.

Bulk phase electrolyte (M)	Anacystis			Spinach		
		$FeCN^{3-}$			$FeCN^{3-}$	
	Ψ_o (mV)	C_o (nM)	P700	Ψ_o (mV)	C_o (nM)	P700
(1-1) 0.01	-141	0.09	$1.5 \cdot 10^{-3}$	-86	49	1.8
" 0.10	-84	62	1.1	-38	12500	450
" 0.20	-68	390	7	-28	39500	1400
(2-2) 0.02	-62	780	13.9	-35	17600	630

Displayed values were calculated on the basis of the following magnitudes: Surface charge density, Anacytis thylakoids -0.09 C/m^2; spinach thylakoids -0.03 C/m^2 (K. Kalosaka and G. C. Papageorgiou, unpublished experiments). Chl content of samples 10 µg/ml, equivalent to 56 nM P700 in Anacystis and 28 nM P700 in spinach. Bulk phase concentration of $FeCN^{3-}$ 1 mM. It is assumed that Chl: P700 equals 200:1 in cyanobacteria and 400:1 in spinach.

thetic O_2 evolution. Maximal effect is elicited by 100-350 mM monovalent, 20-30 mM divalent and 2 mM trivalent cations. This valence hierarchy is in qualitative agreement with the Gouy-Chapman and Boltzmann equations (Figure 1) and in general agreement with the results of other workers[17,19,51,52].

(2) Wide difference exist among cations of the same valence, with regard to the maximal stimulation of the $FeCN^{3-}$ Hill reaction and the concentration required for that.

(3) Cations stimulate photoinduced electron transport to the positively charged acceptor methylviologen^{2+}, but to a lesser extent than for the $FeCN^{3-}$ Hill reaction. Here, concentrations for maximal effect are neary the same for cations of the same valence.

(4) At monovalent cation concentrations exceeding 300-400 mM, and divalent cation concentrations exceeding 30-40 mM, the $FeCN^{3-}$

Hill reaction becomes progressively inhibited.

It is obvious that the simple electrostatic screening model is inadequate to account for observations 2, 3, and 4. Additional contributing factors may include: specific cation binding; chaotropic and antichaotropic effects of anions; second or higher order effects, such as conformational changes, changes in the surface topography of membranes, membrane adhesion, etc.

INTERACTIONS OF CATIONS WITH THE INNER THYLAKOID SURFACE

A suitable marker for the inner thylakoid surface is P700, the reaction center chromophore of photosystem I. Photooxidation of P700 is reported by absorption loss at 700 nm, reduction by the recovery of the absorption loss. In cyanobacteria, electron donor to P700 is Cyt c-550 and to a leeser degree the Cu-protein plastocyanin[52]. These proteins are negatively charged and easily dissociable from the membrane. In subthylakoid fragments, cations stimulate electron donation from plastocyanin to P700 and the process appears to be under electrostatic control. Analogous experiments with intact thylakoids have not been reported.

Employing aged Anacystis permeaplasts, we examined whether cations added to the suspension medium could accelerate the dark recovery of the P700 absorption loss, that is speed up electron donation to oxidized P700 by the endogenous donor. Significant acceleration was observed only when KCl (2 mM) and the K^+ ionophore valinomycin(3 μM) were present simultaneously. The acceleration depended on the concentration of KCl in a way suggesting that its cause was a positive shift of the inner surface potential as a result of the imported K^+ ions. Valinomycin, or KCl, added alone had no effect on the dark reduction of P700 by the endogenous donors. We may conclude, then, that without the ionophore, K^+ cannot penetrate through the thylakoid membrane.

SUMMARY

Ion-permeable cells (permeaplasts) of the cyanobacterium Anacystis nidulans were prepared enzymatically and were characterized with respect to several structural and functional indices. The permeaplasts contain intact, ion-impermeable thylakoids and are photosynthetically active. Employing these cells, we investigated the effects of cations, acting either on the outer, or on the inner thylakoid membrane surface, on photoinduced electron exchanges with anionic donors (Cyt $\underline{c}$-550, plastocyanin; innersurface), or anionic acceprors (FeCN^{3-}; outer surface). Cations accelerate such exchanges by accumulating near the solution-membrane interfaces and screening the negative surface charge of membranes. Electrostatic screening, however, is not the only contributing factor, and other electrolyte-linked influences must be invoked in order to interpret the experimental observations.

ACKNOWLEDGEMENTS

Experiments referred to in this report were performed with the cooperation of K. Kalosaka, T. Lagoyanni, and G. Sotiropoulou.

REFERENCES

1. R. EMERSON, W. ARNOLD, J. Gen. Physiol., 16, 191 (1932)
2. R. EMERSON, Ann. Rev. Plant Physiol., 9, 1 (1958)
3. R. EMERSON, E. RABINOWITCH, Plant Physiol., 35, 477 (1960)
4. L. N. M. DUYSENS, J. AMESZ, B. M. KAMP, Nature, 190, 510 (1961)
5. W. STOECKENIUS, D. M. ENGELMAN, J. Cell Biol., 42, 613 (1969)
6. S. J. SINGER, G. L. NICOLSON, Science, 175, 720 (1972)
7. D. BRANTON, D. W. DEAMER, Membrane Structure (Springer-Verlag, Berlin and New York, 1972)
8. P. MITCHELL, Chemiosmotic Coupling and Energy Transduction (Glynn Res., Bodmin, Cornwall, England, 1968)
9. S. MCLAUGHLIN, in Current Topics in Membranes and Transport, edited by F. Bronner and A. Kleinzeller (Academic Press, New York, 1977), Vol. 9, pp. 71-144.
10. J. BARBER, Ann. Rev. Plant Physiol, 33, 261 (1982)

11. K. MATSUURA, K. MASAMOTO, S. ITOH, M. NISHIMURA, Biochim. Biophys. Acta, 547, 91 (1985)
12. G. C. PAPAGEORGIOU, K. KALOSAKA, T. LAGOYANNI, G. SOTIROPOU-LOU, in New Methods in Membrane Research and Biological Energy Transduction, edited by L. Packer (Plenum Press, London, in press).
13. K. MASAMOTO, S. ITOH, M. NISHIMURA, Biochim. Biophys. Acta, 591 (1980)
14. M. EISENBERG, T. GRESALFI, T. RICCIO, S. MCLAUGHLIN, Biochemistry, 18, 5213 (1979)
15. J. BARBER, Biochim. Biophys. Acta, 594, 253 (1980)
16. R. P. RAND, Ann. Rev. Biophys. Bioengineer., 10, 277 (1981)
17. A. BINDER, E. TEL-OR, M. AVRON, Eur. J. Biochem.,67, 187 (1976)
18. S. ITOH, Biochim. Biophys. Acta, 548, 596 (1979)
19. G. C. PAPAGEORGIOU, T. LAGOYANNI, Biochim. Biophys. Acta (1985) in press
20. S. ITOH, Biochim. Biophys. Acta, 548, 579 (1979)
21. W. LOCKAU, Eur. J. Biochem., 94, 365 (1979)
22. S. LIEN, A. SAN PIETRO, Arch. Biochem. Biophys., 194, 128 (1979)
23. N. TAMURA, S. ITOH, Y. YAMAMOTO, M. NISHIMURA, Plant and Cell Physiol., 22, 603 (1981)
24. W. HAEHNEL, A. PROPPER, H. KRA, Biochim. Biophys. Acta, 593, 384 (1980)
25. D. WALZ, Biochim. Biophys. Acta., 505, 279 (1979)
26. R. G. PICCIONI, D. C. MAUZERALL, Biochim. Biophys. Acta, 504, 384 (1978)
27. J. J. BRAND, FEBS Lett., 103, 114 (1979)
28. C. M. C. YU, J. J. BRAND, Biochim. Biophys. Acta, 591, 483 (1980)
29. J. J. BRAND, P. K. MOHANTY, D. C. FOCK, FEBS Lett., 155, 120 (1983)
30. R. A. WAVARE, P. K. MOHANTY, Photobiochem. Photobiophys., 3, 327 (1982)
31. R. A. WAVARE, P. K. MOHANTY, Indian J. Biochem. Biophys, 20, 301 (1983)
32. A. N. GLAZER, Biochim. Biophys. Acta, 768, 29 (1984)
33. G. SOTIROPOULOU, T. LAGOYANNI, G. C. PAPAGEORGIOU, in Advances in Photosynthesis Research, edited by C. Sybesma (M. Nijhoff/Dr. W. Junk Publishers, The Hague, 1984), Vol. 2, pp. 663-666.
34. K. KALOSAKA, G. C. PAPGEORGIOU, as Ref. 33, pp. 707-710.
35. J. B. THOMAS, W. DE ROVER, Biochim. Biophys. Acta, 16, 391 (1955)

36. W. W. FREDRICKS, A. T. JAGENDORF, Arch. Biochem. Biophys., 104, 39 (1964)
37. W. A. SUSOR, D. W. KROGMANN, Biochim. Biophys. Acta, 88, 11 (1964)
38. Y. FUJITA, J. MYERS, Arch. Biochem. Biophys., 113, 730 (1966)
39. E. WAX, W. LOCKAU, Z. Naturforsch.,35c, 98 (1980)
40. D. I. ARNON, B. D. MCSWAIN, H. Y. TSUJIMOTO, K. WADA, Biochim. Biophys. Acta., 357, 231 (1974)
41. H. L. CRESPI, S. E. MANDEVILLE, J. J. KATZ, Biochem. Biophys. Res. Commun., 9, 569 (1962)
42. J. BIGGINS, Plant Physiol., 42, 1447 (1967)
43. H. SPILLER, P. BOGER, Meth. Enzymol., 66, 446 (1980)
44. B. GERHARDT, A. TREBST, Z. Naturforsh., 20B, 879 (1965)
45. B. WARD, J. MYERS, Plant Physiol., 50, 547 (1972)
46. G. C. PAPAGEORGIOU, Biochim. Biophys. Acta, 461, 379 (1977)
47. G. C. PAPAGEORGIOU, H. TZANI, J. Appl. Biochem., 2, 230 (1980)
48. C. J. LUDLOW, R. B. PARK, Plant Physiol., 50, 547 (1969)
49. S. J. ROBINSON, C. S. DEROO, C. F. YOCUM, Plant Physiol., 70, 154 (1982)
50. G. C. PAPAGEORGIOU, T. LAGOYANNI, Biochim. Biophys. Acta, 724, 323 (1983)
51. C. S. DEROO, C. F. YOCUM, Biochem. Biophys. Res. Commun., 100, 1025 (1981)
52. R. A. WAVARE, P. MOHANTY, Photobiochem. Photobiophys., 6, 189 (1983)

4.6 Picosecond Processes of Photosynthesis in Laser Absorption Studies

A. Yu. BORISOV and A. P. RAZJIVIN

Department of Photosynthesis, A. N. Belozersky Laboratory, M. V. Lomonosov Moscow State University, 119899 Moscow, USSR

and

R. V. DANIELIUS and R. J. ROTOMSKIS

Laser Research Center of V. Kapsukas, Vilnius University, 232734 Vilnius, USSR

The demand for new energy sources has stimulated extended studies of sunlight energy conversion processes in photosynthesis. The primary photosynthetic processes are at present the basic field of research. These processes are characterized by high quantum efficiency, an important factor for designings artificial sunlight energy converters. Studying primary photosynthetic processes enables us to understand and model sun energy conversion.

The universally accepted scheme of photosynthetic primary processes includes: (i) absorption of a light quantum and further transfer of the electronically excited state (excitation) via pigment molecules of light harvesting antenna (LHA); (ii) excitation trapping by the reaction center (RC) and transformation of this energy into the energy of separated charges of different signs in RC.

The charge separation processes in RCs were studied almost exclusively by means of absorption laser picosecond spectroscopy (LPS) on isolated RC preparations[1-5]. Absorption LPS appeared to be the most appropriate for investigating these processes: RCs exhibit large photoinduced absorption changes while RC fluorescence is too weak for registration with an appropriate spectral and time resolution[6]. On the other hand, many important aspects of excitation migration over LHA may be solved with the help of fluorescence. The intensity of fluorescence is proportional to the concentration of excited molecules. This makes it possible to observe excitation energy distribution over LHA as well as excitation migration via light harvesting pigments if ultrashort light pulses are used. The experimental data obrained in the 1960's by the phase-fluorometer method show that photosynthetic objects are characterized by long-lived fluorescence[7]. But subsequent works showed that the main part of light harvesting pigment has a short-lived fluorescence with the time constant ~ 30-100 ps [8,9]. But these data, obtained with the modified phase-fluorometer method, had to be verified by more direct and informative methods.

The appearance of picosecond lasers and adequate registration instruments made direct measurements possible. The first picosecond fluorescence investigations indicated short-lived fluorescence [10-12]. Yet subsequent measurements caused some doubts about the validity of the results: when the sample is excited by strong laser light pulses, non-linear effects may appear[13-15]. Some papers dealing with the problems of non-linear effect influence on natural photosynthesis were published[16].

As noted above, primary light energy conversion in photosynthesis comprises the two main stages: the excitation transfer via LHA pigments toward RCs and the trapping and stabilization of this energy in RCs (electron transfer via RC molecules). Significant progress has been made in separate investigation of these two stages, but the problem of their coupling has received little attention so far. This is largely due to the absence of a unified approach to the investigation of these two stages. The low quantum yield of RC fluorescence ($\lesssim 10^{-3}$), its spectral similarity to LHA fluorescence, on one hand, and the relatively low sensitivity of the absorption LPS method, on the other, hindered the coupling problem investigation by means of absorption and fluorescence LPS. The former difficulty is apparently of principal nature, while the latter one was overcome to some extent when a picosecond difference absorption spectrometer with the sensitivity better than 10^{-3} A-units and with selective excitation was designed[17,18]. Using absorption picosecond spectroscopy makes it possible to follow the time course of photoinduced absorption changes due to antenna molecules as well as the appearance of signals as a result of charge separation in RCs. Thereby we can discriminate between the kinetics of energy transfer from LHA to RCs and the multicomponent kinetics of excitation deactivation in LHA, and thus study the parameters of the process.

To investigate primary photosynthesis processes by picosecond absorption spectroscopy, one has to fulfill certain essential requirements: (i) selective excitation of photosynthetic preparation in visible and in near infrared spectral regions; (ii) high sensitivity for measuring small absorption changes in samples with the absorbance ~ 1 A-unit; (iii) sufficiently low picosecond excitation pulses to exclude non-linear processes which may interfere with experimental results as strong laser pulses are used. The above listed requirements were fulfilled in a difference spectrophotometer with a picosecond time resolution which was developed on the basis of parametric picosecond light oscillators[19,20].

Figure 1 presents the block scheme of a picosecond laser difference spectrophotometer. YAG:Nd^{3+} laser with passive mode-locking was used as a driving generator. A single pulse selected from a pulse train is amplified in a two-stage amplifier, then three beams are formed. The first beam, after its frequency has been

400 LASERS IN ATOMIC, MOLECULAR AND NUCLEAR PHYSICS

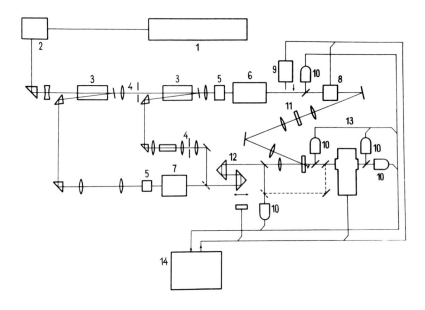

FIGURE 1 A block diagram of picosecond absorption spectrometer.
1 - picosecond YAG:Nd^{3+} laser, 2 - monopulse extraction scheme, 3 - YAG:Nd^{3+} two-stage amplifier, 4 - diaphragm, 5 - second harmonic generator, 6 - picosecond parametric oscillator of excitation channal, 7 - picosecond parametric oscillator of probing channel, 8 - attenuator (tuning optical filter), 9 - electromechanical shutter, 10 - photodetector, 11 - opal glass, 12 - time delay line, 13 - monochromator, 14 - registration and control arrangement with a micro-computer.

doubled, pumps an optical parametric oscillator (OPO) on KDP crystals, the second beam pumps OPO on the LiNbO$_3$ crystal. The wavelength of excitation and probing beams is tuned continuously and independently in a 0.8-1.5 μm spectral range, the pulse energy being 1-2 mJ for KDP OPO, and 0.7-2 μm and 0.05-0.1 mJ for LiNbO$_3$ OPO. The first beam is used for sample excitation, the second - for absorption probing. An electromechanic shutter for measuring the zero level and an attenuator (a tuning optical filter) were put into the excitation beam. A picosecond continuum source on D$_2$O pumped by the third beam of the main radiation was used for beam probing at difference spectrum measurements in a wide range. OPO was used to measure the dependences of absorption changes on excitation intensity and the kinetics of absorption changes. The application of considerable narrow-band radiation made it possible to dispense with a spectral device in the registration channel, and to use photodiodes as photoreceivers. As a result, a sencitivity of $> 10^{-3}$ units could be obtained, the absorption being ~ 1 unit. The error level of the measurements was 2-3% if absorption changes were $5 \cdot 10^{-2}$ units. When probing was carried out by continuum, a typical measurement error was of $2-5 \cdot 10^{-3}$ units.

The time resolution of the spectrometer is up to 10 ps when a single exponential process is recorded. When the relaxation process is more complicated, computer simulation is needed. The initial processing of the information and the operation on the spectrometer were carried out by a microcomputer "Electronica D3-28" through an individual interface.

We might just as well discuss the advantages of selective excitation and probing and, in particular, high sensitivity needed for photosynthetic studies. First, excitation in the long wavelength bacteriochlorophyll (BChl) band of LHA excludes all problems connected with energy transfer from additional pigments (carotenoids) to BChl (the efficiency of this process is ~ 0.3 and may vary). Second, one can achieve uniform excitation of the sample by excitation wavelength tuning. It is important because the relaxation of excitation in LHA depends on excitation intensity. For example, one can use chromatophore samples with high optical density at the absorption peak (up to 10 A-units) which makes it possible to obtain a high amplitude of the RC signal (up to 0.3 A-units). But this will be useful only if the selective probing at the band wings is available too. The high values of the ratio between the photoinduced RC signal and the initial absorption on the same wavelength are registered near 800 nm in the absorption band

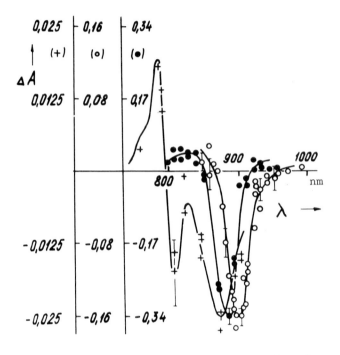

FIGURE 2 Difference spectra of photo-induced absorption changes in Rhodospirillum rubrum chromatophores at high (●) and low (o) intensities of exciting 900 nm light pulses (> 10^{16} and < 10^{15} photon/cm^2, respectively). The probing pulse coincides in time with the arrival of the exciting one. The curve (+) carried out at low (< 10^{15} photon/cm^2) intensity of 915 nm picosecond pulses with 450 ps time delay for probing pulse relative to exciting pulse. Absorbance of chromatophores in a 1 mm cell at 880 nm was ≥ 1 A-unit.

of RC monomeric BChl of R. rubrum and R. sphaeroides chromatophores. The results given below show absorption changes due to oxidation of ~ 10% RCs in chromatophores and corresponding to an excitation intensity of about 1 photon per 300-400 BChl molecules.

Purple bacteria are the traditional object for photosynthetic studies, for their photocynthetic apparatus is simpler than that of higher plants and algae. Membrane preparations containing photosynthetic apparatus (chromatophores) may be extracted from purple bacteria cells by biochemical methods. These preparations are

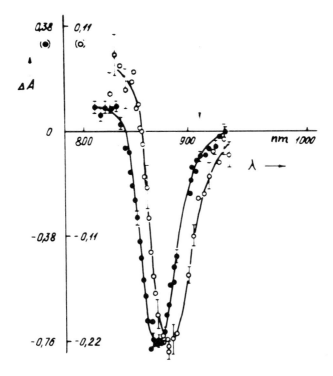

FIGURE 3 Difference spectra of photo-induced absoption changes in R. sphaeroides R-26 chromatophores at high (●) and low (○) intensities of exciting 840 nm pulses (>10^{16} and <10^{15} photon/cm^2, respectively). The probing pulse coincides in time with exciting one. Absorbance of the sample on the exciting wavelength was about 0.5 A units.

more adequate for optical investigations than suspensions of bacteria cells. Chromatophore membranes contain the pigment-protein complexes of RCs and LHA. Each pigment-protein complex contains some pigment molecules whose spatial arrangement is determined by special packing of the protein component[23,24]. The RC isolated from some purple bacteria may serve as an example of pigment-protein complex[23]. In turn, the LHA is assembled from dozed pigment-protein complexes of one or more types, different from RC[24-27].

We investigated chromatophore preparations from purple photosynthetic bacteria Rhodospirillum rubrum, Rhodopseudomonas spha-

404 LASERS IN ATOMIC, MOLECULAR AND NUCLEAR PHYSICS

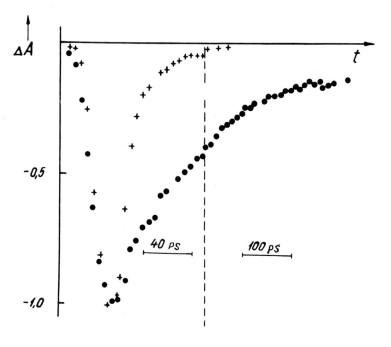

FIGURE 4a The kinetics of absorbance changes of B^o and B^m signals for complex B890 from C. minutissimum chromatophore. Curve (+) recorded at 873 nm under 840 nm excitation ($4 \cdot 10^{16}$ photon/cm^2); curve (o) recorded at 900 nm under 840 excitation ($3 \cdot 10^{15}$ photon/cm^2); sample absorbance at 885 nm was about 1.5 A unit.

aeroides, strain R-26, Rhodopseudomonas viridis, and the pigment-protein comlex B890 from Chromatium minutissimum. These 4 species represent sufficiently fully the entire suborder of purple bacteria. That is why the obtained picture of excitation transfer processes via LHA and excitation trapping by RCs in chromatophores, may be characteristic of the suborder of purple basteria as a whole.

Control experiments show a sufficiently high quality of chromatophore preparations. The number of LHA BChl molecules was about 30-35 per RC. The absolute quantum yield of RC oxidation in chromatophores due to continuous illumination was estimated at 0.87± 0.15. With the aid of the relative method of quantum yield estima-

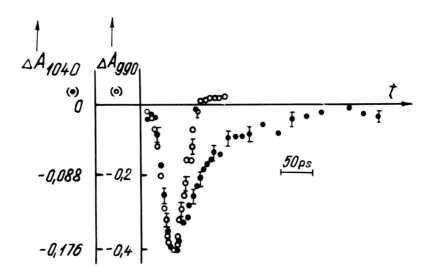

FIGURE 4b The kinetics of absorbance changes of B_o and B_m signals for R. viridis chromatophores. Both curves were registered by 930 nm exciting pulses with high (●) and low (o) intensity (10^{14} and 10^{16} photon/cm^2, resp.).

tion[21], which excludes the influence of destroyed and initially defected chromatophores, we obtained the value 0.90±0.03.

It is accepted that the long-wavelength pigment-protein complex of LHA in chromatophores is represented only by one pigment spectral form [24,25,28]. In our experiments it was shown that spectral changes in LHA long-wavelength absorption band of purple bacteria chromatophores are of complex nature (Figures 2,3). These changes were interpreted as the manifestation of two spectral forms of LHA[29-31]. The following scheme of energy transfer from LHA to RC in bacterial photosynthesis as presented here is based on an analysis of spectral (Figures 2,3) and kinetic (Figure 4) data, as well as exciting pulse intensity dependences (Figures 5,6)[32-34]:

$B_o \longrightarrow B_m \longrightarrow RC$

where B_o - the bulk BChl of LHA, B_m - the minor fraction of LHA localized in the vicinity of RC.

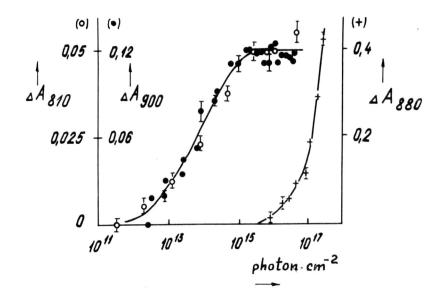

FIGURE 5 Dependence of absorption changes of R. rubrum chromatophores on the energy of 915 exciting pulses (light curves): curves (+) and (o) were measured at 880 nm (in B_o band) and 900 nm (in B_m band), respectively; probing pulse coincided in time with the exciting one in both cases; curve (o) that corresponds to P800 band shift due to RC oxidation, was measured at 810 nm with 450 ps time delay for probing pulse relative to the exciting one.

The minor components of LHA, whose absorption bands are shifted of longer wavelengths compared to the RC absorption band, have been considered in works on plant photosynthesis (the components with a fluorescence maximum at 720-740 nm)[35]. It is accepted that photoinduced excited states begin to leak to these forms at low temperatures due to inactivation of RCs[36].

The long wavelength bleaching band in the difference picosecond spectrum of chromatophores was interpreted as a result of the transition of 3-5 highly interacting BChl molecules to an excited state. Moreover, it was proposed[33,38] that all pigment molecules of the minor form B_m comprise a structure similar to an associate

LASER STUDY OF PHOTOSYNTHESIS

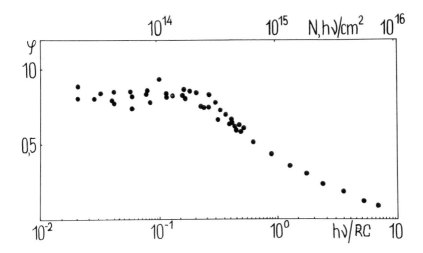

FIGURE 6 Dependence of the absolute quantum yield of RC photo-oxidation in R. rubrum chromatophores on intensity of exciting pulses (for absorbed quanta only).

of dye molecules and that this associate is incorporated into a single pigment-protein complex. If this assumption is valid, such a pigment protein complex, in principle, may be purified from chromatophores of purple bacteria. Although the analysis of the LHA protein composition has yielded no minor polypeptides[42], the LHA heterogeneity is indicated by the studies of BChl oxidation in RC deficient R. rubrum mutant[39], by the low-temperature absorption spectra measurements[40] and by the fluorescence polarization spectra of chromatophores[41].

At the present time it is not clear if the absorption changes found on the long wavelength side of the absorption band of LHA testify to LHA structural heterogeneity. For example, in[37] it was suggested that as the antenna BChl dimer absorbs a light quantum, a mixing of the singlet excited state and the charge transfer state takes place in this dimer. The difference spectrum of the minor form corresponds to the absorption of the charge transfer state while the spectrum of the major form - to the absorption of the singlet excited state. The CT-state level is ~ 30 meV lower than the S_1 level. The logical suggestion is that the energy transfer to RC via antenna molecules may be connected with some kind of charge migration instead of singlet excited states migration. But this model fails to explain the behaviour of the light curves (Figure 5).

Yet there may be more involved interpretations of the difference spectrum B_m as well. It has been shown that this spectrum can be well simulated by a bleaching of the main absorption band and by a short-wavelength shift of the same band. The absorption changes, which may be desribed in this way, are registered in some other objects with strong intermolecular interaction - dye aggregates[43], and polycrystals of Chl a[44]. The absorption band shift may be of electrochromic nature as a result of charge separation in antenna, as suggested in [33,34]. Evidence for such interpretation is the large amplitude of B_m bleaching. A cross-section of B_m molecule absorption is to be 3-fold higher than that of BChl dimers of LHA or RC if we want to obtain such an amplitude of B_m bleaching.

According to new data obtained by absorption picosecond spectroscopy, the scheme of excitation transfer and trapping in chromatophores from photosynthetic bacteria differs from the traditional one. It is quite difficult to determine the influence of annihilation processes on the kinetics of absorption changes of LHA bacteriochlorophylls. The point is that an increase in exciting pulse intensity results in absorption band bleaching (Figure 4). But this bleaching is not manifested in low intensity excitation due to its rapid relaxation as compared with exciting and probing pulse duration (~ 20 ps). The B_m bleaching decay time hardly changes (Figure 6) for the broad exciting intensity range (10^{14} - 10^{16} photon/cm^2) when the kinetics is on a fixed wavelength where only the B_m signal appears. This "independence" may occur if the annigilation processes are very rapid and can be accomplished at the stage of excitation transfer $B_o \longrightarrow B_m$. We believe that with the development of an universal laser spectrometer to take parallel absorption and fluorescent measurements in identical experimental conditions, it would become possible to make progress in solving this problem. On the other hand, we hope that the use of difference spectrophotometers with subpicosecond time resolution will open up new opportunities for a better understanding of energy migration and conversion in photosynthetic objects.

REFERENCES

1. T. L. NETZEL, P. M. RENTZEPIS, J. LEIGH, Science, 182, 238 (1973)
2. C. G. ROCKLEY, M. W. WINDSOR, R. J. COGDELL, W. W. PARSON, Proc. Natl. Acad. Sci. USA, 72, 2251 (1975)
3. P. L. DUTTON, K. J. KAUFMAN, B. CHANCE, P. M. RENTZEPIS, FEBS Letters, 60, 275 (1976)
4. S. L. SHAPIRO, A. J. CAMPILLO, in Ultrashort Light Pulses, edited by S. L. SHAPIRO (Springer Verlag, Berlin, 1977).
5. S. A. AKHMANOV, A. Yu. BORISOV, R. V. DANIELIUS, A. S. PISKARSKAS, V. D. SAMUILOV, A. P. RAZJIVIN, Pisma v JETF, 26, 655 (1977) (in Russian)
6. R. K. CLAYTON, Photosynthesis. Physical Mechanisms and Chemical Patterns (Cambridge Univ. Press, Cambridge, London, New York, New Rochelle, Melbourne, Sydney, 1980).
7. A. Yu. BORISOV, in Photosynthesis in Relation to Model Systems, edited by J. Barber (Biomedical Press, Elsevier/North Holland, London, 1979), pp. 1-26.
8. A. Yu. BORISOV, M. D. IL'INA, Biokhimiya, 36, 822 (1971) (in Russian)
9. A. Yu. BORISOV, V. I. GODIK, Bioenergetics, 3, 211 (1972)
10. M. SEIBERT, R. R. ALFANO, S. L. SHAPIRO, Biochim. Biophys. Acta, 292, 493 (1973)
11. V. Z. PASCHENKO, A. B. RUBIN, Kvantovaya Electronika, 2, 1336 (1975) (in Russian)
12. V. H. KOLLMAN, S. L. SHAPIRO, A. J. CAMPILLO, Biochim. Biophys. Res. Communs., 63, 917 (1975)
13. A. J. CAMPILLO, V. H. KOLLMAN, S. L. SHAPIRO, Science, 193, 227 (1976)

14. D. MAUZERALL, Biophys. J., 16, 87 (1976)
15. G. E. SWENBERG, N. E. GEACINTOV, M. POPE,
    Biophys. J., 16, 1447 (1976)
16. J. BRETON, N. E. GEACINTOV, Biochim. Biophys. Acta, 594,
    1 (1980)
17. R. DANIELIUS, A. PISKARSKAS, V. SIRUTKAITIS, A. STABINIS,
    J. YASEVICHUTE, Parametric Light Oscillators and Picosecond
    Spectroscopy (Mokslas, Vilnius, 1983), pp. 138-152 (in Russ.)
18. A. S. PISKARSKAS, in Laser Applications in Atomic, Molecular,
    and Nuclear Physics (MIR Publ., Moscow, 1979), pp. 249-313
    (in Russian)
19. R. A. GADONAS, R. V. DANIELIUS, A. S. PISKARSKAS,
    Kvantovaya Elektronika, 8, 669 (1981) (in Russian)
20. R. DANIELIUS, A. PISKARSKAS, V. SIRUTKAITIS,
    Kvantovaya Elektronica, 9, 2491 (1982) (in Russian)
21. A. Yu. BORISOV, V. I. GODIK, S. G. FETISOVA,
    Molekularnaya Biologiya, 8, 458 (1974) (in Russian)
22. V. C. REMSEN, in The Photosynthetic Bacteria, edited by R. K.
    Clayton, W. R. Sistrom (Plenum Press, New York, London,
    1978), pp. 31-60.
23. C. GINGRAS, in The Photosynthetic Bacteria, edited by R. K.
    Clayton, W. R. Sistrom (Plenum Press, New York, London,
    1978), pp. 119- 131.
24. J. P. THORNBER, R. J. COGDELL, B. K. PIERSON, R. E. SEFTOR,
    J. Cell Biochem., 23, 159 (1983)
25. R. PICOREL, G. BELANGER, G. GINGRAS, Biochemistry, 22,
    2491 (1983)
26. K. SAUER, L. A. AUSTIN, Biochemistry, 17, 2011 (1978)
27. R. J. COGDELL, J. P. THORNBER, FEBS Lett., 122, 1, (1980)
28. R. J. COGDELL, J. G. LINDSAY, J. VALENTINE, J. DURANT,
    FEBS Lett., 150, 151 (1982)
29. A. Yu. BORISOV, R. A. GADONAS, R. V. DANIELIUS, A. S. PISK-
    ARSKAS, A. P. RAZJIVIN, S. G. KHARCHENKO, Doklady AN SSSR,
    264, 980 (1982) (in Russian)
30. J. A. ABDOURAKHAMNOV, R. V. DANIELIUS, A. S. PISKARSKAS,
    A. P. RAZJIVIN, R. J. ROTOMSKIS, in Abstracts of 6 Int.
    Congress on Photosynthesis, Brussels, 1, 244 (1983)
31. R. V. DANIELIUS, V. V. KRASAUSKAS, R. J. ROTOMSKIS, in
    Studies in Spectroscopy and Quantum Electronics, VI (V. Kap-
    sukas Vilnius Univ. Press, Vilnius, 1983), p. 95 (in Russ.).
32. A. P. RAZJIVIN, R. V. DANIELIUS, R. A. GADONAS, A. Yu. BORI-
    SOV, A. S. PISKARSKAS, FEBS Lett., 143, 40 (1982)
33. A. Yu. BORISOV. R. A. GADONAS, R. V. DANIELIUS, V. S. KOZLOV-
    SKI, A. S. PISKARSKAS, A. P. RAZJIVIN, S. G. KHARCHENKO,
    Doklady AN SSSR, 266, 482 (1982) (in Russian)
34. R. V. DANIELIUS, A. P. RAZJIVIN, Izvestiya AN SSSR, ser. fiz.,
    48, 466 (1984) (in Russian)

35. W. L. BUTLER, C. J. TREDWELL, R. MALKIN, J. BARBER, Biochim. Biophys. Acta, 545, 309 (1979)
36. V. Z. PASCHENKO, S. S. VASIL'EV, V. N. KORVATOVSKI, V. B. TUSOV, G. P. KUHARSKIN, A. B. RUBIN, Doklady AN SSSR, 273, 1252 (1983) (in Russian)
37. E. GAIZAUSKAS, G. TRINKUNAS, L. VALKUNAS, in Advance in Photosynthesis Research, edited by C. Sybesma (Martinus Nijhoff/ Dr. Junk publishers, Hague, 1984), Vol. 1, pp. 49-52.
38. A. Yu. Borisov, R. A. GADONAS, R. V. DANIELIUS, A. S. PISKARSKAS, A. P. RAZJIVIN, FEBS Lett., 138, 25 (1981)
39. J. GOMEZ, P. PICOREL, J. M. RAMIREZ, P. PEREZ, R. R. DEL CAMPO, Photochem. Photobiol., 35, 399 (1982)
40. F. F. LITVIN, B. A. GULYAEV, Izvestiya AN SSSR, ser. biol., 43 (1970) (in Russian)
41. H. KRAMER, Structural Aspects of Energy Transfer in Photosynthesis (Thesis, Leiden, 1984), p. 92.
42. R. A. BRUNISHOLZ, R. SUTER, H. ZUBER, Hoppe-Seyler's Z. Physiol. Chem., 365, 675 (1984)
43. R. A. GADONAS, R. V. DANIELIUS, A. S. PISKARSKAS, S. RENCH, Izvestiya AN SSSR, ser. fiz., 47, 2445 (1983) (in Russian)
44. R. Danielius, R. GADONAS, P. MALIJ, Chem. Phys. Lett., (1985) (in press)

4.7 The Structure, Function and Assembly of the Light-Harvesting Antenna of Photosynthetic Purple Bacteria

R. VAN GRONDELLE

Department of Biophysics, Physics Laboratory of the Free University, De Boelelaan 1081, 1081 HV Amsterdam, The Netherlands

1. INTRODUCTION

The photosynthetic apparatus of photosynthetic purple bacteria such as Rhodopseudomonas sphaeroides and Rhodopseudomonas capsulata contains at least two types of major light-harvesting pigment-protein complexes: B800-850 and B875. Species such as Rhodospirillum rubrum only contain the latter.

The B800-850 comlex shows absorption at 800 nm, due to BChl 800 and at 850 nm due to BChl 850. It contains about one carotenoid per two BChl molecules.

The B875 complex shows absorption at 875 nm due to BChl 875 and contains one carotenoid per BChl molecule (B875 of Rps. sphaeroides).

The B800-850 comlex is found at the periphery of the light-harvesting antenna. B875 forms a central network that interconnects several reaction centers (the "lake model" for energy transfer). The dominating direction of excitation energy transfer is from carotenoid to BChl 800, BChl 850 or BChl 875 and then among the different BChl's finally to BChl 875.

Once in the B875 network, the excitation is transferred among a large number of identical BChl 875 molecules, a process which

413

can be described as a random walk that the excitation performs on the B875 network, until a reaction center is reached where a charge separation may be initiated.

Three concepts are of fundamental importance for the understanding of this process. Firstly, the excitation may be lost during the transfer process. Secondly, upon arrival at the reaction center, the excitation is not necessarily trapped; it may escape from the trap and resume the hopping process. Thirdly, if an excitation finds a closed or inactive trap, it may escape and, due to the fact that the B875 network connects several reaction centers, try elsewhere.

The emissions of BChl 800, BChl 850, and BChl 875 can be observed at all temperatures and these can be used to study the losses that occur in each pool of antenna molecules separately. In addition absorption and linear- and circular-dichroism spectra are available.

Using lithium dodecyl sulfate polyacrylamide gel electrophoresis (LDS-PAGE), a separation of the two antenna complexes in a relatively intact state is obrained[1,2]. In addition several "intermediate" complexes are separated that contain various ratios of BChl 800 and BChl 850.

The amplitude of BChl 800 absorption in all these complexes is rather low, but incubation with LDAO restores this completely.

In this paper I will shortly deal with the spectroscopic properties of these preparations, discuss the transfer and trapping of excitations in the intact system and, finally, make a few comments about the mechanism of association of these comlexes to a fuctioning antenna.

2. THE B800-850 COMPLEX

Figure 1 shows a model for the pigment organization in the B800-850 comlex based on a combination of spectroscopic and biochemical

LASER STUDY OF PHOTOSYNTHESIS

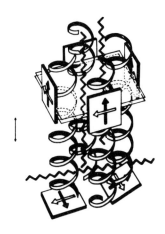

FIGURE 1 Schematic model of the B800-850 complex. The upper square boxes are the porphyrin heads of BChl 850, the lower ones of BChl 800. Open arrows: Q_Y transitions, solid arrows: Q_X transitions. Zig-zag lines are carotenoids (spheroidene) and the spirals represent the α-helical sections of the constituent peptides. The plane of the membrane is horisontal; the vertical bar represents 5 Å. From ref. 6, with permission.

data[3-6]. A minimum unit of B800-850 must contain at least 4 BChl 850, 2 BChl 800, and 3 carotenoid molecules. These pigment molecules are non-covalently bound to two idendical pairs of proteins (α and β), each of which contains a central hydrophobic stretch of amino acids, that is supposed to the α-helical and transmembrane[7,8]. Each apoprotein contains a central histidine at about the same position, which may be the binding site for BChl 850. Two of the four subunits (β) contain a second histidine at about 20-25 Å, measured along the α-helix, from the first, at the hydrophobic/hydrophylic interface. These may be the binding sites for the two BChl 800's.

The BChl 850's have their porphyrin heads arranged in such a way that the Q_Y's are all parallel and the Q_X's are all perpendicular to the membrane plane. In Figure 1 the four BChl 850 Q_Y's form a circularly degenerate oscillator to account for the observed fluorescence polarization.

If each of the four porphyrin heads shows a small displacement along the normal of the membrane plane, the four BChl 850 Q_Y's may form a left-handed helix, and this induces sufficient rotational strength to account for the observed circular dichroism.

The two BChl 800 molecules have their porphyrin heads almost in the plane of the membrane with the Q_X's and Q_Y's mutually perpendicular. The BChl 800 Q_X's may be tilted out of the membrane plane but at most 25°. The distance between the centers of the two BChl 800 porphyrin heads must be less than 19 Å. From the observed BChl 800 fluorescence yield at low temperatures the BChl 800 - BChl 850 dipole-dipole distance may be estimated to be smaller than 21 Å, if transfer to only one BChl 850 takes place, or, less than 24 Å if two equivalent BChl 850's are involved. The latter value appears to give the best agreement with the structural data.

The carotenoid content of the B800-850 comlex is heterogeneous. About one-third of the carotenoids transfers its excitation exclusively to BChl 800, while the remaining two thirds transfer their excitation energy to BChl 850. The former carotenoid is oriented more or less parallel, the latter perpendicular to the membrane plane. The BChl 800 carotenoid shows a red shifted absorption.

At room temperature energy transfer between the BChl 800's and BChl 850's is fast enough to establish a thermal equilibrium between the excitation densities on both types of BChl molecules. At low temperatures (4 K) the rate of energy transfer from BChl 800 to BChl 850 will be about $3 \cdot 10^{12}$ s^{-1} to account for the observed losses. The low temperature spectroscopic data do not give any indication for a further heterogeneity of the BChl 800 absorption band.

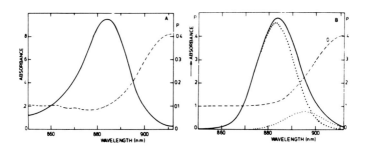

FIGURE 2 Measured (A) and simulated (B) absorption (———) and fluorescence polarization (- - - - -) spectra of chromatophores of Rps. sphaeroides R26. The dashed curves in B represent the major, λ_{max} = 883 nm, and minor, λ_{max} = 896 nm, BChl species present in the B875 antenna. The spectra were recorded at 4 K. From ref. 11, with permission.

3. THE B875 COMPLEX

The minimum unit of B875 probably contains about 6 BChl molecules[9]. At room temperature the polarization of the fluorescence increases somewhat upon excitation in the red wing of the long wavelength absorption band[10,11]. This effect is dramatically enhanced at low temperatures.

In intact chromatophores after cooling below 77 K similar effects are observed. (See Figure 2.) However, at room temperature the polarization of the emission is constant (p ≃ 0.12) upon excitation over the whole long wavelength absorption band[11]. To explain this observation it was proposed that each B875 hexamer contains one special BChl molecule, BChl 896, that at low temperatures is responsible for all the emission. At room temperature a thermal equilibrium will exist between the excitation densities on BChl 875 and BChl 896 and this will lead to a decrease in the observed fluorescence polarization in isolated complexes, due to strongly overlapping emission bands. In the intact membrane, at room temperarute energy transfer between different BChl 896 molecules will occur, which leads to the complete absence of these effects.

We note that the proposed BChl 896 appears similar to the "minor form" BChl introduced by Borisov c.s.[12,13]. However, the properties that these authors claim for their special antenna appear in disagreement with those of BChl 896 as discussed above. From these data no evidence can be obtained that supports the hypothesis that BChl 896 forms a special antenna that connects the bulk BChl 875 with the reaction center.

4. TRAPPING, LOSS, AND ANNIHILATION OF EXCITATIONS IN PHOTOSYNTHETIC SYSTEMS

In the intact photosynthetic system the BChl 875 molecules form a network (more or less homogeneous) that connects several reaction centers. At moderate and high picosecond pulse intensities ($\gtrsim 10^3$ photons/cm^2 per pulse) more than one excited BChl 875 molecule may be generated in this network, or domain, and due to extensive excitation transfer two excitations may collide and annihilate at least one of the pair according to the following scheme.

$$BChl^* + BChl^* \xrightarrow{k_2} BChl^* + BChl$$

The "rate constant" k_2 includes the hopping of both excitations among the BChl 875 molecules and the actual probability of annihilation upon collision. The annihilation competes with the trapping and loss processes at moderate and high pulse intensity and it can be shown that a study of the yield of trapping and losses allows for a determination of the various rate constants involved in the process[14,15]: the rate of energy transfer between nearest neighbors, k_h, the rate of trapping by active and inactive reaction centers, k_t^o and k_t^c respectively and the rate of annihilation upon collision, k_a. Moreover estimates about the number of connected photosynthetic units (the "domain size") can be obtained.

To do so we must make a simplified model for the B875 network.

LASER STUDY OF PHOTOSYNTHESIS   419

In the following it is assumed that the domain may be represented by a square lattice. Each domain contains λ reaction centers, that are distributed regularly. The number of antenna molecules per reaction center is N and energy transfer takes place to nearest neighbors only. The reaction center occupies only a single lattice point and the rates of excitation transfer to and from the reaction center are assumed to be equal.

We then calculate the probabilities of trapping by an open, f_t^o, or a closed trap, f_t^c, for a single excitation:

$$f_t^{o,c} = [N(1-z)[G_N(o;z) + \eta_t^{o,c}/(1-\eta_t^{o,c})]]^{-1}\Big|_{z=1-\varepsilon}$$

where $\varepsilon = k_\ell/(k_\ell + 4k_h)$ and $1-\eta_t^{o,c} = k_t^{o,c}/(4k_h + k_\ell + k_t^{o,c})$ with k_ℓ the rate of loss processes, including fluorescence, for each antenna molecule $G_N(o;z)$ is the Greens function of the lattice representing the photosynthetic unit.

Very similarly we fing for the probability of annihilation, f_a, of one excitation of a pair of excitations in a domain, in the absence of trapping:

$$f_a = [N_D(1-z)[G_{N_D}(o,z) + \eta_a/(1-\eta_a)]]^{-1}\Big|_{z=1-\varepsilon}$$

where N_D is the number of pigment molecules per domain ($N_D = \lambda N$) and η_a is the probability that the excitations will escape annihilation upon collision:

$$1-\eta_a = k_a/(2k_\ell + 8k_h + k_a)$$

G_{N_D} is the Greens function of the lattice representing the domain. We note the occurance of N_D in the expression for f_a.

Using these probabilities $f_t^{o,c}$ and f_a we can define a set of effective rate constants for trapping by open/closed traps, $k_t^{o,c}$, and for annihilation, k_2:

$$k_t^{o,c} = k_\ell \cdot f_t^{o,c}/(1 - f_t^{o,c})$$

$$k_2 = k_\ell \cdot 2f_a/(1 - f_a)$$

We will be able to describe the competition between annihilation and trapping after one additional assumption. This concerns the fact that even if all the reaction centers in a domain are open before the pulse, a fraction of them will become closed during the pulse. For the description of trapping in such a mixture of open and closed traps no exact theory is available and therefore we make the following approximations: if a fraction x of the traps in a domain is closed, the rate of trapping in such a domain is given by:

$$k_1(x) = (1-x)k_1^o + xk_1^c$$

It is possible to show that although this expression is not exact, it is nevertheless a good approximation[15].

To take into account the fluctuations occurring in each domain and moreover the fact that there are many domains in a photosynthetic system, we use a Pauli-Master equation[14-16]. This leads to an expression for the total probability of loss per excitation, $U_n(z)$, and the total fraction of traps closed, $V_n(z)$, where n is the average number of open reaction centers in a domain before the pulse is given ($0 \leqslant n \leqslant \lambda$) and z is the average number of excitations genereated per domain.

Figure 3 shows an experiment with chromatophores of the purple bacterium R. rubrum[15]. The excitation wavelength was 532 nm and the pulse width about 30 ps. The fluorescence was detected at 900 nm. The fluorescence yield vs pulse intensity was measured with either all the traps initially closed (state $P875^+$) or open. Moreover the fluorescence yield induced by a weak Xenon flash, fired about 1 ms after the picosecond laser pulse was measured and in principle this yields the fraction of traps closed by the intense laser pulse.

LASER STUDY OF PHOTOSYNTHESIS 421

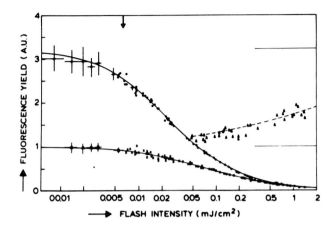

FIGURE 3 Fluorescence yield as a function of the pulse intensity in R. rubrum chromatophores, with all reaction centers in the closed state by continuous background illumination; x, with all the reaction centers open before the pulse. The fluorescence detected with a weak xenon flash 1 ms after the laser pulse in the case where the reaction centers are initially all open is shown by the open triangles (Δ). The arrow indicates the intensity of the laser flash where there is on the average on excitation per domain. From ref. 15, with permission.

The fits shown in Figure 3 were done for $\lambda \simeq 14\text{-}17$, $N_D \simeq 700\text{-}800$, $k_\ell = 5 \cdot 10^8 \text{ s}^{-1}$, $k_1^o = 9.5 \cdot 10^9 \text{ s}^{-1}$.

It is calculated from the random walk expressions that these values correspond to

$$k_h > 10^{12} \text{ s}^{-1}$$

$$k_t^o = (4\text{-}6) \cdot 10^{11} \text{ s}^{-1}$$

$$k_t^c = 1.4 \cdot 10^{11} \text{ s}^{-1}$$

$$k_a > 5 \cdot 10^{12} \text{ s}^{-1}$$

If $k_a \gg k_h$ is assumed, or perfect annihilation upon a collision of two excitations, it follows that: $k_h = (1\text{-}2) \cdot 10^{12} \text{ s}^{-1}$.

Using the Förster expression for excitation energy transfer between two neighboring antenna molecules, it is found that the "average" BChl 875 - BChl 875 dipole-dipole distance is about 13-15 Å. It may be noted that these numbers are in close agreement with those given be Freiberg et al. (see these proceedings) from an analysis of the fluorescence lifetimes.

We finally remark that very similar results were obtained with chromatophores of Rhodopseudomonas capsulata.

5. THE ASSEMBLY OF THE LIGHT-HARVESTING ANTENNA OF RHODOPSEUDOMONAS SPHAEROIDES

Aerobically grown cells of Rps. sphaeroides will develop a photosynthetic light-harvesting system if the oxygen tension is low enough. This enables one to study the membrane bound pigment-protein complex over a wide concentration range compared to e.g. membrane area. Moreover, using sucrose density centrifugation, a pigmented band can be separated (UPB) that is probably a precursor of a true chromatophore[17,18]. Excitation annihilation and fluorescence emission spectra were studied in a number of these preparations at different stages of development[19]. The largest contrast was found between 0hr UPB/0hr chromatophores (0hr reflects the fact that these membranes were isolated almost immediately after interrupting the O_2-flow) and 40hr chromatophores.

In all preparations energy transfer from all the pigments to BChl 875 was good, although the 0hr UPB had an increased BChl 850 fluorescence level[19,20]. The fluorescence yield versus pulse intensity curves for the 0hr preparations were shifted to about 6-fold higher intensities (see Figure 4) implying the existence of much smaller domains, approximately 100-200 BChl 875 molecules (3-6 connected reaction centers). The 40hr preparation gave a domain size of about 1000-2000 connected BChl molecules. Intermediate preparations gave intermediate results. In general UPB fractions have smaller domains than chromatophores.

Thus, even at the extremely early stages of development,

LASER STUDY OF PHOTOSYNTHESIS 423

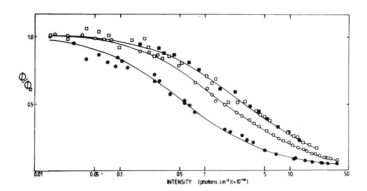

FIGURE 4 Fluorescence yield as a function of flash intensity for membranes isolated from Rps. sphaeroides. The excitation flash was a 30 ps, 532 nm pulse from a frequency doubled, modelocked Nd^{3+}-YAG laser. The fluorescence of all preparations was detected at 900 nm. The absorbance for all samples at 532 nm was 0.1. Φ/Φ_o was normalized to 1.0 at low intensities. The symbols for the curves are:
□ 0hr UPB; ■ 0hr ICM; ○ 21hr UPB; ● 40hr ICM.
From ref. 19, with permission.

several reaction centers are connected. There is about a five to ten-fold increase in domain size, and an apparent increase in the energy transfer efficiency during development.

REFERENCES

1. R. M. BROGLIE, C. N. HUNTER, P. DELEPELAIRE, R. A. NIEDERMAN, N.-H. CHUA, R. K. CLAYTON, Proc. Natl. Acad. Sci. USA, 77, 87 (1980)
2. C. N. HUNTER, R. A. NIEDERMAN, R. K. CLAYTON, in Proc. of the Vth Int. Congr. on Photosynthesis, edited by G. Akoyunoglou (Balaban Int. Science Services, Philadelphia, 1981), Vol. 3, pp. 539-545.
3. J. BRETON, A. VERMEGLIO, M. GARRIGOS, G. PAILLOTIN, in Proc. of the Vth Int. Congr. on Photosynthesis, edited by G. Akoyunoglou (Balaban Int. Science Services, Philadelphia, 1981), Vol. 3, pp. 445-459.
4. J. D. BOLT, K. SAUER, Biochim. Biophys. Acta, 546, 54 (1979)
5. R. VAN GRONDELLE, H. J. M. KRAMER, C. P. RIJGERSBERG, Biochim. Biophys. Acta, 682, 208 (1982)

6. H. J. M. KRAMER, R. VAN GRONDELLE, C. N. HUNTER,
   W. H. J. WESTERHUIS, J. AMESZ, Biochim. Biophys. Acta, 765,
   156 (1984)
7. M. H. TADROS, F. SUTER, G. DREWS, H. ZUBER,
   Eur. J. of Biochem., 129, 533 (1983)
8. H. ZUBER, R. A. BRUNISHOLZ, G. FRANK, P. FUGLISTALLER,
   W. SIDLER, R. THEILER, in Proc. of the Workshop on Molecular
   Structure and Function of Light-Harvesting Pigment - Protein
   Complexes and Photosynthetic Reaction Centers, Zürich,
   Switzerland, 1983, pp. 56-58.
9. R. VAN GRONDELLE, C. N. HUNTER, J. G. C. BAKKER, H. J. M.
   KRAMER, Biochim. Biophys. Acta. 723, 30 (1983)
10. J. D. BOLT, C. N. HUNTER, R. A. NIEDERMAN, K. SAUER,
    Photochem. Photobiol., 34, 653 (1981)
11. H. J. M. KRAMER, J. D. PENNOYER, R. VAN GRONDELLE,
    W. H. J. WESTERHUIS, R. A. NIEDERMAN, J. AMESZ,
    Biochim. Biophys., accepted for publication.
12. A. Yu. BORISOV, R. A. GADONAS, R. V. DANIELIUS,
    A. S. PISKARSKAS, A. P. RAZJIVIN, FEBS Lett., 138, 25 (1982)
13. A. P. RAZJIVIN, R. V. DANIELIUS, R. A. GADONAS, A. Yu. BORISOV, A. S. PISKARSKAS, FEBS Lett., 143, 40 (1982)
14. W. Th. F. DEN HOLLANDER, J. G. C. BAKKER, R. VAN GRONDELLE,
    Biochim. Biophys. Acta, 725, 492 (1983)
15. J. G. C. BAKKER, R. VAN GRONDELLE, W. Th. F. DEN HOLLANDER,
    Biochim. Biophys. Acta, 725, 508 (1983)
16. G. PAILLOTIN, C. E. SWENBERG, J. BRETON, N. E. CEACINTOV,
    Biophys. J., 25, 513 (1979)
17. R. A. NIEDERMAN, C. N. HUNTER, G. S. INAMINE, D. E. MALLON,
    Proc. Vth Int. Congr. on Photosynthesis, edited by G. A.
    Akoyunoglou (Balaban Int. Science Services, Philadelphia,
    1981), Vol. 5, pp. 663-674.
18. C. N. HUNTER, J. D. PENNOYER, R. A. NIEDERMAN, in Cell Function and Differentiation, part B, Biogenesis of Energy
    Transducing Membranes and Membrane and Protein Energetics
    (Alan R. Liss Inc., New York, 1982), pp. 257-265.
19. C. N. HUNTER, H. J. M. KRAMER, R. VAN GRONDELLE,
    Biochim. Biophys. Acta, 1984, accepted for publication.
20. C. N. HUNTER, R. VAN GRONDELLE, N. G. HOLMES, O. T. G. JONES,
    R. A. NIEDERMAN, Photochem. Photobiol., 30, 313 (1979)

5. ULTRAFAST PROCESSES AND TECHNIQUES

5.1 Generation, Propagation and Compression of Femtosecond Light Pulses

BERND WILHELMI

Sektion Physik der Friedrich-Schiller-Universität Jena, DDR-6900 Jena, German Democratic Republic

1. INTRODUCTION

Considerable progress has taken place in the last 16 years in the generation of ultrashort light pulses[1.0]. Until 1965 the duration of the shortest light pulses remained at about 1 ns despite of refined electronic techniques. Picosecond light pulses were first generated in 1965 by passive modelocking of a ruby laser[1.1] and one year later the same method was successfully applied to Nd-glass lasers[1.2], where pulse durations of only some picoseconds were achieved. Since that time several methods have been developed in order to modelocke various lasers and to generate ever shorter light pulses at wavelengths from the UV to the IR. In 1981 the first pulses with a duration of less than 0.1 picoseconds or 100 femtoseconds were generated by improvements of the passively mode-locked dye laser[1.3] using the colliding pulse ring laser configuration. Utilizing an additional dispersive element in such ring resonators for pulse compression, pulses as short as 50 femtoseconds were generated[1.4-1.9]. The shortest light pulses were obtained by amplifying the ultrashort light pulses from a passively modelocked dye ring laser and by passing the amplified pulses through a nonlinear optical element, which produces a chirp (frequency sweep), and through an appropriate dispersive device that compresses the chirped pulses.

The rapid development of femtosecond lasers and femtosecond light pulse diagnostics stimulated an enormous progress in the whole field of ultrafast measuring techniques. Ultrashort light pulses permit novel investigations of extremely rapidly proceeding physical and chemical phenomena on the femtosecond time scale. On the one hand, ultrashort light pulses allow new and deep insights into nature and in particular into the temporal evolution of some of the most fundamental processes in materials, most of which occur on a picosecond or subpicosecond time scale. Among the fundamental processes that have been measured are the free decay of molecular vibrations, electronic relaxation processes in small and large molecules as well as in solids, phonon decay and energy migration in solids, charge transfer processes and other nonradiative transfer processes. On the other hand the interaction of light pulses with condensed matter influences shape and duration of the light pulses. Moreover femtosecond technology provided new capabilities for the manipulation of photophysical and photochemical processes and for making extremely fast electro-optical components, e.g. switches, modulators and receivers.

In this report we describe after some preliminary remarks the compression of light pulses by propagating them through various types of optical media and through dispersive linear optical devices. On this basis we can explain the formation of pulses with frequency sweep and the compression of such phasemodulated light pulses. Finally we explain the generation of very short light pulses by passively modelocking and intracavity pulse conpression in laser resonators which contain resonant and nonresonant optical samples.

1.1. Preliminary Remarks on Ultrashort Light Pulses, Definitions and Notations

ULTRAFAST PROCESSES AND TECHNIQUE 427

According to Fourier's theorem the generation of ultrashot light pulses is connected with the emission of light in a comparatively broad spectral range where certain relations exist among the phases of all the Fourier components of the electric field.

Fourier Description of Light Pulses

Electric field strength E(t) and its Fourier transform $\underset{\sim}{E}(\omega)$

$$E(t) = \frac{1}{2\pi} \int_{-\infty}^{\infty} d\omega \underset{\sim}{E}(\omega) e^{i\omega t} = E^{(-)}(t) + E^{(+)}(t) \qquad (1.1)$$

where

$$E^{(-)}(t) = \frac{1}{2\pi} \int_{0}^{\infty} d\omega \underset{\sim}{E}(-\omega) e^{-i\omega t}$$

$$E^{(+)}(t) = \frac{1}{2\pi} \int_{0}^{\infty} d\omega \underset{\sim}{E}(\omega) e^{i\omega t}$$

$$E^{(-)}(t) = E^{(+)}(t)^* \qquad \text{complex analytical signal}$$

$$E(t) = E^{(-)}(t) + c.c.$$

Reverse transformations

$$\underset{\sim}{E}(\omega) = \int_{-\infty}^{\infty} dt\, E(t) e^{-i\omega t}$$

$$\underset{\sim}{E}^{(-)}(\omega) = \int_{-\infty}^{\infty} dt\, E^{(-)}(t) e^{-i\omega t} = \begin{cases} \underset{\sim}{E}(\omega) & \text{for } \omega \leq 0 \\ 0 & \text{for } \omega > 0 \end{cases} \qquad (1.2)$$

Light Pulse of Carrier Frequency ω_L Time Domain:

$$E(t) = \frac{1}{2} \mathcal{E}(t) e^{+i\omega_L t} + c.c. \qquad (1.3)$$

$$\mathcal{E}(t) = |\mathcal{E}(t)| e^{i\phi(t)} \qquad \text{complex temporal amplitude} \qquad (1.4)$$

(slowly varying pulse envelope)

$|\mathcal{E}(t)| = A(t)$ temporal amplitude

$\phi(t)$ temporal phase

$$J(t) = c\varepsilon_o \overline{E^2(t)}^T = 2c\varepsilon_o E^{(-)}(t) E^{(+)}(t) \quad \text{instantaneous intensity}$$
(instantaneous power per area) (1.5)

$$J(t) = c\frac{\varepsilon_o}{2} A^2(t)$$

$I(t) = J(t)/\hbar\omega_L$ instantanious photon flux density (1.6)

τ_L pulse duration (FW HM of I(t))

Frequency Domain:

$$\underset{\sim}{E}(\omega) = a(\omega) e^{+i\varphi(\omega)} \tag{1.7}$$

$a(\omega)$ spectral amplitude

$\varphi(\omega)$ spectral phase

Pulse energy (per area) $\mathcal{E}_L$

$$\mathcal{E}_L = \int_{-\infty}^{\infty} dt\, J(t) = 2c\varepsilon_0 \int_{-\infty}^{\infty} dt\, E^{(-)}(t) E^{(+)}(t) = \frac{\varepsilon_0 c}{2} \int_{-\infty}^{\infty} dt\, A^2(t) \tag{1.8}$$

Parseval's Theorem:

$$\mathcal{E}_L = \int_0^{\infty} d\omega\, \tilde{J}(\omega) \tag{1.9}$$

$\tilde{J}(\omega) = \frac{2\varepsilon_0 c}{2\pi} |\underset{\sim}{E}(\omega)|^2 \equiv \frac{\varepsilon_0 c}{\pi} a^2(\omega)$ spectral intensity of the light pulse (recorded by a spectrometer without time resolution)   (1.10)

$\Delta\omega$ spectral pulse width (FWHM of $\tilde{J}(\omega)$).

Time-bandwidth Product

$$\gamma \equiv \frac{\Delta\omega}{2} \tau_L \geqslant c_B \tag{1.11}$$

c_B constant of the order of unity whose value depends on the shape of the intensity profile

$\gamma = c_B$ for bandwidth-limited (transform-limited) pulses (shortest pulses for given bandwidth)

In linear optics $J(t)$ and $\tilde{J}(\omega)$ appear as the directly measurable parameters of light pulses. There exist no unique relation between $J(t)$ and $\tilde{J}(\omega)$ (because of the phases in the field strengths, which are lost in building the square of the modulus).

Light Pulses with Frequency Sweep ("Chirped" Pulses)

Example: Gaussian laser pulse with linear frequency sweep (see Fig. 1.1)

ULTRAFAST PROCESSES AND TECHNIQUE 429

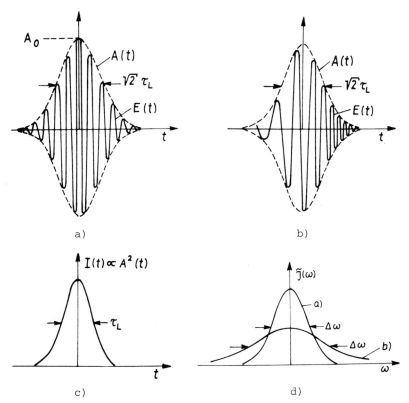

FIGURE 1.1
a) electric field strength of a bandwidth-limited pulse with Gaussian profile
b) electric field strength of a "chirped" pulse with Gaussian profile
c) instantaneous intensity of the pulses from a) and b)
d) spectral intensity of the Gaussian pulses from a) and b)

Electric field strength

$$E(t) = \frac{1}{2} A_o e^{i\phi_o} e^{-\gamma t^2 + i\beta t^2} e^{i\omega_L t} + \text{c.c.} \quad (1.12)$$

$\tau_L = \sqrt{\frac{2 \ln 2}{\gamma}}$ pulse duration (FWHN of $I(t)$)

βt^2: special phase modulation that describes a frequency shift linearly increasing in time (linear frequency sweep or "chirp")

430 LASERS IN ATOMIC, MOLECULAR AND NUCLEAR PHYSICS

The half width $\Delta\omega$ of the spectral intensity is obtained as

$$\Delta\omega = 2\sqrt{2\ln 2 \left(\frac{\gamma^2 + \beta^2}{\gamma^2}\right)} \qquad (1.13)$$

Under the condition $\phi(t) = \phi_0 + \beta t^2 = \text{const}$ ($\beta = 0$) we obtain

$$\wp = c_B^{Gauss} = 0.441 \qquad (1.14)$$

$\wp$ increases with increasing β^2. Pulses with constant phase are denoted as bandwidth-limited or Fourier-limited.

2. PROPAGATION OF LIGHT PULSES THROUGH NONRESONANT OPTICAL SAMPLES

2.1. Basic Equations

We start from the wave equation

$$-\frac{\partial^2}{\partial z^2} E + \frac{1}{c^2}\frac{\partial^2}{\partial t^2} E = -\mu_0 \frac{\partial^2}{\partial t^2} P \qquad (2.1a)$$

where $P = P^L + P^{NL}$ (2.1b)

is the polarization, which is composed of the linear term P^L and the nonlinear term P^{NL}. (Note that we restrict our consideration to plane wave propagating in the z direction).

In the frequency domain we use the relation

$$\underline{P}^L(\omega) = \varepsilon_0 \underline{\chi}(\omega) \underline{E}(\omega) =$$

$$= \varepsilon_0 \left[\underline{\chi}(\omega_L) + \underline{\chi}'(\omega_L)(\omega - \omega_L) + \frac{1}{2}\underline{\chi}''(\omega_L)(\omega - \omega_L)^2 + \ldots\right] E(\omega) \quad (2.2a)$$

between linear polarization and electric fieldstrength where we expanded the linear susceptibility at the center frequency ω_L. The expansion coefficients are given by the wave number

$$k_L = \frac{\omega_L}{c}\sqrt{1 + \underline{\chi}(\omega_L)} = \frac{\omega_L}{c} n_0(\omega_L) \qquad (2.2b)$$

the group velocity

$$v_L = \left(\frac{\partial \omega}{\partial k}\right)_L \qquad (2.2c)$$

and the reciprocal group velocity dispersion

$$k_L'' = \left(\frac{\partial^2 k}{\partial \omega^2}\right)_L = \frac{\lambda_L^3}{2\pi c^3}\left(\frac{d^2 n}{d\lambda^2}\right)_L \qquad (2.2d)$$

Discribing fieldstrength and polarization by the concept of carrier frequencies and pulse envelopes according to

$$E(t,z) = \frac{1}{2}\bar{E}(t,z)e^{i(\omega_L t - k_L z)} + c.c. \qquad (2.3a)$$

$$P^{L,NL}(t,z) = \frac{1}{2}\bar{P}^{L,NL}(t,z)e^{i(\omega_L t - k_L z)} + c.c. \qquad (2.3b)$$

we obtain for the Fourier-transform of (2.2a)

$$\bar{P}^L(t,z) = \varepsilon_o \chi(\omega_L)\bar{E} - i\varepsilon_o \chi'(\omega_L)\frac{\partial}{\partial t}\bar{E} - \frac{1}{2}\varepsilon_o \chi''(\omega_L)\frac{\partial^2}{\partial t^2}\bar{E} + \ldots (2.3c)$$

which is substituted in the wave equation (2.1a). Provided both the slowly varying envelope approximation (SVEA) and the neglect of the dispersion of the nonlinear source term are justified we obtain

$$\left(\frac{\partial}{\partial z} + \frac{1}{v_L}\frac{\partial}{\partial t}\right)\bar{E} - \frac{ik''}{2}\frac{\partial^2 \bar{E}}{\partial t^2} = -\frac{i\mu_o \omega_L^2}{2k_L}\bar{P}^{NL} \qquad (2.4)$$

We consider now the lowest-order nonlinear effect by which the wave of carrier frequency ω_L influences its own propagation in the sample. For nonresonant interaction the third-order nonlinearity is represented by

$$\bar{P}^{(3)}(t,z) = 2\varepsilon_o n_o \tilde{n}_2 |\bar{E}(t,z)|^2 \bar{E}(t,z) \qquad (2.5a)$$

where

$$n(t,z) = n_o + \tilde{n}_2 \bar{E}(t,z)^2 \quad \text{and} \quad n_2|\bar{E}|^2 \ll n_o \qquad (2.5b)$$

Using the abbreviation

$$\tilde{\varkappa} = \frac{k_L \tilde{n}_2}{n_o} \qquad (2.6a)$$

and the coordinate transformation

$$\eta = t - z/v_L, \quad \xi = z \qquad (2.6b)$$

in (2.4) we obtain

$$i\frac{\partial}{\partial \zeta}\bar{E} = -\frac{k''}{2}\frac{\partial^2}{\partial \eta^2}\bar{E} + \tilde{\mathcal{H}}|\bar{E}|^2\bar{E} \qquad (2.7)$$

This equation, which is often referred to as the nonlinear Schrödinger equation, decribes the distortion of light pulses under the action of group velocity dispersion and of nonresonant, nonlinear polarization. Note that one can employ this equation not only when dealing with plane waves, for which we derived it, but also when calculating wave propagation in polarization preserving single mode fibers; in this case we have only to substitute the bulk parameters k'' and $\tilde{\mathcal{H}}$ by the adequate fiber parameters[2.1].

Let us first study two limiting cases, namely that of dispersive linear optical samples and that of nondispersive nonlinear optical samples.

2.2 Dispersive Linear Optical Samples

With neglect of the nonlinear term in (2.7) the general solution is

$$\bar{E}_L(\zeta,\eta) = \frac{1-ik''/|k''|}{2\sqrt{\pi|k''|\zeta}} \int_{-\infty}^{\infty} d\eta' \bar{E}_{LO}(\eta') e^{\frac{i(\eta-\eta')^2}{2k''\zeta}} \qquad (2.8)$$

where $\bar{E}_{LO}(t) \equiv \bar{E}_L(\zeta=0, \eta \equiv t)$ is the field strength envelope at the boundary. For the particular case of a phase modulated Gaussian entrance pulse with field strength

$$E_{LO}(t) = \frac{1}{2}|\bar{E}_{mo}|e^{-2\ln2(t/\tau_{LO})^2} e^{+i(\beta_o t^2 + \varphi_{LO})} e^{+i\omega_L t} + c.c. \qquad (2.9a)$$

and with instantaneous frequency

$$\omega(t) = \omega_L + \frac{d}{dt}\Delta\varphi(t) = \omega_L + 2\beta_o t \qquad (2.9b)$$

(i.e. a pulse with linear frequency sweep, constant chirp) we obtain from (2.8)

$$\bar{E}_L(\xi,\eta) = |\bar{E}_{mo}|\exp[-2\ln2(\eta/\tau_L)_e^2 - i2\ln2(\eta/\tau_L)^2[4\ln2k''\xi/\tau_{Lo}^2 + \beta_o\tau_{Lo}^2(1+2k''\beta_o\xi)/2\ln2]]/\sqrt{2k''\xi i[2\ln2/\tau_{Lo}^2 - i\beta_o]+1}$$

where the pulse duration τ_L evolves as

$$\frac{\tau_L(\xi)}{\tau_{Lo}} = \sqrt{1 + 2\beta_o k'' L_\beta\left[1-(1-\xi/L_\beta)^2\right]} \quad (2.9c)$$

and where

$$L_\beta = -\frac{2\beta_o\tau_{Lo}^4 k''}{(4\ln2k'')^2 + (2\beta_o\tau_{Lo}^2 k'')^2} \quad (2.9d)$$

is a characteristic length. (Note the space-time analogy in comparing (2.9) with results of diffraction theory. This analogy was first investigated by Akhmanov and Khokhlov[2.2].) For $L_\beta > 0$, which requires $k''\cdot\beta_o < 0$ the pulse is shortened on its path from $\xi = 0$ to $\xi = L_\beta$ (see. Fig. 2.1). At this point the pulse is bandwidth-limited, and for $\xi > L_\beta$ the pulse duration increases. Thus chirped entrance pulses can be compressed by linear optical means (down chirp requires $k'' > 0$ and up chirp $k'' < 0$). At $\xi = L_\beta$ the pulses are bandwidth-limited. (Note that only the linear frequency sweep has been compensated. Nonlinear frequency sweep needs more complicated dispersive elements in order to be compensated.) Bandwidth limited Gaussian input pulses double their duration over a propagation length, the so-called dispersion length,

$$L_D \approx 0.6\frac{\tau_{Lo}^2}{k''} \quad (2.9e)$$

The total electric field strength is given by

$$E(\xi,\eta) = \frac{1}{2}\bar{E}(\xi,\eta)e^{i(\omega_L t - k_L z)} + c.c. =$$
$$= \frac{1}{2}|\bar{E}_{mo}|\sqrt{\frac{\tau_{Lo}}{\tau_L}}e^{-2\ln2(\eta/\tau_L)^2}e^{i\tilde{\omega}(\xi,\eta)} + c.c. \quad (2,10a)$$

434 LASERS IN ATOMIC, MOLECULAR AND NUCLEAR PHYSICS

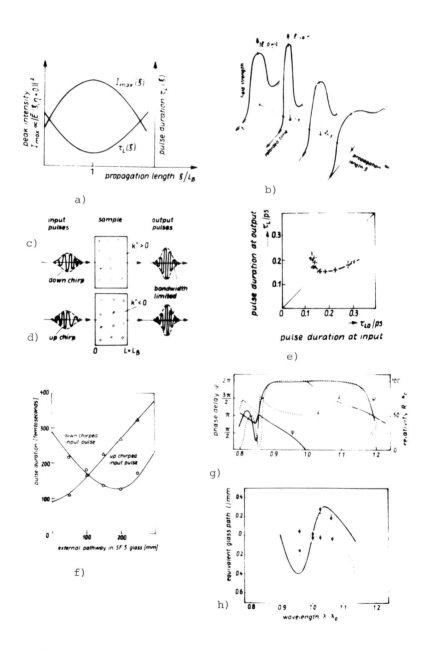

FIGURE 2.1 Pulse shaping in dispersive media

where $\tilde{\omega}(\xi,\eta) = \omega_L + 2\beta(\xi)\eta$ (2.10b)

is the instantaneous frequency

$$\beta(\xi) = \frac{(2\ln 2)^2 k''(\xi - L_\beta)}{\tau_L^2(\xi)\,\tau_L^2(L_\beta)}$$ (2.10c)

is the chirp parameter.

Fig 2.1a shows the dependence of pulse duration τ_L and maximum intensity I_{max} on the pathlength ξ in the sample. Fig. 2.1b presents the pulse evolution in space and time and Fig. 2.1c,d demonstrate the generation of bandwidth-limited pulses for down-chirped and up-chirped input pulses. Fig. 2.1e presents the shaping of down-chirped femtosecond pulses ($\lambda = 0.61\,\mu$m) in a BK5 glass sample if 17 cm length[1.4]. The long entrance pulses ($\tau_{Lo} > 0.17$ ps) experience compression, whereas very short pulse become lengthened because of $\xi > L_\beta$. Fig. 2.1f gives experimental results on the compression of down-chirped femtosecond pulses where $\tau_{Lo} = 260$ fs in BK7 glass[1.8]. We observe qualitatively the same dependence on the pathlength as depicted in Fig. 2.1a.

Up to now we considered chirp production and compensation in spatially homogeneous, linear optical samples, e.g. in glass rods. Furthermore we explained that grating devices may be used instead of such camples. Let us now discuss the chirp generation by dielectric multilayer mirror. The electric field $\underline{E}_r(\omega)$ reflected from a mirror is connected to the incident wave $\underline{E}_i(\omega)$ by

$$\underline{E}_r(\omega) = \underline{E}_i(\omega)\sqrt{R(\omega)}\,e^{i\varphi(\omega)}$$

where $R(\omega)$ is the reflectivity and $\varphi(\omega)$ the phase shift. Assuming a dielectric multilayer mirror with the optical thickness of all layers being $\lambda_o/4$ (λ_o resonance wavelength of the mirror, reflectivity and phase shift have been calculated for several numbers of layers and several angles of incidence[2.3, 2.4]. Fig. 2.1g shows results that refer to a standard multilayer stack of nineteen $\lambda_o/4$ layers (dashed lines) and to a stack of eightteen $\lambda_o/4$ layers covered by one $\lambda_o/2$ layer (full lines). Note that the phase shift φ depends nonlinearly on the wavelength λ. The chirp is here caused (or varied) by a nonlinear frequency dependence of the phase. The chirp originating from reflection at the multilayer stack is equivalent to the chirp produced in passing through dispersive material of length l if

$$(d^2\varphi/d\omega^2)_L = (d^2k/d\omega^2)_L \, l$$

Thus we obtain for the "equivalent glass pathlength" l on the wavelength scale

$$l = \frac{1}{2\pi}[2\frac{d\varphi}{d\lambda} + \lambda\frac{d^2\varphi}{d\lambda^2}][\frac{d^2n}{d\lambda^2}]^{-1}$$

Using $d^2n/d\lambda^2$ for quartz glass SQ1 and the calculated phase shift $\varphi(\lambda)$ as depicted in Fig. 2.1g we get the equivalent pathlengths l of the mirrors shown in Fig. 2.1h.

From these figures it is obvious that the laser mirrors produce up chirp (l < 0) or down chirp (l > 0) in dependence on the layer system used and on the laser wavelength λ with respect to the resonance wavelength λ_o of the mirror. For mirrors with the optical thickness of all layers being $\lambda_o/4$ the chirp is relatively small near the mirror resonance λ_o while it increases rapidly at the edge of the reflectivity band. Independently on the number of layers the blue side of the relectivity band produces up chirp, the red side down chirp. The chirp increases with increasing number of layers. For instance, for a number of layers equal to 13 and 23 at λ/λ_o =0.9 the equivalent glass length l is ~ 0.2 mm and ~ 0.6 mm, respectvely. The operation at one of the two edges of the reflectivity band, which may be employed to obtain sufficient transmission, is connected with a strong variation of losses and phase with wavelength. More advantageously output coupling can be achieved by reducing the number of layers or by increasing the thickness of the first layer from $\lambda_o/4$ to $\lambda_o/2$. In the latter case Fig. 2.1g shows that the variation of l vs λ/λ_o becomes stronger near the mirror resonance λ_o.

To check the computed results experimentally the equivalent glass length l has been measured for two sets of similar high-quality mirrors with various values of λ_o, which were all manufactured under comparable conditions especially for this purpose[2.4]. With the aim to measure the equivalent glass path we used an intracavity measurement technique, where the chirp of light pulses can be compensated for by variation of the length of a glass path of length $L = L_o + l$. The accuracy of this intracavity method is about 10 μm.

2.3 Dispersionfree Nonlinear Optical Samples

With neglect of group velocity dispersion ($k'' = 0$)

$$\bar{E}(\xi,\eta) = \bar{E}_o(\eta) e^{-i\tilde{\mathcal{H}}|E_o(\eta)|^2 \xi} \tag{2.11a}$$

is a solution of (2.7). Since $\tilde{\mathcal{H}}$ is real in loss free media, the modulus of the field strength amplitude remains constant whereas the temporal phase of the pulse changes in passing through the sample. The phase changes significantly on propagation lengths of the order of

$$L_{NL} = 1/(\tilde{\mathcal{H}}|E_{om}|^2) = \tilde{n}_o/(n_2 k_L |\bar{E}_{om}|^2) = n_o/n_2 J_m k_L \tag{2.11b}$$

Near the pulse maximum we obtain with

$$|\bar{E}_o|^2 = |\bar{E}|_m^2 + \frac{1}{2}(\frac{\partial^2}{\partial \eta^2}|\bar{E}_o|^2)_m \eta^2 \tag{2.12}$$

an output pulse with linear frequency sweep (constant chirp) where the chirp parameter β is given by

$$\beta = \frac{1}{2}\tilde{n}_2(-\frac{\partial^2}{\partial \eta^2}|E_o|^2)_m \xi \tag{2.13}$$

This means that the nonlinearity gives rise to a frequency chirp, which linearly increases with the path length ξ in the sample.

438 LASERS IN ATOMIC, MOLECULAR AND NUCLEAR PHYSICS

In most cases there holds $\tilde{n}_2 > 0$ for the electronic contribution to the nonlinearity, and hence positive values of β result, which corresponds to an up chirp, or, in other words to an increase of frequency with increasing time. From the consideration in 2.2 it can be inferred that it is possible to eliminate such a chirp and to achieve a shortening when allowing the pulses the pass through a linear optical sample with $k" < 0$. We see that in this way pulse compression can be achieved in arrangements of separate nonlinear and linear optical elements[2.5-2.7]. (Note that (2.12) is only an approximation of the pulse shape near the maximum. The actual pulse shape causes contributions to the instantaneous frequency that depend nonlinearly on time (see Fig. 2.2a). As explained in 2.2 such contributions can not easily compensated by using linear optical samples with group velocity dispersion. Let us consider, e.g., a Gaussian shaped pulse. Near the maximum we obtain approximately a constant up-chirp, whereas the chirp varies significantly and even charges its sign in the wings. Passing such a phase-modulated pulse through a linear optical sample with $k" < 0$ (or an equivalent grating device) its central part experiences compression and in addition secondary pulses appear at the position of the wings of the input pulse, see Fig. 2.2b).

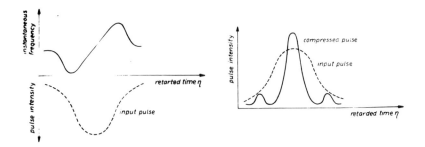

FIGURE 2.2 Chirp generation by phase modulation (a) and compression of compressed pulses

ULTRAFAST PROCESSES AND TECHNIQUE 439

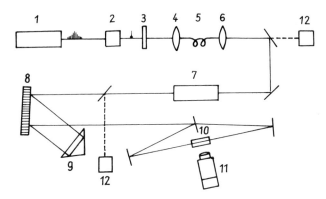

FIGURE 2.3a   Experimental set-up
                1 Nd phosphate glass laser;   8 gtating;
                2 single pulse selection;   9 prism;
                3 neutral filter;   10 TPF set-up;
                4,6 microobjectives   11 OMA2 vidicon;
                5 graded core index fiber;   12 spectrograph.
                7 amplifier;

Let us describe an experiment (see Fig. 2.3a), where pulses from a passively modelocked Nd-glass laser (τ_L = 5 ps, $\Delta \nu/c$ = 5 cm^{-1}, $\tilde{\mathcal{E}}_L$ = 2 μJ) have been used as input pulses$^{2.8}$. The input pulses are almost bandwidth-limited ($\tau_L \Delta \nu$ = 0.75). By use of a microscope objective (f = 35 mm) the pulses are coupled into a short piece (L = 40 cm) of graded-core index fiber. Fibers offer the advantage of small beam diameter in the whole sample. We employed graded-core index fibers (GCF) instead of single-mode fibers (SMF) because the GCF supplies the advantage of comparatively large core diameter, which allows to couple into the fiber about ten times the energy that would have been possible when using SMF. In our case of a rather short fiber and wavelength near the zero-dispersion wavelength (k" = 0), which is about 1.3 μm in typical silica fibers, we can neglect the influence of group velocity dispersion. In the fiber there arise chirped pulses due to the nonlinear effect as described in 2.2. The chirped pulses are then amplified. A bandwidthlimited pulse will always be broadened in time by passage through an optical element with a finite bandwidth (e.g. dilter, amplifier), whereas a chirped pulse can be shortened as it has been experimentally demonstrated by Ippen and Shank$^{2.17}$. Thus due to the high amplification necessary in our case to compensate losses and to provide suitable energies for measurements the effective bandwidth of the amplifier (about 5 nm) will already lead to weakly shortened pulses. It should be noted however that the finite amplifier bandwidth limits the usable fiber length since only part of the broadened pulse spectrum will be amplified.

The amplified pulse is then compressed in a dispersive delay line consisting of a holographic grating with 651 lines per mm and a right angle prism. The grating was blazed for light of a wavelength of 1020 nm, the angle of incidence was 2° and the first order diffraction angle amounted to 40°. The delay length could be varied between 12.5 and 95 cm. A decrease of the pulse duration with increasing delay length could be measured up to an optimum of the delay length L, beyond this value the pulse broadens again due to GVD (Fig. 2.3b).

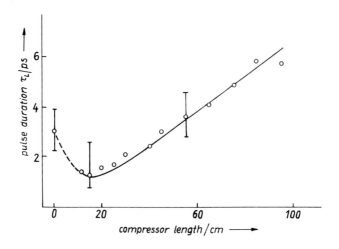

FIGURE 2.3b Duration of compressed pulses versus delay length

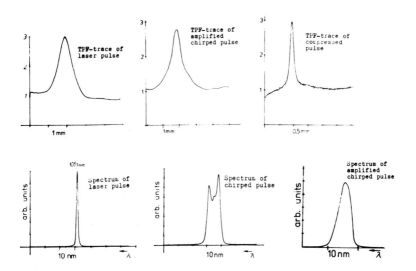

FIGURE 2.3c TPF traces and spectra of pulses

A typical TPF-trace of a compressed pulse is depicted in Fig. 2.3c, having a length of 0.14 mm, measured by imaging onto the OMA-vidicon. The spatial resolution of this system was measured to be 0.1 mm as it was obtained by registrating the image of a keen edge. Under consideration of the spatial resolution and of the refractive index of the solution RH 6G in ethylenglycol (n = 1.435) a Gaussian shaped pulse corresponding to the TPF-trace would have a duration of 0.7 ps. Note that satellite pulses, which may appear in pulse compression, have here been supressed by frequency filtering during amplification. (This becomes apparent from the pulse spectra taken in front of and after the amplifier.)

2.4 Dispersive Nonlinear Optical Samples

In many experimental situations we can neither neglect the nonlinearity (as in 2.2) nor the group velocity dispersion (as in 2.3) and thus we have to solve the full "nonlinear Schrödinger equation" (2.7). In dependence on the fact whether $k''n_2$ is positive or negative we obtain defferent types of solution.

2.4.1 Formation of "Rectangular" Pulses with Linear Frequency Sweep

In many experiments conditions are somewhat different from those assumed in our previous consideration in order to achieve optimum compression. In addition to the central part of the pulse where the frequency sweep can be approximated by a linear function of η, the pulse in composed of wings where the chirp exhibits another functional dependence on η. Therefore in dispersionfree nonlinear optical samples the wing parts will not be compressed together with the central part but satellite pulses occur before and after the main pulse. Care must be taken to minimze the power content of these satellites and to achieve maximum compression. For this purpose the combined action of linear dispersion and nonlinearity in the optical fiber has to be taken into account[2.14]. Let at the pulse wavelength the dispersion parameter k" be positive; in connection with $\tilde{n}_2$ being positive too this means that the pulse becomes longer in passing through the nonlinear dispersive medium, see Fig. 2.4b. After the pulse has passed an appropriate distance[2.14]

$$\xi \gg 0.6\, \tau_{Lo} / \sqrt{\widetilde{æ} k'' \bar{E}_{mo}} \tag{2.14a}$$

in the sample it attains a nearly rectangular temporal shape with the pulse duration

$$\tau_{Lr} \approx 2.9 \sqrt{k_o k'' \tilde{n}_2 / n_o}\; \bar{E}_o\, \xi \tag{2.14b}$$

the field strength

$$\bar{E}_r = 0.6 \sqrt{\bar{E}_o \tau_{Lo} / (\sqrt{\widetilde{æ} k'''}\, \xi\,)} \tag{2.14c}$$

and the full frequency sweep

$$\Delta\omega_r = 1.4 \sqrt{\frac{\widetilde{æ}}{k''}}\, \bar{E}_{mo}\left(1 - 0.28 \frac{\tau_{Lo}}{\sqrt{\widetilde{æ} k'' \bar{E}_{mo}}\, \xi}\right) \approx 1.4 \sqrt{\frac{\widetilde{æ}}{k''}}\, \bar{E}_{mo} \tag{2.14d}$$

where the instantaneous frequency increases linearly with time over nearly the whole pulse length. (For very long path lengths,

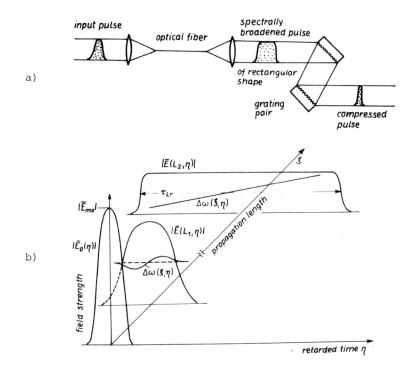

FIGURE 2.4 a) Schematic of the experimental set-up
b) Evolution of pulse shape and frequency sweep

$\xi > 6\tau_{Lo}/\sqrt{\tilde{\mathcal{H}}k''}\,\bar{E}_{mo}$, the chirp does not increase significantly and the only effect is a further lengthening of the pulse). Most of the pulse energy can afterwards be compressed into one very short single pulse, where the final duration is

$$\tau_L \approx 2\sqrt{k''/\tilde{\mathcal{H}}}\,/\,\bar{E}_{mo} \qquad (2.15)$$

This means that the ultimate pulse duration decreases with increasing fieldstrength of the input pulse and is independent of the duration of the latter pulse. In other words, the compression factor τ_{Lo}/τ_L increases with the pulse duration of the input pulse. 95% of the maximum attainable compression factor can already be obtained at

TABLE 2.1 Comprizes some experimental results

	input pulse		fiber		output pulse	compression factor		reference
	τ_{L0} [fs]	peak power [W]	core diameter [μm]	length [m]	τ_L [fs]	exp.	theor.	2.14
Shank	90	$7 \cdot 10^3$	3.3	0.15	30	3	3	2.9
Ippen	65	10^5		0.008	16	4	10	1.10
Grischkowski	$6 \cdot 10^3$	100	4	30	600	10	11	2.10
Nikolaus	$5.4 \cdot 10^3$	$0.6 \cdot 10^3$	4	30	450	12	28	2.11
Nakatsuka	$5.5 \cdot 10^3$	10	4	70	$1.5 \cdot 10^3$	3.7	3	2.12
Johnson	$33 \cdot 10^3$	240	3.8	105	410	80	700	2.13
Damm	$5 \cdot 10^3$	$4 \cdot 10^5$	50 (GCF)	0.4	700	7		2.8

$$\xi = 5.6 \tau_{Lo} / \sqrt{\tilde{\mathcal{H}} k''} \bar{E}_{mo} \qquad (2.16)$$

Using (2.15) we can now estimate the ultimately achievable pulse duratuon. In typical single-mode fibers (for wavelengths of about 600 nm) we have $\tilde{n}_2 \approx 1.5 \cdot 10^{-22}$ m^2/V^2, $\tilde{\mathcal{H}} \approx 6.5 \cdot 10^{-16}$ m/V^2 and $k'' \approx 6.5 \cdot 10^{-26}$ s^{2}m^{-1}. The maximum possible fieldstrength $\bar{E}_{mo}$ of the input pulse is estimated to be about 10^{10} V/m, i.t., about one tenth of the inneratomic fieldstrength, near which the expansion of the nonlinear polarization breaks down. For higher fieldstrength the phase modulation becomes so complicated that compression is impossible (this fieldstrength corresponds to the maximum intensity of 10^{13} W/cm^2). Using these data we obtain for the ultimately possible pulse duration $(\tau_L)_{min} \approx 2$ fs. This value roughly corresponds to one cycle of the light wave. But with these fieldstrength the application of (2.7) is already questionable. Table 2.1 comprizes some experimental results.

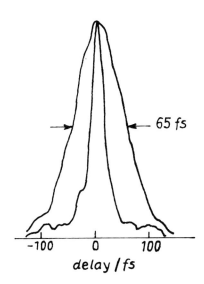

FIGURE 2.5 SHG-autocorrelation trace of the input pulse and the compressed pulse$^{1.10}$

Fig. 2.5 presents the autocorrelation trace of the shortest optical pulse ever obtained until now (τ_L = 16 fs)[1.10]. It consists of only 8 cycles of the optical field.

2.4.2 Formation of Solitons, Soliton Lasers

Let us now discuss interesting solutions of (2.7) for $k''\tilde{n}_2 < 0$. With $\tilde{n}_2 > 0$ this requires $k'' < 0$, which occurs in many transparent materials at long wavelengths due to the influence of IR absorption bands (see Rig. 2.6).

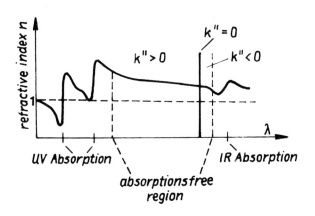

FIGURE 2.6 Refractive index versus wavelength

Now the chirp generated by the nonlinear process can simultaneously be compensated for by the linear optical process. In particular, in appropriate materials bandwidth-limited input pulses can even shortened under certain conditions. Moreover pulses can even propagate over long path length without changing their shape. The propagation of sech-shaped laser pulses has been investigated where the pulse envelope is given as[2.15]

$$\bar{E}_o(\eta) = \frac{a}{\tau}\sqrt{k''/\tilde{x}}\,\text{sech}(\eta/\tau), \quad \tau_L = 1.76\,\tau, \qquad (2.17)$$

a is dimensionless amplitude factor. Numerical evaluation of (2.17) reveals the interesting phenomenon that input pulses with a = 1 experience distortionless propagation through the sample. Pulses exhibiting this property are called solitary pulses. For a being a natural number >1 there appear solitary solutions (solitons) of a more complex type. The passage of these pulses is not distortionless throughout, but the pulses reproduce their shape only after the periodicity length

$$\xi_o = \frac{\pi}{2} |\tau^2/k''| \qquad (2.18)$$

which is of the order of magnitude of L_D (2.9e). Within the periodicity interval the pulses are split up, which is shown in Fig. 2.6.

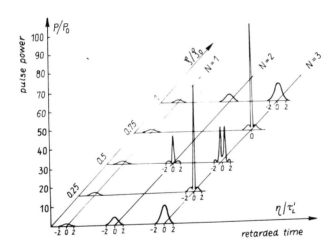

FIGURE 2.7 Soliton propagation

These calculations could be experimentally confirmed by Mollenauer, Stolen and Gordon[2.15]. Recently Mollenauer and Stolen[2.16] constructed a soliton laser where a single mode fiber is used in an additional feed back loop (see Fig. 2.8a).

The soliton laser consists of a synchronously pumped color center laser, which is tunable from 1.4 through 1.6 μm. Some part of the laser output is coupled into a polarization-preserving single

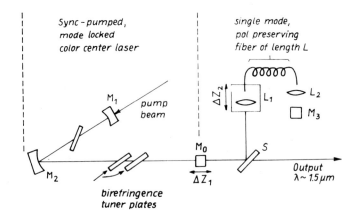

FIGURE 2.8a Scematic of a soliton laser

mode fiber from where it is retroreflected into the laser cavity. If the length of this feedback loop is an integral multiple of the main cavity length, the reflected pulse will coincide with the pulses already present in the cavity. Without feedback loop the laser pulse duration is about 8 ps. With feedback loop the pulses become shorter. When switching on the laser, the initially broad pulses are narrowed in passing the fiber. The narrowed pulses are coupled back into the main cavity and force the laser to produce shorter pulses; finally the pulses become soliton-shaped, and the profile of the laser output pulses coincides with that of the re-injected pulses. The experiment demonstrate (see the relationships between pulse duration, fiberlength and peak power in Fig. 2.8b) that the (N = 2) solitons are generated. The shortest pulses obtained so far by such lasers have a duration of about 200 fs.

3. PROPAGATION OF LIGHT PULSES THROUGH RESONANT OPTICAL SAMPLES

3.1 Exact Resonance

In exact resonance and for τ_L being long compared with the phase decay time τ_{21} the light pulse experiences no phase modulation when passing the sample. Fig. 3.1 gives the result of pulse shaping

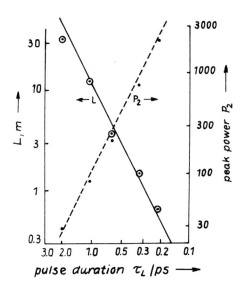

FIGURE 2.8b Fiber length L and peak power P_2 at input to fiber versus pulse duration τ_L. Straight lines: theoretical results for N = 2 solitons; points: experimental one.

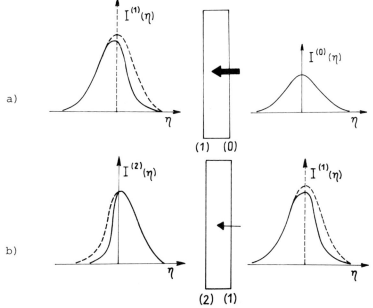

FIGURE 3.1 Temporal profile of the light pulse at the entrance and the exit of the gain medium (a) and the saturable absorber (b).

under these conditions$^{1.0}$.

3.2 Near Resonance

For deviations of the carrier frequency ω_L of the light pulse from ω_{21} and broad pulse spectra the light pulses experience phase modulation when passing absorbers or amplifiers.

Under suitable conditions linear frequency sweep can be obtained in the center of the pulse. The main effect can be explained by saturation of dispersion, which originates from saturation of absorption (see Fig. 3.2)

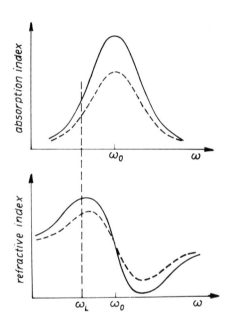

FIGURE 3.2 Idealized absorption profile and the correspunding dispersion curve before (—) and after (---) the passage of the pulse

We calculate such frequency chirp following references 3.1 and 3.2 (see Fig. 3.3). The chirp depends on the detuning $(\omega_L - \omega_{21})$, on the saturation and on the ratio τ_L/τ_{21}. Such chirped pulses can be compressed as explained in Section 2.

ULTRAFAST PROCESSES AND TECHNIQUE   451

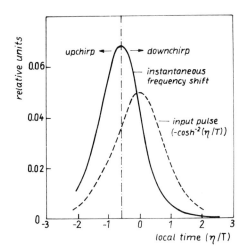

FIGURE 3.3   Chirp arising from saturable absorber

3.3 Chirp Generation and Chirp Compression in Passively Modelocked Lasers

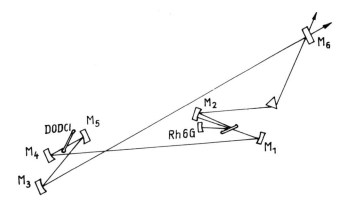

FIGURE 3.4   Scheme of a ring laser

Fig. 3.4 shows the scheme of a ring laser used for passive modelocking. On each round trip the pulses pass through the amplifying sample, the saturable absorber, possible frequency filters and dispersive samples and it is reflected at several mirrors. These optical samples effect the modulus and the phase of the fieldstrength of the light pulses as explained in Section 2 and Subsections 3.1 and 3.2 (cf. [1.3]-[1.10]). In [3.2] and [3.3] these effects have been calculated. The minimum pulse duration is obtained when the chirp that arises by passing the pulse through the saturable absorber and the amplifier and reflecting it at mirrors is compensated by a glass sample of appropriate length. (The glass pathlength is varied by changing the position of one or two prisms in the cavity (Two prisms are used to compensate for the angular dispersion of the prisms.).) Thus the generation of very short and bandwidthlimited light pulses can be explained. The shortest pulses appear, when the pathlength needed for compensation is maximum (see Fig. 3.5).

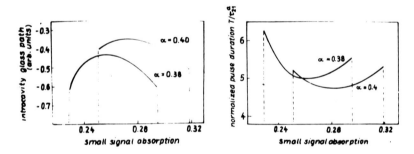

FIGURE 3.5 Intracavity path length of glass needed for chirp compensation (a) and pulse duration (b) versus small signal absorption of the saturable absorber (α small signal gain of the amplifier, $T = \mathcal{T}_L/1.76$)

Fig. 3.6 shows some experimental results on intracavity chirp compensation$^{3.2}$. By use of extracavity pulse compression (cf. Section 2) it can be checked wether linear frequency sweep appears in the output pulses or not. These measurements indicate that the pulses have no linear frequency sweep and are almost bandwidth-limited for optimum intracavity chirp compensation. By undercompensation and overcompensation of chirp inside the cavity down--chirped pulses and up-chirped pulses can be obtained, respective-

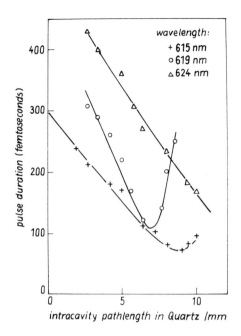

FIGURE 3.6 Intracavity chirp compensation

ly. (Note that the light pulses, which were used as the input pulses in Fig. 2.f had been generated in this way.)

Finally we would like to emphasize that the light pulses propagating inside the cavity have soliton-like properties. The field-strength reproduces itself after each round trip with respect to modulus and phase. The treatment of the CPM regime (colliding pulse modelocking) in such set-ups, where the two counterpropagating pulses meet in the saturable absorber, show that these pulses even remain stable in collisions, which is typical for solitons[3.2]. Note that these pulses are obtained by the simultaneous action of absorptive and dispersive effects in contrast to the generation of solitons in nonresonant media (cf. Section 2.) and in purely absorptive media where $\tilde{\tau}_L < \tilde{\tau}_{21}$.

454 LASERS IN ATOMIC, MOLECULAR AND NUCLEAR PHYSICS

REFERENCES

1.0 J. HERRMANN, B. WILHELMI, Laser for Ultrashort Light Pulses (German, Berlin, 1984).
1.1 H. W. MOCKER, R. J. COLLINS, Appl. Phys. Lett., 7, 270 (1965)
1.2 A. J. DEMARIA, D. A. STETSER, H. MEYNAU, Appl. Phys. Lett., 8, 174 (1966)
1.3 R. L. FORK, B. I. GREENE, C. V. SHANK, Appl. Phys. Lett., 38, 197 (1981)
1.4 W. DIETEL, E.DÖPEL, D. KÜHLKE, B. WILHELMI, Opt. Comm., 43, 433 (1982)
1.5 J.C. DIELS, W. DIETEL, E. VAN STRYLAND, Lasers 82 (New Orleans, 1982)
1.6 J. J. FONTAINE, W. DIETEL, J. C. DIELS, IEEE J., QE 19, 1467 (1983)
1.7 D.KÜHLKE, W. RUDOLPH, B. WILHELMI, IEEE J., QE 19, 526 (1983)
1.8 J. C.DIELS, J. J. FONTAINE, I. C. MCMICHAEL, B. WILHWLMI, W. DIETEL, D. KÜHLKE, W. RUDOLPH, Kvant. Elektr., 10, 2398 (1983)
1.9 J. C. DIELS, W. DIETEL, E. DÖPEL, W. RUDOLPH, B. WILHELMI, paper submitted to Lasers '84 (San Francisco, 1984)
1.10 J. G. FUJIMOTO, A. M.WEINER, E. P. IPPEN, Appl. Phys. Lett., in press, paper at IQEC, Anaheim 1984
2.1 H. P. WEBER, paper in this volume
2.2 S. A. AKHAMOV, paper in this volume
2.3 W. DIETEL, E. DÖPEL, W. RUDOLPH, P. WILHELMI, in Lasers '83 (San Francisco, 1983)
2.4 W. DIETEL, E. DÖPEL, K. HEHL, W. RUDOLPH, E.SCHMIDT, Opt. Comm., 50,179 (1984)
2.5 E. B. TREACY, Phys. Lett., $28A$, 34 (1968)
2.6 T. K. GUSTAFSON, J. P. TARAN, H. A. HAUS, J. R. LIFSCHITZ, P. L. KELLEY, Phys, Rev., 177, 1196 (1969)
2.7 A.LAUBEREAU, Phys Lett., $29A$, 539 (1969)
2.8 T. DAMM, M. KASCHKE, F. NOACK, B. WILHELMI, to be published
2.9 C. V.SHANK, R. L. FORK, R. T. YEN, R. H. STOLEN, W. J. TOMLINSON, Appl. Phys. Lett., 40, 761 (1982)
2.10 D.GRISCHKOWSKY, A. C. BALANT, Appl. Phys. Lett., 41, 1 (1982)
2.11 B. NIKOLAUS, D. GRISCHKOWSKY, Appl. Phys. Lett., 42, 1 (1983) and Appl. Phys. Lett., 43, 228 (1983)
2.12 H. NAKATSUKA, D. GRISCHKOWSKY, A. C.BALANT, Phys. Rev. Lett., 47, 1910 (1981)
2.13 A. M. JOHNSON,R. H. STOLEN, W. M. SIMPSON, paper at Ultrafast Phenomena (Monterey, 1984)
2.14 B. MEINEL, Opt. Comm., 47, 343 (1983)
2.15 L. F. MOLLENAUER, R. H. STOLEN, J. P. GORDON, Phys. Rev. Lett., 45, 1095 (1980)

2.16 L. F. MOLLENAUER, R. H. STOLEN, paper at
 Ultrafast Phenomena (Monterey, 1984)
2.17 E. IPPEN, C. V. SHANK, Appl. Phys. Lett., 27, 488 (9175)
3.1 W. RUDOLPH, B. WILHELMI, Opt. Comm., 49, 371 (1984)
3.2 W. DIETEL, E. DÖPEL, W. RUDOLPH, B. WILHELMI, D. C. DIELS,
 J. J. FONTAINE, paper presented at UPS III, Minsk 1983,
 Isvest. Akad. Nauk, 48, 480 (1984)
3.3 W. RUDOLPH, B. WILHELMI, Appl. Phys., B35, 37 (1984)

5.2 Short Pulses in Optical Fibers

H. P. WEBER, W. HODEL and B. VALK

Institute of Applied Physics, University of Berne, 3012 Berne, Switzerland

> The combined action of group-velocity dispersion (GVD) and self phase modulation (SPM) leads to characteristic modifications of psec pulses in glass fibers. The various contributions and their effects are discussed and compared with experiments.

1. INTRODUCTION

Propagation of short pulses in single mode fibers is an ideal subject to study optical nonlinear phenomena because:

(a) The beam cross-section is known and unchanged along the fiber[1]. The optical waves can thus be considered as plane waves with constant waist diameter.

(b) The small core dimensions lead to high intensities in the fiber and due to the low losses[2] very long interaction lengths are possible.

Whereas at low intensities only group-velocity dispersion (GVD) modifies propagating pulses nonlinear effects introduce additional perturbations.

In the last few years much work has been done on this subject (see for example[3]). In our paper we will concentrate on the optical Kerr-effect, i.e. the intensity dependence of the refractive

458 LASERS IN ATOMIC, MOLECULAR AND NUCLEAR PHYSICS

index which leads to self-phase modulation (SPM) of an optical pulse. In order to gain an intuitive understanding we will start with a detailed analysis of GVD, then treat SPM in the absence of dispersion and in a third chapter examine the combined action of SPM and GVD on pulse propagation. Dispersive self-phase modulation (DSPM) in the wavelength region of negative GVD can result in the propagation of stable pulse forms known as optical solitons. In the normal dispersion regime it offers the possibility of pulse compression using an external dispersive delay-line. This technique has led to the generation of the shortest pulses to date[4].

2. GROUP- VELOCITY DISPERSION (GVD)

In the following we treat the single mode fiber as a one dimensional object. The functions describing the transverse field distribution of the fundamental mode are only of quantitative significance and will therefore be omitted.

For convenience consider a pulse of duration τ with Gaussian amplitude at the input of the fiber

$$A(t) = A_o \exp(-1/2(t/\tau)^2) \exp(i\omega_o t) \tag{1}$$

where $\omega_o = 2\pi c/\lambda$ is the centre frequency, λ the vacuum centre wavelength and c the speed of light in vacuo. The spectral amplitude of the pulse is given by the Fourier-transform of (1)

$$A(\omega) = (1/2\pi)^{1/2} \int_{-\infty}^{\infty} A(t) \exp(-i\omega t) \, dt \tag{2}$$

and can easily be shown to be

$$A(\omega) = A_o \tau \exp[-1/2(\Delta\omega \tau)^2] \tag{3}$$

with $\Delta\omega = \omega - \omega_o$. After travelling a length L of fiber the pulse will accumulate a phase shift

$$\Delta\phi(\omega) = -k(\omega)L \tag{4}$$

where $k(\omega)$ denotes the propagation constant of the fundamental mode and has to be known as a funcion of frequency to evaluate the integral

$$A_L(t) = (1/2\pi)^{1/2} \int_{-\infty}^{\infty} A(\omega) \exp(-ik(\omega)L)\exp(i\Delta\omega t)d\omega \quad (5)$$

which represents the temporal pulse amplitude at distance L. Due to the frequency dependence of the propagation constant the temporal intensity distribution of the pulse changes with distance whereas the spectrum remains the same.

If pulse distortion due to higher order dispersion is negligible the truncated expansion of $k(\omega)$ up to second order suffices

$$k(\omega) = k(\omega_o) + (\frac{\partial k}{\partial \omega})_{\omega_o}\Delta\omega + \frac{1}{2}(\frac{\partial^2 k}{\partial \omega^2})_{\omega_o}\Delta\omega^2 \quad (6)$$

Inserting (6) into (5) and using the abbreviations

$$k_o = k(\omega_o), \quad k'_o = (\frac{\partial k}{\partial \omega})_{\omega_o} \quad \text{and} \quad k''_o = (\frac{\partial^2 k}{\partial \omega^2})_{\omega_o} \quad (7)$$

one gets for A_L

$$A_L(t^*) = A_o (\tau/\tau_L)^{1/2} \exp(ik_o L)\exp[-1/2(\frac{t^*}{\tau_L})^2]\exp(i\Phi(t^*)) \quad (8)$$

where $\tau_L^2 = \tau^2 + (k''_o L/\tau)^2$ gives the new time duration and $t^* = t - k'_o L$ denotes the new time variable.
The phase function $\Phi(t^*)$ is biven by

$$\Phi(t^*) = \Phi_o + 1/2 \, k''_o L (\frac{t^*}{\tau \tau_L})^2 \quad (9)$$

with $\Phi_o = \frac{1}{2}\mathrm{arctg}(-k''_o L \tau^{-2}) = \mathrm{const.}$

The first term in the expansion (6) therefore leads to an overall phase-shift of the wave and the second term is responsible for the "group-delay" defined as

$$\tau_g = k'_o L = \frac{L}{v_g(\omega_o)} \quad (10)$$

where $v_g(\omega_o) = (k'_o)^{-1}$ is group-velocity of the centre frequency ω_o. The pulse is now represented with respect to a new time variable t^* which is the local time in a coordinate frame moving at the group-velocity $v_g(\omega_o)$. The third term finally is responsible for a variation in the pulse form. The pulse shape at distance L is still Gaussian

$$I_L(t^*) = A_L(t^*)^2 = A_o^2(\frac{\tau}{\tau_L}) \exp[-(\frac{t^*}{\tau_L})^2] \qquad (11)$$

with a duration τ_L that increases with distance:

$$\tau_L = \tau[1 + (k_o''L/\tau^2)]^{1/2} \geq \tau \qquad (12)$$

If the fiber is not operated at the zero dispersion wavelength defined by $k_o'' = 0$, the pulse at distance L is always broadened in time independent of the sign of k_o''. This is due to the square law dependence on the second order term

$$k_o''L = \frac{-L}{v_g^2(\omega_o)} (\frac{\partial v_g}{\partial \omega})_{\omega_o} = (\frac{\partial \tau_g}{\partial \omega})_{\omega_o} \qquad (13)$$

which describes the change in transit time τ_g with change in frequency. Hence pulse reshaping occurs because the group-velocity is wavelength dependent which implies that every spectral component of the pulse travels at a slightly different velocity. This phenomenon is termed group-velocity dispersion (GVD) and is called normal if the group-velocity is a decreasing function of frequency ($k_o'' < 0$) and anomalous ($k_o'' > 0$) otherwise. The despersive spreading is seen to follow a hyperbolic law (12) in analogy to the diffraction of Gaussian laser beams in free space. The amount of spreading is dependent on the strength of dispersion ($|k_o''|$), the length travelled (L) and the linewidth of the input pulse ($\propto 1/\tau^2$), as seen from eq. 12. A characteristic length for dispersion is the distance at which the width of the input pulse has approximately doubled due to GVD. This "dispersive length" is given by[3]

$$z_d \approx 1.4 \, \tau^2/|k_o''| \qquad (14)$$

The phase function of the pulse (9) is quadratic in time. Since

ULTRAFAST PROCESSES AND TECHNIQUE

$$\Delta\omega(t^*) = - \frac{\partial(\Delta\tilde{\Phi}(t^*))}{\partial t^*} \tag{15}$$

we have

$$\Delta\omega(t^*) = - k_o'' L \frac{t^*}{(\tau\tau_L)^2} \tag{16}$$

and the pulse is seen to exhibit a linear chirp in time, i.e. the momentary frequency is either a linearly increasing or decreasing function of time depending on whether the dispersion is normal or anomalous.

We should mention at this point that in a single mode fiber there are two contributions to the total dispersion[5]: the ordinary material dispersion and the waveguide dispersion. The latter expresses the wavelength dependence of the propagation constant due to the geometrical configuration of the waveguide. This fact can be used to shift the zero dispersion point towards longer wavelengths. We should keep in mind that the boundary between normal and anomalous dispersion is actually dependent on the specific fiber used.

In summary GVD can be described by a quadratic transfer function in the frequecy domain, where the sign of the quadratic term is determined by the nature of dispersion (normal or anomalous). Dispersion always broadens a pulse temporally whereas the spectrum remains unaffected. The resulting temporal broadening is dependent on the strength of dispersion, the fiber length and the initial linewidth of the pulse. The output pulse is linearly chirped, i.e. the momentary frequency is a linear function of time which is increasing (decreasing) when dispersion is normal (anomalous).

3. <u>SELF-PHASE MODULATION (SPM)</u>

Assume now that a pulse with a Gaussian envelope (1) is propagating in a single mode fiber in the absence of dispersion. For simplicity the fiber is taken to be polarization maintaining[6].

Due to the optical Kerr-effect the refractive index of the fiber material is a slightly increasing function of intensity, described by

$$n(t) = n_o + n_2 I(t) \qquad (17)$$

where I(t) is the temporal intensity distribution and n_o denotes the refractive index for zero intensity. The nonlinear coefficient n_2 is usually very small ($n_2 = 3.2 \ 10^{-16} \ W^{-1} cm^2$ for silica[7]) so that the induced change in refractive index will have no influence on the guiding properties of the fiber, i.e. self-focussing effects are negligible. The resulting phase modulation after traveling a length L of fiber, however, can have considerable effects on the pulse spectrum as we shall see below. Since the phase modulation is caused by the wave inself the process is termed self-phase modulation (SPM). The phase change as a function of time at distance L is given by

$$\Delta\Phi(t) = kL\delta n(t) \text{ with } \delta n(t) = n_2 I(t) \qquad (18)$$

To calculate $\Delta\Phi$ according to (18) one should take into account the transverse intensity distribution of the fundamental mode[1]. This is done by an avaraging procedure where one sets

$$I(t) = P(t)/A_{eff} \qquad (19)$$

with P(t) the pump power (in watts) as a function of time and A_{eff}, the effective area, is the fiber overlap integral over the fundamental mode field ψ[7].

$$A_{eff} = \langle\psi 2\rangle^2 / \langle\psi 4\rangle \qquad (20)$$

where

$$\langle\psi k\rangle = 2 \int_0^\infty \psi^k(r) \ r \ dr \qquad (21)$$

Close to the cut-off of the second order mode, however, one has $A_{eff} \cong A_{core}$ and the core area is a good approximation to A_{eff}. To allow for losses the length L in equation (18) should be replaced by an effective value as well

$$L_{eff} = \int_o^L \exp(-\alpha l)\, dl \qquad (22)$$

where α is a linear absorption parameter[7]. Bearing these facts in mind we will in the following drop the suffices that denote the effective values for core area and fiber length. The induced phase change is seen to be directly proportional to the product of the intensity times the fiber length which is typical of nonlinear processes. A good measure for the nonlinearity is therefore the value of $\Delta\Phi_{max}$, the maximum phase shift which usually occurs at the centre of the pulse

$$\Delta\Phi_{max} = kL\delta n_{max} = n_2 kL\, I_{max} \qquad (23)$$

If only SPM has to be considered the temporal pulse shape does not change along the fiber, whereas the spectral amplitude at distance L is now given by the Fourier integral

$$A_L(\omega) = (1/2\pi)^{1/2} A_o \int_{-\infty}^{\infty} \exp[-1/2(t/\tau)^2] \exp[i(\Delta\Phi(t) - \Delta\omega t)]\, dt \qquad (24)$$

and the spectral intensity distribution

$$I_L(\omega) = |A_L(\omega)|^2 \qquad (25)$$

must in general be calculated numerically.

Some features, however, are straightforward. Since the frequency-shift at time t is equal to the negative time derivative of the phase-perturbation (18) we now have

$$\Delta\omega(t) \propto -\frac{\partial I(t)}{\partial t} \qquad (26)$$

and the pulse exhibits a nonlinear chirp which is proportional to the time derivative of the intensity invelope. Hence SPM acts in a similar way in the time domain as dispersion does in the frequency domain and as a consiquence one would expect the output pulse to be spectrally broadened. Equation (26) shows that the maximum frequency shift is determined by the maximum slope of the pulse intensity and for a Gaussian pulse we get the estimate

$$\Delta\omega_{max} \approx 0.86 \frac{\Delta\Phi_{max}}{\tau} \approx 0.52\, \Delta\Phi_{max}\, \Delta\omega_i \qquad (27)$$

The spectral broadening is therefore directly proportional to the amount of nonlinearity described by $\Delta\Phi_{max}$ and the initial linewidth (FWHM) of the pulse ($\Delta\omega_i$). A useful parameter is the critical power for SPM, which is the power necessary to approximately double the spectral width of a pulse in a fiber of length z_d (14).

This critical power depends on pulse and fiber properties as follows[3]

$$P_1 = \frac{nc\lambda A_{eff}}{16\pi z_d n_2} 10^{-7} \; W \qquad (28)$$

where n is the refractive index of the fiber core and the nonlinear coefficient n_2 is in electrostatic units ($1.1\; 10^{-13}$ esu for silica). For any reasonable pulse-shape there are in general two different times within the pulse that lead to the same frequency shift (see Figure 1) and it is their mutual phase-relationship that determines the actual form of the output spectrum.

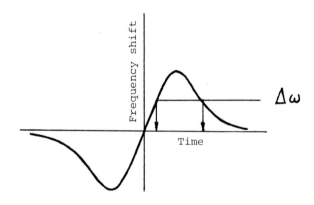

FIGURE 1 The instantaneous frequency as a function of time for a Gaussian shaped pulse. There are in general two different times that lead to the same frequency shift

For maximum phase-shifts $\Delta\Phi_{max} \gtrsim \pi$ this leads to the appearance of interference peaks . Figure 2 shows SPM-spectra belonging to

ULTRAFAST PROCESSES AND TECHNIQUE 465

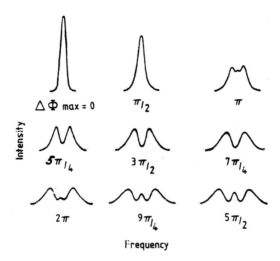

FIGURE 2 SPM spectra of a Gaussian pulse for different values of the maximum phase-shift

different values of $\Delta\Phi_{max}$ as experimentally verified by Stolen et al[7]. Since in the dispersionless case peak intensity and (effective) fiber length are equivalent parameters these spectra can be viewed as either a function of length for a given power or as a function of peak intensity for a fixed fiber length.

In summary SPM is described by a transfer function proportional to the pulse intensity envelope in the time domain. Consequently the temporal pulse shape is not affected by SPM but the corresponding spectra are broadenend and may show significant interference structures. The amount of broadening is directly proportional to the pulse peak-intensity, the (effective) fiber length and the intinal linewidth.

4. DISPERSIVE SELF-PHASE MODULATION (DSPM)

Under realistic conditions(if the fiber is not operated exactly at the zero dispersion wavelength) pulse propagation in optical fibers is subject to the simultaneous action of group-velocity dispersion and self-phase modulation. The amount of temporal broadening was seen to be strongly dependent on the pulse-linewidth (12) which is increasing along the fiber due to SPM. The spectral broadening is in turn dependent on the pulse peak intensity which is reduced by GVD. The combined action of GVD and SPM may thus lead to pulse shapes and spectra quite different from what would be expected if only one process had to be considered. At the beginning of pulse gradually broadens in time and the corresponding spectrum shows the characteristic interference structures. However, after the pulse peak-intensity has been reduced to such an extent that no further significant spectral broadening occurs GVD becomes dominant. The pulse keeps on broadening in time whereas the spectral width remains approximately constant, the interference peaks at the same time becoming less pronounced. These features are illustrated in Figure 3 which shows the temporal and spectral evolution of a pulse along the fiber. The input pulse was taken to be of hyperbolic secant form with amplitude

$$A(z=0,t) = N \operatorname{sech}(t/\tau) \qquad (29)$$

where the amplitude factor N is related to the actual power P normalized to the critical power for SPM P_1 defined earlier (28) through

$$N = (P/P_1)^{1/2} \qquad (30)$$

The pulse is seen to evolve into a flat-topped square-like form which is in fact typical of DSPM pulses.

The data shown in Figure 3 have been calculated in a straightforward way using an algorithm developed in an earlier work on DSPM by Fisher and Bischel[8]. The fiber is cut into segments which alternately act nonlinear and dispersive. The action of GVD and SPM is described using the appropriate transfer function in the frequency and time domain, respectively. A fast Fourier transform is applied at each segment interface in order to represent the field with respect to the correct variable. The number of segments

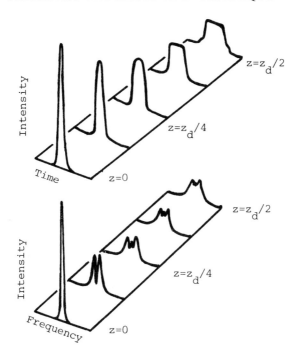

FIGURE 3 Temporal and spectral shapes of a pulse with N = 5 for various values of the normalized length z/z_d

is then increased until the temporal and spectral intensity distributions at the output end of the fiber remain the same. We used this approach since it is quite flexible and physically intuitive.

It is interesting to remark that an input pulse symmetric in time and frequency always leads to a symmetrically broadened spectrum. This fact can be used to gain information on the temporal symmetry of the input pulse by looking at the output spectrum. Figure 4 shows examples of output spectra of pulses whose temporal asymmetry was changed by fine-adjustment of the dye laser that generated them. The autocorrelation functions of all pulses were nearly identical.

An alternative way to describe nonlinear pulse propagation in optical fibers is to solve the nonlinear wave-equation including

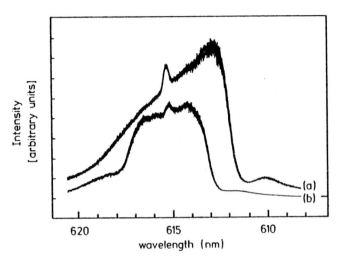

FIGURE 4 Asymmetric output spectra obtained by changing the alignment of the dye laser

dispersion. If we denote the (complex) pulse amplitude function by A the corresponding equation can be put into the following convenient form[3]:

$$i\left[\frac{\partial A}{\partial (z/z_d)}\right] = \frac{\pi}{4}\left[\frac{\partial^2 A}{\partial (t/\tau)^2} + 2|A|^2 A\right] \quad (31)$$

which is the dimensionless nonlinear Schrödinger-equation. The physical length z has been normalized to the dispersive length z_d (14) defined in Chapter 2. GVD is described by the first and SPM by the second term on the right hand side of equation (31) respectively.

From what was said previously we know that GVD leads to linearly chirped pulses whereas the chirp resulting from SPM is proportional to the time derivative of the pulse intensity and in general is a nonlinear function of time. The time derivative of a typical flat-topped pulse as shown in Figure 3 however, is negligible over almost the entire pulse length and one would expect GVD to linearize the chirp of such pulses. That this is indeed the case is illustrated by Figure 5 where the phase of the pulse (29)

ULTRAFAST PROCESSES AND TECHNIQUE 469

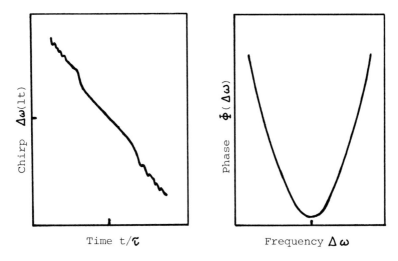

FIGURE 5 The instantaneous frequency as a function of time and the phase as a function of frequency for a pulse with N = 5 at a fiber length $z = z_d/2$

with N = 5 at $z = z_d/2$ is plotted as a function of frequency along with the corresponding chirp[10].

An estimate for the distance at which a DSPM pulse exhibits a linear chirp is provided by[8]

$$z_c \approx 0.66\, \tau/(\lambda_o^{-1} k'' \delta n_{max})^{1/2} \qquad (32)$$

This is the shock distance derived by Fisher and Bischel demanding that the low and high frequency parts of a pulse are separated by one pulse length due to DSPM. Typical DSPM pulses therefore have two very interesting properties: at a characteristic distance z_c, depending on pulse and fiber parameters, they possess a largely broadened spectrum and an almost clean linear chirp over the entire pulse length. This makes them favourite candidates for recompression using a dispersive delayline such as a grating-pair[11]. The action of a grating-pair is conveniently described by its transfer function in frequency space given by

$$\Phi_{gp} = \Phi_o - \beta\omega^2 \qquad (33)$$

where Φ_o is a constant and the coefficient of the quadratic term depends on the actual geometry of the grating configuration and the centre-wavelength λ of the pulse as follows

$$\beta = b\lambda^3/(4\pi c^2 d^2 \cos^2 \gamma) \qquad (34)$$

In this expression b is the centre to centre distance of the grating pair, d their groove spacing and γ the angle between the normal to the input grating and the diffracted beam at λ^{10}. If the input pulse to the grating is described in terms of frequency

$$A_{in} = A_{in}(\omega)\exp(i\Phi(\omega)) \qquad (35)$$

the output pulse is simply

$$A_{out} = A_{in}(\omega)\exp[i(\Phi(\omega) + \Phi_{gp}(\omega))] \qquad (36)$$

So if the phase function $\Phi(\omega)$ of the input pulse is quadratic in ω it can be corrected for by properly adjusting the grating geometry. Ideally the pulse emerging from the grating pair is the shortest possible pulse belonging to the largely broadened input--spectrum and may be much shorter than input pulse to the fiber itself. We again would like to emphasize that both processes treated earlier are of importance: SPM produces the necessary spectral broadening and GVD linearizes the chirp of the pulse so that it can be corrected for using a grating pair.

It is worthwhile noting that the shortest pulses to date (16 fs at $\lambda \cong 0.6 \mu$m which corresponds to ~ 8 optical cycles only) have been realized using this technique[4]. For details concerning pulse compression of DSPM pulses the reader is referred to the article by Tomlinson et al.[10].

In the region of negative GVD ($k_o'' > 0$) the transfer function of dispersion is formally identical to that of a grating-pair and the fiber itself acts as a pulse compressor. This suggests the possibility to exactly balance the influence of SPM and negative GVD by proper choice of the pulse and fiber parameters which would

ULTRAFAST PROCESSES AND TECHNIQUE 471

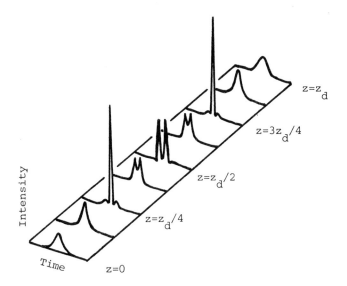

FIGURE 6 Calculated pulse shapes for the N = 3 soliton for different fiber lengths.

result in the propagation of a stable pulse along the fiber. The nonlinear Schrödinger-equation (31) indeed admitts solutions of hyperbolic secant form (29) with astonishing properties provided the amplitude factor N (30) is an integer. This implies that the pulse peak-power is N^2 times the critical power for SPM defined by (28). These solutions are called optical solitons of order N. The fundamental soliton (N=1) is in fact a stationary solution in the sense that it propagates without change in shape. The higher order solitons (N≥2) undergo a complex cycle of narrowing, broadening and even pulse-splitting but are restored to their original shape (temporally and spectrally) at the soliton period z_d (14). Figure 6 shows the temporal evolution of the N=3 soliton. The existence of optical solitons in dispersive fibers was first predicted in a theoretical paper by Hasegawa and Tappert dating from 1973[12] but only recently has been verified experimentally[13].

In summary DSPM is seen to result in temporally and spectrally broadened pulses. A typical DSPM pulse assumes a flat-topped square-like shape. In the beginning of pulse propagation SPM is the dominant effect. The reduction of peak-intensity due to dispersion however, leads to a saturation of the spectral broadening. SPM then has little influence on pulse-reshaping compared to dispersion. GVD tends to linearize the chirp of DSPM pulses which offers the possibility of pulse compresion by means of an external dispersive delay-line (grating-pair). If the dispersion is anomalous the fiber itself acts as pulse compressor. The combination of SPM and negative GVD leads to remarkable propagation features related to optical solitons.

REFERENCES

1. D. GLOGE, Appl. Opt., 10, 2252 (1971)
2. T. MIYA et al., Electron, Lett., 15, 106 (1979)
3. R. H. STOLEN, in New Derections in Guided Wave and Coherent Optics, edited by D. B. Ostrowsky and E. Spitz (NATO ASI Series E 78)
4. J. G. FUJIMOTO et al., Appl. Phys. Lett., 44, 832 (1984)
5. D. GLOGE, Appl. Opt., 10, 2442 (1971)
6. T. OKOSHI, IEEE J. Quant. Electron., QE-17, 879 (1981)
7. R. H. STOLEN and CHINLON LIN, Phys. Rev., A-17, 1448 (1978)
8. R. A. FISHER and W. K. BISCHEL, J. of Appl. Phys., 46, 4921 (1975)
9. B. VALK et al., Opt. Comm., 50, 63 (1984)
10. W. J. TOMLINSON et al., JOSA, B-1, 139 (1984)
11. E. B. TREACY, IEEE, J. Quant. Elrctron., QE-5, 454 (1969)
12. A. HASEGAWA and F. TAPPERT, Appl. Phys. Lett., 23, 142 (1973)
13. L. F. MOLLENAUER at al., Phys. Rev. Lett., 45, 1095 (1980)

5.3 Picosecond Relaxation Processes in Monomeric and Aggregated Dyes Under Solvent Influence

S. K. RENTSCH

Friedrich-Schiller-University of Jena, DDR-6900 Jena, German Democratic Republic

1. INTRODUCTION

The technique of picosecond spectroscopy has been elaborated in the last years to such an extend that spectroscopic measurements in this field become more and more precise. To solve problems in molecular kinetics it is of advantage to employ time resolved fluorescence technique as well as excite-and-probe beam technique. From fluorescence data one gets the time interval in which the relaxation from the S_1 state proceeds. For more detailed studies of relaxation or other kinetic processes with non-fluorescing intermediate states the ps-absorption spectroscopy in a wide spectral region is of high advantage.

For time resolved fluorescence measurements we used time correlated single photon counting with deconvolution down to 40 ps [1]. For higher time resolution up-conversion technique [2] and the streak camera AGAT are available.

For ps-absorption measurements a Nd-YAG spectrometer with a parametric generator [3,4] (Δt = 23 ps) and a Nd-phosphate glass laser spectrometer (Δt = 5 ps) has been elaborated [5].

In the latter spectrometer (Figure 1) a Nd-phosphate glass laser with amplifyer supplies the exciting pulse at the harmonic frequencies (1054, 527, 365, 258 nm) or SRS frequencies generated

474 LASERS IN ATOMIC, MOLECULAR AND NUCLEAR PHYSICS

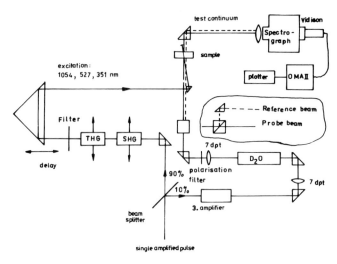

FIGURE 1   Nd-phosphat glass laser spectrometer

by the second harmonic (560 and 620 nm). A section of about 10% from the first harmonic was amplified in a separate amplifier, after that a ps continuum was generated in D_2O. The probe light was split in a probe beam, which tested the excited (or nonexcited) sample, and a reference beam. The relation between the probe beam pulse energies E_{test}/E_{ref} with and without excitation gives transmission like values. The difference of the decadic logarithm of these values is the difference of the optical density with and without excitation ΔD.

The quantity ΔD is essential for the further data processing because it is proportional to the concentration of molecules in various excited or intermediate states. For systems with constant total concentration of molecules ΔD is given quantitatively by

$$\Delta D/z = -\varepsilon_o(\lambda)[c_1(t) + \sum_\alpha c_\alpha(t)] + \varepsilon_1(\lambda)c_1(t) + \sum_\alpha \varepsilon_\alpha(\lambda)c_\alpha(t) - \varepsilon_{1e}(\lambda)c_1(t) - \sum_\alpha \varepsilon_{\alpha e}(\lambda)c_\alpha(t) \quad (1)$$

(c_1 is the concentration in the S_1 state, c_α denotes the concentration of an arbitrary intermediate state, ε_o, ε_1, ε_α are absorption coefficients and ε_{1e}, ε_e gain coefficients). The first term on the right hand side of (1) describes the ground state bleaching, the second and the third account for intermediate state absorption, the fourth and the fifth result from stimulated emission.

2. PICOSECOND SPECTROSCOPIC STUDIES OF POLYMETHINE DYES

With the described experimental technique we investigated polymethine dyes under various conditions. The interest in polymethine dyes came from their practical application as saturable absorbers in laser oscillators and also from their effectiveness as sensitizers of photographic materials. The third point of interest the ability of the dyes to form aggregates in aqueous solution. Optical characteristics of such aggregates exhibits analogies to photo--biological materials. Therefore we hope to be able to use aggregates and their kinetics as models for description of more complicated systems.

Here we report on experimental results of investigations of two polymethine dyes, pseudoisocyanine chloride (PIC) and pinacyanol jodide (PC), in their monomer and dimer form in various solvents and we describe the kinetic behaviour of higher aggregates of PIC.

These molecules exhibit a high mobility in the polymethine chain. It is assumed that torsional vibrations around a bond within the chain effectively couple to modes of the surrounding solvent molecules.

In the theory of radiationless processes the coupling of modes determines the effectivity of energy degradation.

It is well-known that polymethine dyes of the described structure are weakly fluorescent dyes in low viscous solvents. The fluorescence quantum yield and lifetime increase with the solvent viscosity.

The novelty in our experiments with the ps-excite-and-probe

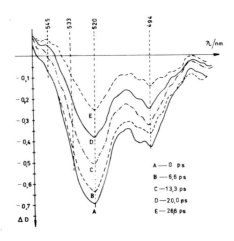

FIGURE 2   ΔD-spectrum of PIC/butanol

spectroscopy was the detection of short living photoisomers. Systematic investigation of the kinetics of isomerization has been performed together with R. Gadonas, R. Danielius, V. Krasauskas and A. Piskarskas of Vilnius university[6-8].

The latest investigation was carried out by our Nd-Phosphate glass laser spectrometer with 5 ps time resolution for dyes exhibiting very fast relaxation processes. In Figure 2 the ΔD-spectrum PIC/butanol after excitation is shown. The solvent bytanol was used because the first study in methanol showed that the time resolution of 5 ps was too small for the observation of photoisomerization in this dye. In Figure 2 a strong bleaching of the $S_o - S_1$-
-absorption was observed, which decrease with a time constant of τ = 22 ps. On the long wavelength side there emerges a new spectrum, the maximum of which lies at (559 ± 5) nm. In the case of photoisomer formation the ΔD-spectrum consist in a superposition of the bleaching - negative ΔD-values, proportional to the concentration of all molecules absent in the ground state - and of positive ΔD values which are proportional to the concentration of photoisomers.

ULTRAFAST PROCESSES AND TECHNIQUE

$$\Delta D(\lambda,t) = -\varepsilon_{01}(\lambda)[c_1(t) + c_{PI}(t)] + \varepsilon_{PI}(\lambda)c_{PI}(t)$$

$$c_o(t) + c_1(t) + c_p(t) = c_{total}$$

(ε_{PI} and c_{PI} are photoisomer absorption coefficient and concentration). A value $\Delta D(t) = 0$ at a fixed wavelength (isosbestic point) cannot be expected.

Photoisomerization on the dye PC was observed in a series of n-alcohols. There a dependence of the S_1-liretime measured by the nearly $\tau \sim \eta$ [9,27]. The determination of the isomerization quantum yield from the $\Delta D(\lambda,t)$ values is not so simple as we thought in our first studies[6]. This is demonstrated in Figure 3. At the maximum bleaching (t = 0) it can be assumed that there exist only exited molecules in the S_1 state but not in other intermediate states, so that $\Delta D(0) = \varepsilon_{01}(t)c_1(0)z$. In an ideal case, if no other absorption exists at the probe wavelength the ΔD-values at delay times greater than the S_1 lifetime, (the values $D(t \gg \tau)$) is due to that part of excited molecules which undergo photoisomerization or other chemical conversion. For the quantum yield of isomerization than holds $\Delta D(t \gg \tau)/\Delta D(0) = c_{PI}/c_1(0) = \Phi_{iso}$.

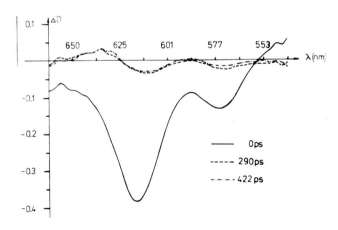

FIGURE 3 ΔD-spectrum of PC/butanol

But the ΔD-spectra of PC (FIGURE 3) show that the photoisomer absorb in a broad spectral region, therefore holds

$$\frac{\Delta D(t \gg \tau)}{\Delta D(0)} = \frac{(\varepsilon_{01} - \varepsilon_{PI}) c_{PI}}{\varepsilon_{01} c_1(0)} < \Phi_{iso}$$

The quantum yield of isomerization can be determined only with a systematic error, because ε_{PI} is unknown. By comparison of measuremets in various solvents the systematic error is constant, if the spectra of the stable form and the photoisomer are non shifted in wavelengths.

The short living species which we called a photoisomer can be characterized by the following spectroscopic data (Table I).

TABLE I Spectroscopic and kinetic data of isomers of polymethine dyes in butanol

Dyes	λ_N(nm)	λ_{PI}(nm)	$\Delta\nu$(cm^{-1})	τ(ps)	Φ_{ISO}
C_2H_5 C_2H_5 /J$^-$	605	635	781	35	0.09
C_2H_5 C_2H_5 /Cl$^-$	523	545	772	22	0.33
H_3C,H_3CN~~~N$^{CH_3}_{CH_3}$/ClO$_4^-$	510	545	1260	36	0.35

Further efforts are necessary to determine the stability and the geometry of this species.

From ns-laser flash experiments the reformation of the instable to the stable form could be measured by testing the decrease of the photoisomer absorption. In PC the reformation time is also viscosity dependent and amounts in methanol to τ_r = 25 ± 5 ns and in butanol to τ_r = 55 ± 15 ns [10]. In PIC no isomer absorption could be detected, that means the reversion to the normal form proceeds in times shorter than the resolution of the excitation pulse of 15 ns.

Results on the geometry of the photoisomer have been obtained by Lau, Werncke et al[11] by means of resonance CARS with nanosecond time resolution. They detected the CARS-spectrum of PC without and with strong excitation. Thereby intensity alteration of lines and new Raman lines indicate a species of other geometry.

All the together confirm the idea of the formation of a photo-isomer in the picosecond time scale in its ground state with altered geometry and a lifetime of a few nanoseconds.

Theoretical concepts have been elaborated to describe the well-known dependence of the lifetime on solvents viscosity. One approach starts from the assumption that the relaxation proceed during the motion of part of a molecule along a configurational co-ordinate in a nearly parabolic potential with a barrier under influence of fluctuational forces[12]. The solvent viscosity is proportional to the drag coefficient of the equation of motion.

Detailed stulies[13] of the relaxation process in a wide range of viscosity showed a complicate relationship between τ and η which contains not only that part where the lifetime τ decrease with decreasing solvent viscosities, but a region of middle viscosity, where τ is nearly independent on solvent viscosity and a low viscosity region, where even an increase of τ with decreasing η is expected.

For the dye PC $\tau(\eta)$ has been measured in a wide range of about three orders of magnitude of viscosity[15]. These results are suitable to prove theoretical results. The theoretical analysis of the experimental data of PC[14] showed that besides the barrier crossing an additional relaxation channel (Franck-Condon transition) has to be taken in consideration, which is domenant at high viscosity.

By picosecond spectroscopy not only lifetime measurements can be performed precisely, but moreover the formation kinetics and the quantum yield of the photoisomer can be determined. These data allow to prove models for the photoisomerization process. We described three models, used in the literature, with rate equations[16]. The solutions of the rate equation can be proved experimentally[17]. A distinction between the models is possible measuring the formation kinetics in the time range $t \leqslant \tau$ with high accuracy.

3. PICOSECOND STUDIES OF MONOMER-DIMER-SYSTEMS

Both dyes, PC and PIC, which we studied as monomer in several alcohols, form dimers in aqueous solution, which exist in a chemical equilibrium with monomers.

From the quantum theoretical description we know that the excited dimer levels originate from monomer levels, split owing to molecular interaction. For so called sandwich-dimers the transition to the higher level is allowed, but to the lower one forbidden. The vibronic coupling yield a broad dimer absorption band with a maximum at shorter wavelength than the monomer maximum and weak shoulder on the long wavelength side.

The dimer spectrum is always superposed with that of the monomers. The pure dimer spectrum is not known for the dyes under study. Recently a dimer spectrum of PIC was calculated by computer simulation[18]. The known monomer spectrum multiplied by a factor was subtracted from the experimental spectra under various conditions. In this way a dimer spectrum was obtained. Picosecond investigation of such a monomer-dimer system implicitely contains the task to measure two superposed ΔD-spectra and their kinetics. Using of special excitation wavelength for selective excitation of only monomers or dimers we obtained ΔD-spectra of both components, because at strong excitation the selectivity is low.

Another way to separate both components is the kinetics. In Figure 4 the ΔD-spectrum at different times after excitation is shown. Both the maxima of monomers and dimers are bleached, but the monomer maximum decays faster than the dimer one. The time constants are different in nearly one order of magnitude.

The monomer component of PC decays with $\tau = 16 \pm 8$ ps whereas

ULTRAFAST PROCESSES AND TECHNIQUE 481

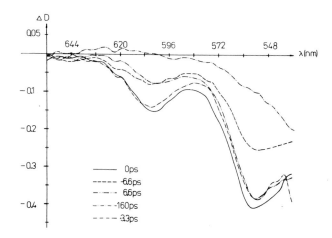

FIGURE 4 ΔD- spectrum of the PC monomer-dimer system in water/methanol

the dimer decays with τ = 100 ± 20 ps along the whole bleached band. The kinetics opened a new way to isolate a dimer spectrum experimentally.

The ΔD-spectrum of the system some time after excitation must be a pure dimer spectrum. In PC we find that the monomers undergo photoisomerization and so a little remainder ΔD-value of monomer bleaching superposed the dimer spectrum at a longer delay time.

Analogous studies of the kinetics of PIC at room temperature [7,26] showed no difference in the relaxation times of monomers and dimers. The reason is the extremly short monomer relaxation time and the strong superposition of both spectra in which only the dimer relaxation could be observed. We performed the study at a temperature of 10°C, where the relaxation processes proceed more slowly. In Figure 5 it can be clearly seen that the ΔD-curve consists of two components, which must be assigned to the monomer and dimer relaxation.

The main result in studing monmer-dimer systems of polymethine dyes consists in the detection of different relaxation kinetics of both components. Dimers always exhibit a longer relaxation time

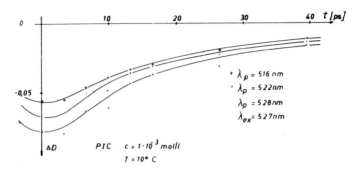

FIGURE 5 Kinetics of the PIC monomer-dimer system

than monomers. The reason is the strong intermolecular coupling of the molecules in the dimer which prevent the molecular torsion around the polymethine chain. In this way the system was rigidized and therefore the radiationless relaxation is less effective in the dimers than in the monomers.

4. PICOSECOND STUDIES OF AGGREGATES

The dye PIC forms aggregates in aqueous solution at a concentration of $c \sim 10^{-3}$ mol/l. The aggregation is accompanied by an alteration of the optical spectrum, which exhibits now the strong and narrow J-band, shifted to longer wavelengths (λ = 573 nm, $\Delta\tilde{\nu}$ = 163 cm^{-1}), but the monomer and dimer spectra decrease in the aggregated solution. The J-band polarization is perpendicular to that of the monomer and dimer bands. By comparison with an investigation of PIC crystals the axis of the J-band polarization was determined to be the b-axis, the direction in which the overlap of the π-electron system of the neighbouring molecules is maximal[19]. The excitonic character of the J-band was recently supported by density matrix descriptions of finite molecular systems[20]. The lineshape of the PIC aggregates was calculated in detail in[21]. Aslangul[20] showed that an aggregate of a number of molecules of N = 3, ..., 100 interact with the radiation dield as a unit. Excitation of a part or a molecule within the aggregate does not exist, so that it is not necessary to speak about energy transfer within the aggregates. In a real system there may exist monomers, dimers and aggregates in a chemical equilibrium so that energy exchange between the components may be possible.

Kinetic studies of aggregates were carried out by fluorescence measurements by Dähne and coworkers[22,23]. We have carried out such work together with the group of A. Piskarskas from Vilnius university[24,25] with the help of excite-and-probe-spectrometers.

The recovery kinetics measured after excitation with λ = 527 nm at probe wavelengths within the monomer and dimer band result in a two exponential decay. The aggregate bleaching first decays fast with 16 ps and than slowly with about 500 ps.

The fast conponent vanishes more and more, as the excitation pulse energy decreases, but the slow component seems to vary between 300 and 600 ps. The difficulty of aggregate studies consist in the measuring of a thin layer (10 μm) of highly viscous aggregated solution, in which the absorbed photons may cause local temperature increase or even destruction of molecules.

Recently Stiel et al[22] reported a strong temperature dependence of the aggregate fluorescence lifetime and quantum yield, measured by weak excitation and single photon counting. The lifetime and the quantum yield decrease strongly with increasing temperature. So it seems to be possible that at strong excitation with ps-pulses local heating causes lifetime shortening of excited aggregates.

Detailed spectral measurements in the range of the J-band after excitation showed that the band itself contains two components of different polarization[9]. Moreover a superposition of transient absorption bands with the aggregate band has been observed[19].

In order to clarify the nature of the aggregate band we used various excitation wavelengths for excitation within the monomer and dimer bands as well as within the aggregate band. Exciting the monomer-dimer band results in the bleaching of the whole absorption band, whereas direct excitation of the J-band results only in the bleaching of the J-band. In this experiment $\Delta D = 10^{-3}$ really could be measured, the absence of bleaching at shorter wavelengths means that the absorption coefficient of the J-band in this spectral range ($\lambda < 573$ nm) is of a factor of about 10^{-3} lower than J-band absorption coefficient. This result agrees with measurements of intensity dependent bleaching[26].

In this experiment[26] bleaching occurs at an excitation intensity of $\lambda = 530$ nm (monomer band) which is more than 10^3 times higher than the J-band (at 573 nm). The aggregate bleaching upon excitation at 530 nm may be caused either by a small absorption cross section of the aggregates itself of by energy transfer from excited monomers to aggregates. If such an energy transfer does

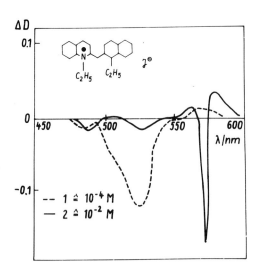

FIGURE 6 ΔD-spectrum of PIC monomers, dimers and aggregates, excited with $\lambda = 530$ nm

exist, one must proceed within a time of about one ps, because no delayed bleaching of the aggregate band was observed.

Summarizing the kinetic data about the PIC aggregate system we found, that monomers and dimers at room temperature exist and relax with their own relaxation time. The J-band relaxation is strongly influenced by temperature and excitation energy. The nature of the extremely short decay at high excitation is not quite clear up to know, superluminescence could not be detected. The recovery time at low excitation energy shows the tendency already observed at dimers. The aggregate is a rigidized system in which the excited state lives longer.

REFERENCES

1. W. BECKER, D. SCHUBERT, H. WABNITZ, Frühjahrsschule Optik, (Jena, 1984)
2. D. SCHUBERT, H WABNITZ, Proc. Conf. Optical Spectroscopy, (Reinhardsbrunn, 1984)
3. H. BERGNER, V. BRÜCKNER, R. GASE, A. SCHLISIO, B. SCHRÖDER, Exp. Techn. Phys., $\underline{30}$, 5. 4/7 (1982)
4. V. BRÜCKNER, this summer scool
5. T. DAMM, M. KASCHKE, M. KRESSER, P. NOAK, S. RENTSCH, W. TRIEBEL, Exp. Techn. Phys., to be published
6. S. RENTSCH, R.DANIELIUS, R.GADONAS, Chem. Phys., $\underline{59}$, 9 (1981)
7. S. RENTSCH, R. DANIELIUS, Chem. Phys. Lett., $\underline{84}$, 416 (1981)
8. S. K. RENTSCH, R. A. GADONAS, A. PISKARSKAS, Chem. Phys. Lett., $\underline{104}$ 2/3, 235 (1984)
9. R. GADONAS, Dissertation (Vilnius, 1984)
10. S. RENTSCH, A. GRANESS, to be published
11. T. J. TSCHOLL, W. WERNCKE, H. J. WEIGMANN, M. PFEIFFER, A. LAU, S. RENTSCH, Proc. Conf. Optical Spectroscopy (Reinhardsbrunn, 1984)
12. M. A. KRAMERS, Physica, $\underline{7}$, 284 (1940)
13. M. KASCHKE, J. KLEINSCHMIDT, B. WILHELMI, Laser Chemistry, in press
14. M. RASCHKE, J. KLEINSCHMIDT, B. WILHELMI, Chem. Phys. Lett., $\underline{106}$, 428 (1984)
15. J. C. MIALOCG, P. GONJON, M. ARVIS, J. Chem. Phys., $\underline{76}$, 1067 (1979)
16. S. RENTSCH, B. WILHELMI, J. Mol. Struct., $\underline{114}$, 1 (1984)

17. S. K. RENTSCH, Chem. Phys., 69, 81 (1982)
18. B. KOPAINSKY, J. K. HALLERMEIER, W. KAISER, Chem. Phys. Lett., 83, 498 (1981)
19. S. MAKIO, N. KANAMURA, J. TANAKA, Proc. Conf. Relaxation of Elementary Excitations (Susoni, Japan, 1979), p. 180
20. G. ASLANGUL, P. KOTTIS, Adv. Chem. Phys., 41, 322-476 (1981)
21. E. W. KNAPP, Chem. Phys., 85, 73 (1984)
22. H. STIEL, K. TEUCHNER, W. FREYER, S. DÄHNE, J. Mol. Structure, 114, 251 (1984)
23. F. FINK, E. KLOSE, K. TEUCHNER, S. DÄHNE, Chem. Phys. Lett., 45, 548 (1977)
24. R. DANIELIUS, R.GADONAS, V. SIRUTKAITIS, K. TEUCHNER, S. DÄHNE, S. RENTSCH, Proc. UPS 80 p. 270

5.4 Picosecond Spectroscopy in Study of the Photoinduced Isomerization of Hexafluorobenzene

J. L. SUIJKER, A. H. HUIZER and C. A. G. O. VARMA

Gorlaeus Laboratories, Department of Chemistry, University of Leiden, P.O. Box 9502, 2300 RA Leiden, The Netherlands

There is a continuing interest in the chemical and physical consequences of substitution of the hydrogen atoms of benzene (C_6H_6) by fluorine[1-9], because no systematic explanation exists as yet for the large effects caused by such substitution. When all hydrogen atoms are replaced by fluorine the molecular behaviour does not resemble that of benzene. In the present context, for instance the near absence of fluorescence and the total absence of phosphorescence upon excitation of hexafluorobenzene (C_6F_6) within its first electronic absorption band, are of interest. From the chemical point of view the formation of the relatively stable Dewar isomer, shown in Fig. 1, formed after UV excitation of C_6F_6, is relevant. The investigations reported here were aimed at an elucidation of the factors responsible for the photophysical and photochemical behaviour of hexafluorobenzene.

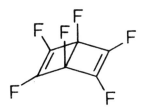

FIGURE 1 Dewar-hexafluorobenzene

In 1967 D. Phillips reported a study on the pressure dependence of the fluorescence quantum yield Φ_f of C_6F_6[10]. The pressure dependence of Φ_f in the case of C_6F_6 could not be explained by the kinetic model introduced by E.M. Anderson and G.B. Kistiakowsky[11], as used in their interpretation of the pressure dependence of Φ_f for gaseous C_6H_6. In order to compare the collisional deactivation processes for C_6F_6 and C_6H_6 we reinvestigated both sets of data by means of a simple two level relaxation model in which we have taken the statistics of the collisions in the gas into account[12]. In our model, illustrated in Fig. 2, the observed pressure dependence for both C_6F_6 and C_6H_6 can be simulated by a single parameter α, which represents the probability to induce a transition from the upper to the lower level in a single collision. Whereas for benzene α is close to unity it is very small for hexafluorobenzene, from which we conclude that the collisional deactivation efficiency is decreased enormously by fluorine substitution. The scheme shown in Fig. 2 applies to the case of excitation within the first UV absorption band. It considers two emissive states $|a\rangle$ and $|b\rangle$ and collision induced relaxation from $|a\rangle$ to $|b\rangle$.

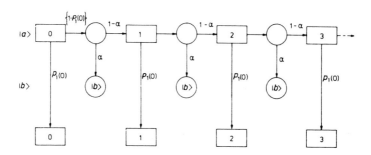

FIGURE 2 Scheme indicating competition between unperturbed decay of initial state $|a\rangle$ and collision-induced transition from state $|a\rangle$ to state $|b\rangle$. The ith numbered rectangle in the upper part indicates that a selected molecule A_s is still in state $|a\rangle$ after its ith collision. The ith rectangle in the lower part indicates that A_s disappears from $|a\rangle$ between its nth and $(n + 1)$th collision.

Circles represent binary collisions involving A_s.
Probabilities for spontaneous and induced transitions
are $P_1(0)$ and α, respectively.

In Figures 3 and 4 it is shown that the pressure dependence of the fluorescence quantum yield of benzene and hexafluorobenzene may be simulated accurately by choosing the value of appropriately.

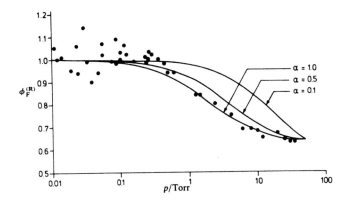

FIGURE 3 Experimental points from ref. (11), representing the relative total fluorescence quantum yield $\Phi_F^{(R)}$ of C_6H_6 as a function of pressure. The curves are single-parameter fits according to the scheme in Fig.2 (λ_{ex} = 254 nm).

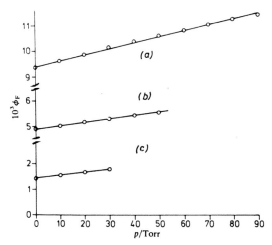

FIGURE 4 Lines according to ref. (10), representing experimental

behaviour of total fluorescence quantum yield ϕ_F of C_6F_6, excited at different wavelengths. The points are obtained by adjusting α according to the scheme in Fig. 2.
(a) $\alpha = 0.13$, $\lambda_{ex} = 270$ nm; (b) $\alpha = 0.05$, $\lambda_{ex} = 265$ nm; (c) $\alpha = 0.05$, $\lambda_{ex} = 254$ nm.

In his article Phillips also reported the first, indirect, observation of the lowest triplet state of hexafluorobenxene. He studied the quenching of the biacetyl phosphorescence by C_6F_6 molecule, causing a transition of the latter to its lowest triplet state. From sensitization experiments of the isomerization of cis-buten-2 by electronically excited C_6F_6, he concluded that both the observed triplet quantum yield and the triplet lifetime are more than an order of magnitude smaller than the corresponding values in the case of C_6H_6. This brought us to look more directly for the triplet state of C_6F_6, but we did not succeed in observing any phosphorescence nor any triplet ESR signal from our pure samples of C_6F_6 at low temperature after UV excitation.

In 1967 I. Haller[13] reported a study on the kinetics and mechanism of the photochemical valence isomerization of C_6F_6 to its Dewar isomer. The experiments were conducted in the gas phase and the sample was excited with UV light of several wavelengths. He determined the quantum yields of isomerization as a function of gas pressure for all the excitation wavelengths involved. From triplet quenching and triplet sensitization experiments he concluded that the triplet of hexafluorobenzene was not involved in the isomerization to the Dewar valence isomer, and he was the first to suggest that the isomer is formed from the first excited singlet state, via the unstable biradical BR shown in Fig. 5.

In an attempt to shed more light on the mechanism of the isomerization to the Dewar form we initiated a study of the metastable species arising from excitation of C_6F_6 in its first electronic absorption band by means of picosecond time-resolved absorption

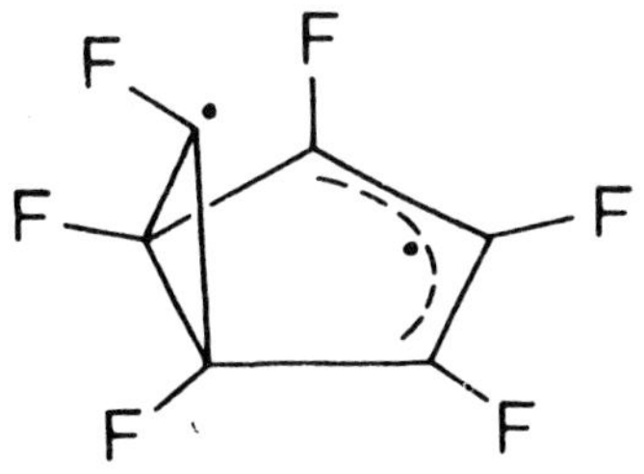

ULTRAFAST PROCESSES AND TECHNIQUE 491

FIGURE 5 The biradical BR

and emission spectroscopy. In these experiments a gaseous or liquid sample is excited with a 265 nm laser pulse of about 10 or 25 ps duration. In the absorption experiments the excited region is then probed by a second pulse after a well defined time delay [14]. In the fluorescence experiments the induced emission from the sample is studied in real time by means of a fast photodetector coupled to a 500 MHz transient digitizer[15] or by means of a streak camera system. A number of metastable species were detected through the transient optical absorptions shown in Fig. 6[17].

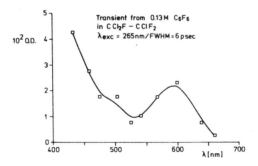

FIGURE 6 Transient absorptions observed at time delay of 50 ps.

The difference in the growth of the absorption at 430 and at 600nm shown in Fig. 7 indicates the formation of several species. The transient absorption around 430 nm is attributed to the cation and anion of hexafluorobenzene, based on the known spectra of these ions. They are formed monophotonically via electron transfer with-

492 LASERS IN ATOMIC, MOLECULAR AND NUCLEAR PHYSICS

in a pair hexafluorobenzene molecules. The absorption band around 600 nm is attributed to the intermediate BR, based on a semi-empirical quantum mechanical calculation (CNDO/S) of its spectrum.

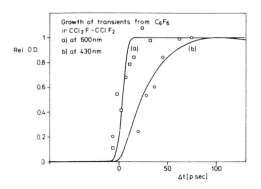

FIGURE 7 Growth of the transient absorption at different wavelengths.

We have discovered UV laser induced fluorescence from hexafluorobenzene in the visible to near infrared wavelength region. A similar red fluorescence could also be observed in the case of other monocyclic aromatic molecules such as benzene and pyridine[15]. These fluorescence bands are shown in Fig. 8.

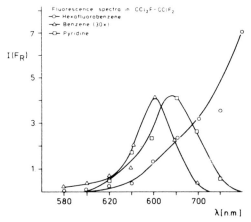

FIGURE 8 Fluorescence in the visible and near-infrared regions observed when solutions of benzene, pyridine and

ULTRAFAST PROCESSES AND TECHNIQUE 493

hexafluorobenzene in $Cl_2FC-CClF_2$ are excited with a
266 nm laser pulse, having a width of 25 ps.

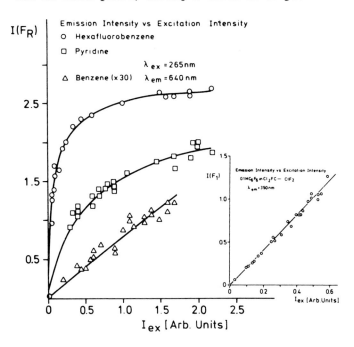

FIGURE 9 Variation of the intensity $I(F_R)$ of the fluorescence of
benzene, pyridine and hexafluorobenzene in $Cl_2FC-CClF_2$
at 640 nm as a function of the integrated energy I_{ex} in
the excitation pulse. The insert in the figure shows the
variation of the intensity of the ordinary fluorescence
of hexafluorobenzene in $Cl_2FC-CClF_2$ at 360 nm as a function of the integrated energy in the excitation pulse.
The largest absolute value of the integrated exctation
energy is the same as in the main part of the figure.

Fig. 9 shows that the intensity of the red fluorescence varies linearly with the integrated energy I_{ex} of the excitation pulse in the case of benzene, but in the case of hexafluorobenzene and pyridine there is saturation at high values of I_{ex}. Since the normal fluorescence of hexafluorobenzene varies linearly with I_{ex} under identical excitation conditions, the saturation does not arise from depletion of the ground state population. The saturation is probably a consequence of absorption of laser photons by the emitting species. From the linear behaviour in the case of benzene we infer that the red fluorescence is induced monophotonically in all the three cases. We have taken advantage of the fact, that the lifetime of the species emitting the red fluorescence of hexafluorobenzene of shorter than that of the metastable species BR, to prove that the red emission arises from an electronically excited state BR* of BR. The evidence is provided in Fig. 10.

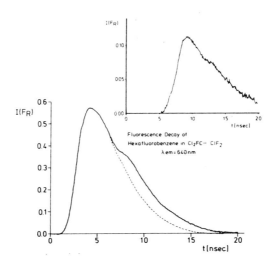

FIGURE 10 Decay of the red fluorescence of hexafluorobenzene in $Cl_2FC-CClF_2$ at 640 nm when a second delayed pulse is absorbed in the transient absorption band at 600 nm (solid curve) and when the latter is absent (dashed curve). The insert in the figure represents the difference between the solid and dashed curves.

It follows from Fig. 7 that the growth in the population of BR comes to an end within less than 10 ps after preparation of the primary excited state S_p of hexafluorobenzene. From this and from the lifetime of BR*, namely τ_R = 3.2 ns, it follows that BR is not formed from BR*, at least not predominantly. The deactivation of

ULTRAFAST PROCESSES AND TECHNIQUE 495

the primary prepared state of hexafluorobenzene may be summarized
as in the following scheme:

$$S_p \begin{cases} \rightsquigarrow BR^* \xrightarrow{h\nu_R} BR, \\ \rightsquigarrow BR \xrightarrow{h\nu_1} \text{Dewar form,} \\ \rightsquigarrow S_1 \xrightarrow{} S_0. \end{cases}$$

When hexafluorobenzene in the gaseous state is excited with
the UV laser pulse, the red fluorescence may also be detected. In
a gas mixture of 740 Torr nitrogen and 17 Torr hexafluorobenzene
at 296 K the lifetime τ_R = 3.2 ns. By comparing this with the value τ_R = 3.2 ns for the liquid solution, we conclude that the lifetime of the state BR* of C_6F_6 is not strongly affected by collisions. It is unlikely that BR and BR* may have to be identified
with molecular aggregates like dimers, excimers or solute-solvent
exciplexes of hexafluorobenzene, because we find that the ratio of
the quantum yields of the red fluorescence in the vapour phase and
in the liquid solution is as high as 0.75 whereas the ratio of the
concentrations C_6F_6 is 0.14.

The decay of the fluorescence at 360 nm of hexafluorobenzene
in the vapour phase and in liquid solutions has been studied with
a carefully corrected and calibrated streak camera[16]. This fluorescence band is emitted by a short living species S_s and a long
living species S_1, which are formed from the primary excited state
S_p. The decay of the fluorescence at 360 nm shown in Fig. 11 is
different from what has been reprted previously[18]. The decay profile is independent of wavelength. We have not observed a time
delay in the growth of the fluorescence at 360 nm. In the case of
pure gaseous hexafluorobenzene at 17 Torr and 296 K the lifetimes
τ_s and τ_1 of the states S_s and S_1 respectively are τ_s = 520±50 ps
and τ_1 = 1820±50 ps. Based on gas kinetics the former value may be
considered as unperturbed by collisions. When 740 Torr nitrogen is

496 LASERS IN ATOMIC, MOLECULAR AND NUCLEAR PHYSICS

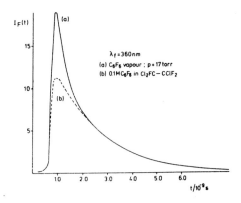

FIGURE 11 Decay of the fluorescence of hexafluorobenzene transmitted through an interference filter with peak transmittance at 360 nm and a bandwidth of 10 nm. The fluorescence is excited with a 266 nm laser pulse of 25 ps FWHM. Curve (a): pure C_6F_6 in the vapour phase at 296 K. Curve (b): 0.1 molar liquid solution of C_6F_6 in $Cl_2FC-CClF_2$. Time scale in ns.

added to the 17 Torr of hexafluorobenzene vapour, the value of τ_s is reduced to τ_s = 280±50 ps, while τ_1 remains unchanged, namely = 1800±50 ps. This behaviour justifies the relaxation model given in Fig. 2.

If 20 Torr trifluoroethanol (TFE) is added to 17 Torr of hexafluorobenzene we find τ_s = 260±50 ps and τ_1 = 1690±60 ps. In the liquid solution of 0.1 M of hexafluorobenzene in $Cl_2FC-CClF_2$ the fluorescence decay has become single exponential with a lifetime τ_1 = 2060±90 ps which is practically equal to that of the long lived component of the fluorescence of C_6F_6 in the pure gas. In the case of a liquid solution of 0.1 M hezafluorobenzene in TFE the fluorescence decays mono-exponentially with a lifetime τ_1 = 1060±60 ps. These observations indicate that TFE quenches the long lived component of the UV fluorescence. Nevertheless, no stable reaction products are found gaschromato-graphically and the UV absorption spectrum of gaseous or liquid samples with C_6F_6 and TFE remain unchanged upon continuous UV irradiation even after a period of 24 hours. By comparing the lifetime of state S_1 in the solvents $Cl_2FC-CClF_2$ and TFE, we conclude that this state does not lead to the Dewar isomer, because otherwise the quantum yield of the isomer would only by reduced by 50% in changing from the first to the second solvent.

The action of TFE in the relaxation of electronically excited C_6F_6 will be considered now in relation with the photoinduced isomerization to the Dewar form. Continuous UV irradiation of hexafluorobenzene vapour (17 Torr) in equilibrium with a drop in the liquid phase and with 740 Torr of nitrogen resulted in a 6.7% yield in the Dewar isomer after a period of 90 minutes. About 50% of hexafluorobenzene at an initial concentration of 0.01 M in $Cl_2FC-CClF_2$ disappears after 90 minutes of continuous UV irradiation and 4.9% of it is then converted to the Dewar form and the other part which is lost is transformed into polymeric material. The quenching of the photoinduced conversion of hexafluorobenzene to its Dewar form and to polymers by TFE must involve the scavenging of an intermediate in these reactions arising as a secondary product of the excitation. We identify the scavenged intermediate with the species BR, causing the transient absorption at 600 nm. We have studied the attenuation of the transient optical density $D(\lambda, \Delta t)$ at 600nm, by TFE in the solution, by keeping the concentration of C_6F_6 constant and by varying the amount of TFE and measuring then the optical density at Δt = 0, 1, and 2 ns after the primary excitation. The transient absorption at 600 nm is still observable if TFE is used as the solvent for C_6F_6. The behaviour of $D(\lambda, \Delta t)$ as a function of the concentration C_Q of TFE is shown in Fig. 12 for λ = 600 nm and Δt = 2 ns.

The figure reveals an asymtotic approach of $D(\lambda, \Delta t)$ to a nonzero value when C_Q tends to infinity. We may explain this behav-

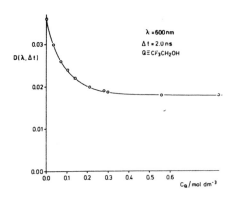

FIGURE 12 Quenching of the transient absorption at 600 nm induced by excitation of a 0.1 molar liquid solution of hexafluorobenzene in $Cl_2FC-CClF_2$ with a 266 nm laser pulse of 25 ps FWHM. The quencher Q is 2,2,2-trifluoroethanol.

iour by assuming that two species contribute to the transient absorption at 600 nm, namely BR and an unknown Z, which is formed simultaneously with BR from the primary excited state. The explanation implies that the transient absorption at 600 nm of the solution of C_6F_6 in TFE is entirely due to Z. The species Z can not be transformed into the Dewar isomer of C_6F_6, because this is not formed when TFE is used as the solvent. Previously we have excluded the possibilities that the absorption at 600 nm involves a transition form the lowest excited singlet state S_1 to a higher singlet state or from the lowest triplet state T_0 to a higher triplet state of hexafluorobenzene[17]. The same reasoning excludes therefore the identification of Z with S_1 or with T_0. The nature of Z remains unknown.

In 1976 R. Pottier and G.P. Semeluk reported a near and far UV spectroscopic study of the fluorinated benzenes[19]. They observed a loss of vibronic structure in the first electronic absorption band which increases with the number of fluorine atoms. The Stokes-shift of the fluorescence increases also with the number of fluorine atoms. We have tried to find vibrational structure in

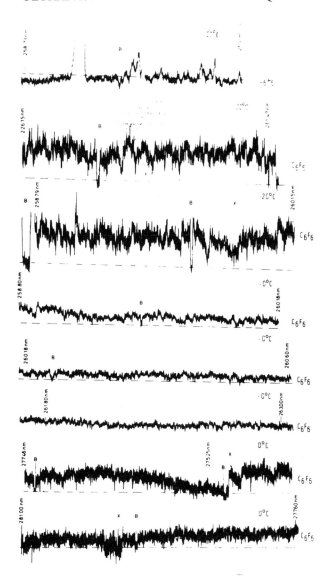

FIGURE 13 Excitation spectrum of the fluorescence of C_6F_6 in a supersonic beam observed at 320 nm. The vapour of a solidified sample was equilibrated with helium at a pressure of 1900 Torr before expansion. The temperature of the solid sample, either -20°C or 0°C, is indicated. The (X) marks the wavelength where the crystal for doubling the frequency of the dye laser was readjusted

or interchanged. The (B) marks where the excitation was interrupted to check the height of the baseline. The upper track illustrates that the signal to noise ratio for the excitation of the fluorescence of C_6H_6 is much larger than in the case of C_6F_6 under identical conditions.

in the first UV absorption band and in the UV fluorescence band of hexafluorobenzene by determining the spectra of the molecule in a supersonic molecular beam (Fig. 13) or in low temperature solid matrices (Fig. 14).

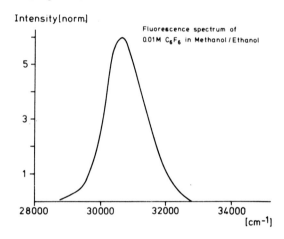

FIGURE 14 UV fluorescence band of hexafluorobenzene in a solid matrix at 4.2 K. The band is three times wider when the molecule is dissolved en a 1:1000 ratio in an argon matrix at 12.4 K.

Neither of the two bands reveal vibrational structure. Based on our observations that the primary excited state S_p decays completely, radiationless and yields in less than 10 ps at least the five products S_s, S_1, BR*, BR and Z, we conclude that the absence of vibrational structure in the absorption band is a consequence of the lifetime of state S_p. The lifetime of S_p is probably less than 10^{-13} s [20]. The lifetime of 2 ns for the fluorescence indicates that the absence of structure in the fluorescence band is not due to lifetime broadening. The absence of structure in this band may be explained by a large difference in equilibrium molecular geometries in initial and final electronic states involved in the emission and by an additional inhomogenous broadening arising from a conformational equilibrium of emitting molecules.

REFERENCES

1. W. HACK, W. LANGEL, Il Nuovo Cimento, 63B, 207 (1981).
2. Y. KOBAYASHI, I. KUMADAKI, Acc. Chem. Res., 14, 76 (1981).
3. R. C. SHARP, E. YABLONOVITCH, N. BLOEMBERGEN, J. Chem. Phys., 76, 2147 (1982).
4. D. V. O'CONNOR, M SUMITANI, J. M. MORRIS, K. YOSHIHARA, Chem. Phys. Letters, 93, 350 (1982).
5. J. ALMLÖF, K. FAEGRI, J.Chem. Phys., 79, 2284 (1983).
6. V. E. BONDEYBEY, J. H. ENGLISH, T. A. MILLER, J. Phys. Chem., 87, 1300 (1983).
7. G. DUJARDIN, S. LEACH, Chem. Phys. Letters, 96, 337 (1983).
8. C. J. MO L, G. R. PARKER, A. KUPERMANN, J. Chem. Phys., 80, 4800 (1984).
9. R. MOORE, F. E. DOANY, E. J. HEILWEIL, R. M. HOCHSTRASSER, J. Phys. Chem., 88, 876 (1984).
10. D. PHILLIPS, J. Chem Phys., 46, 4679 (1967).
11. E. M. ANDERSON, E. B. KISTIAKOWSKY, J. Chem. Phys., 51, 182 (1969).
12. J. L. G. SUIJKER, C. A. G. O. VARMA, A. H. HUIZER, J. Chem. Soc., Faraday Trans. 2, 78, 1945 (1982).
13. I. HALLER, J. Chem, Phys., 47, 1117 (1967).
14. C. A. G. O. VARMA, F. L. PLANTENGA, A. H. HUIZER, J. P. ZWART, PH. BERGWERF, J. P. M. VAN DER PLOEG, J. Photochem., 24, 133(1984).
15. J. L. G. SUIJKER, C. A. G. O. VARMA, A. H. HUIZER, Chem. Phys. Letters, 107, 496 (1984).
16. J. L. G. SUIJKER, M. W. VAN TOL, A. H. HUIZER, C. A. G. O. VARMA, to by published.
17. J. L. G. SUIJKER, C. A. G. O. VARMA, Chem. Phys. Lett., 97, 513, (1983)
18. D. V. O'CONNOR, M. SUMITANI, J. M. MORRIS, K. YOSHIHARA, Chem. Phys. Letters, 93, 50 (1983).
19. R. POTTIER, G. P. SEMELUK, Can. J. Spec., 21, 83 (1967).
20. J. L. G. SUIJKER, C. A. G. O. VARMA, A. H. HUIZER, J. Mol. Struct., 114, 269 (1984).

5.5 The Picosecond System in Jyväskylä

J. KORPPI-TOMMOLA

Department of Chemistry, University of Jyväskylä, SF-40100 Jyväskylä

A synchronously pumped argon ion laser picosecond system to measure absorption recovery times and fluorescence lifetimes of organic dye molecules in solution is described. For the former measurements conventional excite and delayed probe techniques is used while for the latter single photon counting system is used.

The time resolution of the system is now some 15 ps for the absorption recovery and some 50 ps for fluorescence lifetime experiments. A unique, low cost data aquistition system has been developed for both measuring techniques. The system comprises of a microcomputer (TRS-80), an 75 µs (12 bit) A/D converter and a direct transfer line to a Univac 1100/70 computer. The software includes programmes to drive a steppermotor, to record data by signal averaging, to display data in digital and graphical form, to send data from the microcomputer to the main frame computer and finally to deconvolve the experimental data.

The picosecond laser system

The laser system is based on a Coherent INNOVA 15 W mode-locked argon ion laser which synchronously pumps a resonator made of a dye laser (CR 599-04) and a cavity dumper in an extended configuration. The average power at repetition rates of 4 MHz is roughly

504 LASERS IN ATOMIC, MOLECULAR AND NUCLEAR PHYSICS

40 mW of cavity dumped radiation with a pumping power of 700 mW of the mode-locked 514 nm light. The tuning range using rhodamine 6G is form 570 nm to 645 nm. Laser dyes and corresponding optics are at our use continuously from 550nm to 950 nm.

For frequency doubling of the rhodamine 6G wavelengths a temperature tuned 1.5 cm ADA crystal is used while in the red part of the spectrum an angle tuned KDP crystal is used. In this manner we can cover the UV region from 275 nm to 450 nm.

The experimental set up for the autocorrelation and absorption recovery measurements is shown in Figure 1. The excitation beam is chopped at 2000 Hz and the corresponding signal is recorded on a lock-in-amplifier. Further data manipulation is described below.

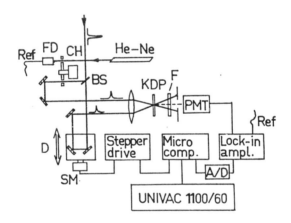

FIGURE 1

Our single photon counting system comprises of a fast (500 ps rise time) photodiode, of two Ortec 583 constant fraction discriminations (CDF) and Ortec 467 time to analog (TAC) converter. For detection of the signal a RCA 8575 photomultiplier tube(PMT) is used in for UV and blue fluorescence while a Hamamatsu R928 tube is used for visible light. Our multichannel analyzer is Nokia LP 4900

ULTRAFAST PROCESSES AND TECHNIQUE 505

with 2048 channels and with a 1200 baud/s I/O interface to our microcomputer. The instrumental function with identical inputs for both CFD:s is about 30 ps, while the resolution with a RCA 8575 PMT in the system is 260 ps. In theory our kinetic resolution is about 50 ps.

The PICO program

To record and autocorrelation trace and an absorption recovery signal a movable, steppermotor controlled optical delay line was constructed. Microcomputer controlled delay line moves at a maximum speed of 11000 steps/s and uses 3200 steps/full turn to advance 1 mm. The total length of the delay line is slightly over 2 ns.

A mixed basic and assembly lanquage program baggage PICO accelerates and decelerates the steppermotor and collects a 1500 point data set for each scan independent of the dealy used. To accomplish the last operation the user defines a summing and a delay index suitable for his use. The summing index defines the number

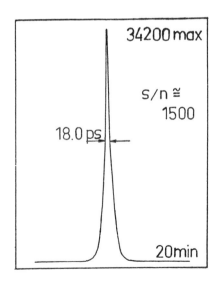

FIGURE 2

of adjacent signal values which are accumulated to make a single observation point. The delay index defines the number of steps, which are skipped during the scan. With these arrangements up to 24 bit data values can be obtained.

Figure 2 shows an autocorrelation trace of one of the picosecond pulses obtained from the system. The recorded data can be displayed graphically and numerically on the microcomputer screen before transferring by a PICO subprogram via a 2400 baud/s direct line to main frame computer. The data can also be stored on floppy disks of the microcomputer. The measuring program allows us to increase the S/N ratio of the system in theory up by a factor of ten as compared to direct analog signal recording for absorption recovery measurements. Good kinetic data are needed especially when studying multiexponential and long lived decay systems.

Our intention is to improve the time resolution of our absorption recovery system further by having two identical delay lines and two lock-in-amplifiers in the system. Recording the autocorrelation trace and the absorption recovery simultaneously allows for deconvolution of the kinetic data. It is estimated that this method will extend our time resolution to the femtosecond region.

The deconvolution programmes

Two programmes have been installed on a Univac 1100/70 computer to estimate kinetic results from recorded absorption recovery and single photon counting data. These programmes have originally been developed at Ris in Denmark to study slow positron decays from metal surfaces. One of these programmes FLUORES allows for searching the instrumental response function directly from the kinetic data. Up to five Gaussian quess functions to allow for asymmetric instrumental functions can be used. The 80 K program performs a point by point least square deconvolution to produce the origin

ULTRAFAST PROCESSES AND TECHNIQUE

of the spectrum, half-widths and locations of the Gaussians and the preliminary lifetimes up to five exponentials.

A program called FLUORFIT is finally used to get the refined kinetic data either using the response function obtained from FLUORES of from the experiment. The program allows for fixing the intensities and lifetimes of the varios exponential components. An option to choose and/or fix the background, the origin and the total area of the spectrum are also provided. A further advantage is to change the number of exponentials after the instrumental corrections have been made.

Figure 3 shows our instrumental response function of our system and a preliminary result of the fluorescence lifetime of Phodamine B in solution.

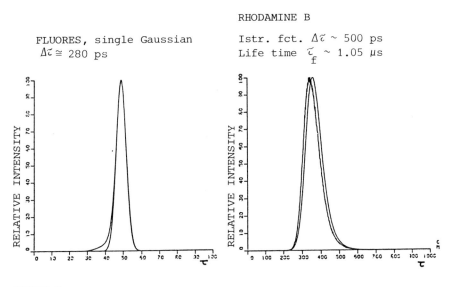

FIGURE 3

6. NEW LASER METHODS AND SOURCES

6.1 Classical, Semiclassical and Quantum Echo

V. P. CHEBOTAYEV and B. Ya. DUBETSKY

Institute of Thermophysics, Siberian Branch of the USSR Academy of Sciences, Novosibirsk, USSR

The mechanism of appearance of an echo in the ensemble with an inhomogeneously broadened line is discussed. It has been shown that the echo arises due to the phase jump of the nonlinear part of oscillations and is not associated with an assumption of the quantum nature of system excitation. The echo is reported to be experimentally observed in the classical system. The influence of the finite number of oscillators on the echo in classical and quantum systems is taken into account. Due to this factor, additional coherent responses appear in the ensembles with determinate frequencies. In principle, it is possible to observe an echo on one particle with quantized internal degrees of freedom. Qualitatively considered are the regularities typical of echo and their use in the problem of observation of interference of atoms resulting from scattering on a resonance standing wave of an atom with quantized internal and translational degress of freedom.

1. The phenomenon of echo is known to be as follows. When the system with an inhomogeneously broadened line is excited by two pulses, a coherent response arises in the system in a time equal to the delay between pulses T. The response contains an information on ralaxation processes that is usually concealed due to great inhomogeneous broadening. Use of the typical peculiarities of the echo has provided progress in Doppler free spectroscopy, it has become possible to observe the interference of atoms and develop

the interferometry of the subAngström range. There are known numerous other applications of the effect. Traditionally, the echo is related to the interaction of atoms with $\pi/2$ and π-pulses. Another point of view has been developed by us[1] and R. L. Shoemaker[2]. We assign the echo to the phase jump of a dipole moment of an atom. The difference between the approaches is as follows. In the first case an assumption on the quantum nature of system excitation is important, in the second not important. In our approach quantum equations play the secondary part and are used only for the calculation of a dipole moment, the following items are important:

1. Inhomogeneous line broadening, i.e., the existence of an esemble of resonance elements with spread of natural frequencies in some sufficiently broad spectral range $\Delta\omega$.
2. The system nonlinearity by an external action.
3. The principle of superposition.

These conditions can be easily simulated in the classical system where the effects typical of π-pulses are known to be missing. The process of reversible relaxation and an effect of echo may be clearly observed in such a system. And at last, study of the echo in the classical system has enabled us to obtain new results interesting for the echo in the optical range.

2. We have carried out the following experiment[3]. An ensemble of pendulums that were hung on the frame mounted on two piezoceramics was excited by pulses of a periodic external force (see Figure 1). The difference in the positions of the centres of gravity of the pendulums gives an inhomogeneous broadening, i.e. a spread over natural frequencies. Each pendulum is a nonlinear oscillator. The frame displacement is proportional to the sum of deflections of the pendulums from the position of equilibrium. The superposition principle is realized in this way. Figure 2a shows the response of

NEW LASER METHODS AND SOURCES 511

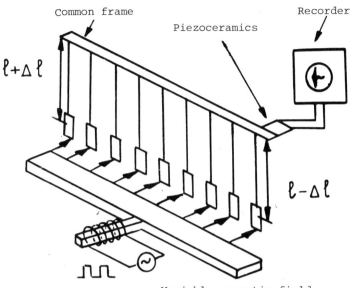

FIGURE 1   Schematic of the experiment

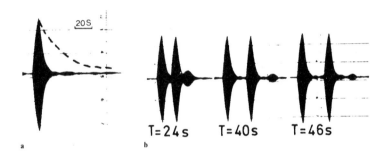

FIGURE 2   a) Variation with time of the oscillations of the ensemble of pendulums (continuous curve) and the envelope of the oscillations of one pendulum (dotted curve) under single-pulse excitation.
b) The same under two-pulse excitation for three values of delay.

the ensemble to a single pulse, the dotted line marks the envelope of the response of one pendulum. It is seen that in the ensemble the response relaxes quicker. This is due to dephasing of the pendulum oscillations. Under the action of the pulse all the pendulums oscillate synchronously at a field frequency ω, then each oscillates at a natural frequency ν. Since the frequencies do not coincide, a phase difference $\Delta\omega t$ arises between the oscillations. Due to this misalignment of the oscillations occurs in a time of the order of an inverse inhomogeneous width $t \geqslant \Delta\omega^{-1}$. It can be visually observed, and it is this fact that results in a more rapid relaxation of the signal in the ensemble.

After the action of the second field pulse the situation changes (see Figure 2b). A coherent response that is an analog of the echo in the classical system appears at a time 2T for different delays. It arises in the following way. Equations of motion of the oscillator are of the form:

$$\ddot{x} + \nu^2(x - ax^3) = 2Gf(t)\cos\omega t$$

where G is the amplitude, f(t) is the field pulse shape, a - characterizes the nonlinearity. Each pulse phases the oscillators at the moment of its action. So after two pulses the linear part of the pendulum oscillations is a superposition of two oscillations:

$$x_1 = a_1 \cos\varphi_1 + a_2 \cos\varphi_2$$

the phases of which increase linearly in time and differ from each other: $\varphi_1 = \nu t$, $\varphi_2 = \nu(t-T)$. Owing to cubic nonlinearity, these phases may be added and subtracted, i.e., under the action of the second pulse the phase jump of the nonlinear part of oscillations (φ_{nl}) occurs respect to the phase φ_1. The following combination of phases is possible:

$$\varphi_{nl}^{(1)} = 2\varphi_2 - \varphi_1 = \nu(t - 2T),$$

the jump $-2\nu T$ corresponds to this combination. This means (see Figure 3) that inversion of the phase φ_1 of the classical oscillator occurs at the moment of the action of the second pulse. Due to a smooth increase of the phase this jump will be further com-

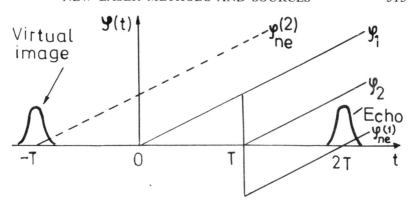

FIGURE 3 Time evolution of the phases of linear (φ_1 and φ_2) and nonlinear ($\varphi_{nl}^{(1)}$ and $\varphi_{nl}^{(2)}$) oscillations.

pensated to the moment 2T. At this moment the nonlinear parts of the oscillations of all oscillators, irrespective ot their natural frequencies, will be in phase. Visually nothing can be seen, the pendulums remain misaligned as they were and only owing to the device that follows the superposition of the oscillations, from the appearance of an echo pulse we can see that the coherence in the system is restored.

In the classical system the phases may be added differently and, in addition to the oscillation that is behind φ_1, a nonlinear oscillation arises with a phase $\varphi_{nl}^{(2)} = 2\varphi_1 - \varphi_2$ that is, on the contrary, ahead the linear one in phase. The phase jump for the nonlinear oscillation will never be compensated in the future, but it was compensated in the past. The related response may be considered to be a virtual image of the first field pulse. It is characteristic that this response is observed in the ensemble of oscillators with an equidistan spectrum: $\nu_k = \nu_o + k\delta$ (k = 1, 2, ..., N; N is the number of oscillators). In fact, if at a certain moment the oscillator phases coincide, then in the time $2\pi/\delta$ the coherence of the system is restored, as for this time the phase variation of each oscillator will be divided by 2π: $\varphi(t + 2\pi/\delta) = \varphi(t) + 2\pi\nu_o/\delta + 2\pi k$, i.e., here a series of coherent response will correspond to each process that phases the oscillator. This is also related to the virtual image. By adding the time $2\pi/\delta$ to the moment of its appearance we can transfer it from the past to the future and observe. We have made the numerical experiment[3]. The square of the sum of oscillator displacements was calculated:

$$P(t) = \left| \sum_{k=1}^{N} x_k \right|^2$$

This quantity is similar to the power of the echo in the response. For the different numbers of oscillators the linear and nonlinear parts of P(t) were calculated:

$$P_1(t) = \frac{1}{2}[NG/(2\omega\Delta\omega)]^2 f_1(t)$$

$$P_{nl}(t) = \frac{1}{2}[NG\varepsilon/(2\omega\Delta\omega)]^2 f_{nl}(t)$$

where $\varepsilon = 3aG^2\tau^3/8V$ is of the order of a relative quantity of nonlinear additions to x_1 arising for the time of the pulse action τ. Figure 4 shows how these quantities vary with an increase of N at $T/\tau = 5$, $\Delta\omega\tau = 15$ and the spectrum equidistant in the range $\omega - \Delta\omega \leqslant \nu \leqslant \omega + \Delta\omega$. It is seen that the linear response is decayed into a set of pulses, the nonlinear one, along with the echo pulse, contains a virtual image displaced in time by $2\pi/\delta$. It is sufficient that the frequencies in the ensemble would form a determinate sequence, that is not necessarily equidistant, for this effect to arise with the great inhomogeneous broadening $\Delta\omega\tau \gg 1$. The time at which the coherent responses are transferred with a given $\Delta\omega$ is linear by the number of elements in the ensemble. With an infinitely large N all the additional pulses will therefore be at infinity and only an echo pulse ramains, i.e., an effect of transfer of coherent response in time results from the allowance for the finite number of elements in the ensemble with determinate frequencies.

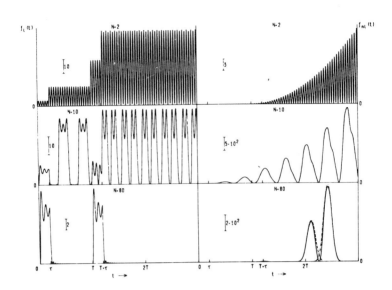

FIGURE 4 Linear and nonlinear responses in the ensemble of oscillators with an equidistant spectrum for different numbers of oscillators N. The dashed and dot-and-dash curves corresponding to the echo and virtual image, respectively, and are plotted by the asymptotic formulas with large N.

The situation changes if the frequencies in the ensemble are random values. It is this case that is borne in mind, as usual, in considerations both of an echo and of other responses in a gas under optical excitation, the value of the response averaged over the ensemble being calculated only. Meanwhile, in such problems as spectroscopy of trapped particles, single atoms and weak flows the allowance for the finite number of elements may turn out to be a topical item. Here the problem itself is put differently. Irrespective to the fact whether the question is about absorption, coherent radiation or, as in our case, about a square of the sum of oscillator displacements $P(t) = (1/2)(NG\tau/\omega)^2 p(t)$, we should consider the response itself to be a random variable and the problem of the theory would be a calculation of the distribution function of this random variable $W_{p(t)}(x)$. In the systems where a principle of superposition is fulfilled, the apparatus of a central limiting theorem is convenient for soving this problem. In the system of classical oscillators outside the region of echo formation only an incoherent noise exists. The following distribution function corresponds to the noise: $W_{p(t)}(x) = (N/\bar{x}) \exp(-Nx/\bar{x})$, where $\bar{x}$ is independent of N, an average value of power is linear in N: $\langle P(t) \rangle = (1/2)N \cdot (G\tau/\omega)^2 \bar{x}$ in full accordance with the fact that without coherence the square of the sum of displacements is reduced to the sum of the squares. In the echo region $|t - 2T| \lesssim \tau$ we have obtained the normal distribution

$$W_{p(t)} = (2\pi)^{-1/2} \sigma^{-1} \exp[-(x-\bar{p})^2/2\sigma^2]$$

with the dispersion the relative value of which decreases as $N^{-1/2}$. As well as an average value of p, the dispersion σ depends on time, the nature of inhomogeneous broadening, the shape of exciting pulses and other factors.

3. It would be interesting to note that the ensemble of the finite number of elements with determinate frequencies may also be realized for an ordinary echo in the optical band. If we use picosecond pulses for excitation and place a field frequency near the ionization threshold of an atom, the whole set of Rydberg levels is excited. We obtain the ensemble of dipoles emiting at the frequency of the transition ω_{jg} between the jth level and the ground state g. Here all the conditions necessary for echo are observed: the inhomogeneous broadening (frequency spread is of the order of an inverse pulse duration and may be greater than an inverse delay between pulses); the superposition principle (we must add the dipoles to the resulting dipole moment of an atom $d(t) = \sum_j d_j(t) \cdot \exp(-i\omega_{jg} t)$, then determine the intensity of emission of an atom that is $I(t) = (4/3)\omega^4 c^{-3} |d(t)|^2$; at last, the nonlinearity

NEW LASER METHODS AND SOURCES 517

arises because of two-quantum processes. It is characteristic that the whole ensemble is concentrated in one atom and, in principle, the echo may be observed in one particle.

In this case the amplitudes of the Rydberg terms are elements of the ensemble. Since they are concentrated in one atom, a coherent interaction sets in between them. If the field pulse excites an atom to one of the terms, then owing to the two-quantum process under the action of the second pulse an atom can pass virtually through the ground state to the other term and emit nonlinearly at the other frequency. The linear oscillations of one element of the ensemble were a source of the nonlinear oscillations of another element. In the case where the field frequency is so close to the ionization threshold that only a small part of the Rydberg spectrum is excited within which the frequencies may be considered to be equidistant, the responses of the type of echo arise not only at the moment 2T but also at the other moments. Actually, let the first pulse excite an atom to the jth level, for the time between the pulses the amplitude phase of this level smoothly changes, under the action of the second pulse a jump-like phase inversion occurs. If an atom is at the same jth level, the situation is similar to that typical of echo, when the phase jump is compensated by the moment 2T. If an atom is on the other j_1 level, the phase $\psi(t)$ varies according to the other law. If with $t > T$ the slope of the dpendence $\psi(t)$ decreases twice, then, evidently, the phase jump will be compensated by the moment 3T and a coherent emission pulse will arise at this moment. It may be shown that owing to the coherent coupling of the terms the pulses of the echo type arise at all distances T_o that are related to the delay T as a rational number. Figure 5 shows the pulses of atom emission calculated by the formula

$$I(t) = 4(\omega |d_{gj}|^2 \tau)^4 E^6 i(t)/(3\hbar^6 c^3 \delta^2)$$

518 LASERS IN ATOMIC, MOLECULAR AND NUCLEAR PHYSICS

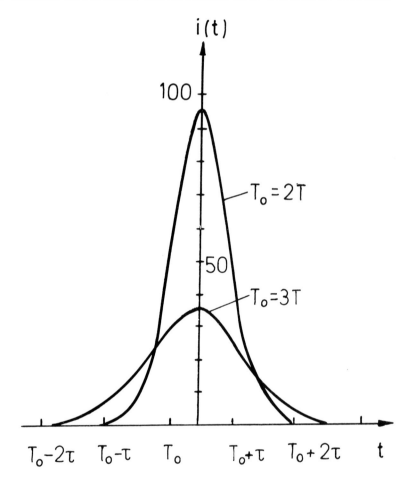

FIGURE 5 Radiation pulses of a Rydberg atom at the moments 2T and 3T

in the assumption that the field is weak, the term summation may be replaced by integration, and the Rydberg spectrum is considered to be equidistant, i.e., it is assumed that $\omega_{j+1,g} = \omega_{j,g} + \delta$, δ = const, d_{gj} = const.

4. When discussing the echo on the Rydberg terms we have passed, in fact, from the classical to the semiclassical echo. Here we mean an ordinary photon echo on the atoms the internal degrees of freedom of which are quantized, the motion of them as the whole

NEW LASER METHODS AND SOURCES

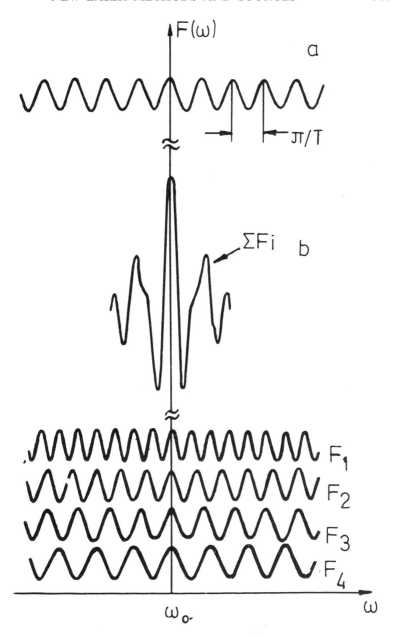

FIGURE 6 Shape of the response $F(\omega)$ in a gas in separated standing-wave fields with a given delay (a) and with variation of the delay and accumulation of the signal from the ensemble of four sinusoids (b)

may be considered to be a classical one. We are interested mostly in applications of the ralated effects in superhigh resolution spectroscopy. These applications are underlay by the effect we predicted in[4] that is the effect of appearance of the response $F(\omega)$ oscillating near the line centre as a function of the field frequency ω with a period inverse to the delay between pulses in a gas on the Doppler broadened transition under the pulse excitation of a standing wave (see Figure 6a). The responses of this type were theoretically studied and often observed experimentlly in one-and two-photon absorption, coherent radiation, in a flow of excited particles for time and spatially separated fields (see references in the survey[5]). The response itself cannot be used in spectroscopy as it is unclear what maximum of the sinusoid $F(\omega)$ should be related to the line centre, the frequency of the atomic transition ω_o. Generation of a sharp maximum requires an ensemble of sinusoids arising at different delays that should be added. In the line centre all the sinusoids are added coherently. Since the period of the sinusoids depends on delay, in the rest points the resulting response decreases and we obtain a resonance with a width inverse to the delay averaged over the ensemble of sinusoids T_o: $\Gamma_R \sim T_o^{-1}$. Spatially separated fields may be used for the averaging. In the system of rest of an atom these fields are pulsed ones, as before, but the delay time depends on the transverse velocity u (T = L/u, where L is the distance between the fields) and when averaging over the transverse velocities u we simultaneously average over the delay times. The width of the arising lines can reach hundreds Hertz. At present the narrowest resonance has been obtained by this methos by R. L. Barger at the 660 nm transition in Ca[6], its halfwidth being Γ_R = 1 kHz. The achieved resolution (Γ_R/ω) amounts to $2 \cdot 10^{-12}$, which is the best resolution obtained in optics.

We would like to emphasize especially that the spatial separation of fields is required for the procedure of averaging of responses over delay times to be realized. There is no difference of this case from that of time separated standing wave pulses. Pulsed fields may also be used for obtaining narrow lines, if at a given frequency a signal is stored at different delay times and then a frequency is tuned. Under spatially unlimited conditions a result of this averaging is[7] the Lorentzian contour with a halfwidth equal to a homogeneous transition halfwidth. If the field is limited, not all pairs of pulses provide an identical contribution but those the delay between which is not greater than the flight time of atoms across the field. In this case the line width seems to be limited by the transit width τ_o^{-1}, where $\tau_0 = a/v_o$ is the flight time of atoms across the field, a is the transverse field size, v_o is the thermal velocity. However there is a velocity spread in gas, and great as the delay T is, there are always the atoms with so small velocity $u \leqslant a/T$, that they do not leave the field for the time between pulses. Owing to these atoms a transit broadening free line will be formed in a gas. This line may be used in two-photon absorption resonances where beam focusing is convenient, however there a transit broadening occurs. In[8] this method is proposed to narrow the line in hydrogen on the 1S-2S transition up to Γ_R = 100 kHz. According to calculation, for the pulse sources with the parameters: wavelength λ = 243 nm, power 10 W, pulse duration $\tau = 10^{-7}$ s, repetition rate 10 Hz, the signal amplitude is 20 times as large as that under continuous excitation earlier calculated[9]. The present method will, possibly, promote the narrowing of a two-photon line in hydrogen that could not yet be made by the other ways.

As far as one-photon lines are concerned, the signal accumulation, with variation of a delay time, is realized experimentally[10] during the investigation of coherent radiation in separated field

FIGURE 7 Lineshape of coherent radiation at the 10.6 μm transition in SF_6

(CRSF) at the 10.6 μm transition in SF_6. Figure 7a shows the CRSF line with the given delay T = 1.45 μs, Figure 7b presents the resonances arising during the variation of the delay from 1.45 to 4.9 μs and the signal accumulation. The lines with a width of up to 10 kHz have been obtained by this method, according to estimations, lines with a width of 500 Hz and less may be obtained.

Here it also seems real to eliminate transit broadening at the expense of an effect of slow atoms. The versions of this effect for continuous excitation were earlier proposed[11-13,15]. Our pulse version is advantageous for two reasons.

A. There is a possibility of compensating the decrease of the resonance amplitude proportional to the number of slow-velocity atoms ΔN which in gas falls as the square of linewidth: $\Delta N/N \sim \Gamma_R^2 \tau_0^2$, where N is the gas density. In a two-photon case the compensation is due to the increase of the power density at the field focusing, in a one-photon case due to the decrease of a pulse duration and the broadening of the Bennet hole up to the thermal velocity. The wave power should be simulataneously largely increased. Here advantageous are the transitions resonant to the radiation from a CO_2 laser that is a sufficiently powerful source for this purpose. In addition to SF_6, transitions in OsO_4, CH_3F, NH_3 and others are also convenient.

B. A line in time separated fields is free of field broadening. This is typical of the method of separated fields. As usual, the transition-saturating power is decreased during the line narrowing. As is shown in[14], experimentally this factor limits the potentialities of narrowing, as with decrease of the source power down to the level of the saturating power the resonance amplitude falls. In case of separated fields the situation is contrary. Here it becomes possible to effectively saturate the transition, no field broadening occurs, as a narrow resonance is observed for the time of flight between the fields, where a strong saturating field is missing at all.

5. Quantum echo. Ref.[16] discusses the possibilities of observing the interference of atoms scattered by a standing wave. On the one hand, the echo peculiarities are used in a purely quantum case, the degrees of freedom, both internal and translational. On the other hand, the need for this peculiarities and for separated fields naturally result from the statement of the problem. In fact, if before the interaction with the field the wave function of an atom was a flat wave with pulse p, due to scattering there appear in it two components with amplitudes ψ_+ and ψ_- corresponding to the change of an atom pulse to a photon pulse $+\hbar \vec{k}$ and $-\hbar \vec{k}$, where $\vec{k}$ is the wave vector of a standing wave. The two waves interfere and a periodic structure appears in the atomic density, the square of the modulus of the wave funstion

$$\rho(z) = |\psi_+ e^{i(\vec{p}+\hbar\vec{k})\vec{r}/\hbar} + \psi_- e^{i(\vec{p}-\hbar\vec{k})\vec{r}/\hbar}|^2 =$$
$$= |\psi_+|^2 + |\psi_-|^2 + 2|\psi_+ \psi_-|\cos(2kz + \arg(\psi_+ \psi_-^*))$$

where the z axis is directed along k. One field is sufficient for the periodic structure to appear, however the amplitude of the density harmonic is proportional to the product of the time passing after the action and of the recoil frequency $\Delta = \hbar k^2/2M$ ($k = |\vec{k}|$, M is the atom mass), the typical frequency responsible for quantum effects. Thus, we can observe the structure in a time $T \gtrsim \Delta^{-1}$. It is the long time in optics and ranges from 10 to 100 μs, for this time atoms can move at a distance $L \sim v_o T \sim 1$ cm. This distance is greater than the optimum field size a. This size is found from the condition that all atoms in a beam would belong to the Bennett hole, i.e. for the time of interaction with the field $\tau = a/u$ they would move along the wave by no more than a wavelength $\lambdabar = 1/k$. Otherwise, for the time of interaction the atoms will cross many nodes of a standing wave, the average field amplitude along the atom trajectory will be small and the interaction effectiveness will decrease. If an atomic beam propagates in the direction perpendicular to the axis z, then for the time τ the particles will move along the wave by a distance $\Delta L \sim u\theta\tau \sim a\theta$, where θ is the angular beam divergence. Then from the relation $\Delta L \lesssim \lambdabar$ we obtain the limitation for the field size $a \lesssim \lambdabar/\theta$, which in optics ($\lambdabar \sim 10^{-5}$ cm) amounts to 10^{-2} cm even with the angular divergence $\theta \sim 10^{-3}$. Thus, the interference can be observed only in the region spatially separated from the exciting field. However at a large space, due to the angular spread in a beam the nodes of the grating (periodic structures) formed by atoms with different velocities will be displaced by a large value $x \sim L\theta$, i.e. the grating will acquire a great additional phase $\varphi \sim L\theta/\lambdabar \sim$ $\sim \theta k v_o/\Delta \sim 100$. Since the averaging over angles is equivalent to that over phases φ, the periodic structures in the spatial distribution of atoms in the beam will be smeared at such a great distance from the exciting field. Consequently, the large phase φ should be compensated for the effect of atom interfernce to be observed. The typical peculiarities of the echo are required here; an atomic beam is excited by the second wave separated at a large distance. At the nonlinear interaction the phase jump of the wave function of an atom occurs, at certain distances it exactly compensates a large phase φ, and it becomes possible to observe a clear interference pattern.

REFERENCES

1. V. P. CHEBOTAYEV, in Coherence in Spectroscopy and Modern Physics, edited by F. T. Arecchi, R. Bonifacio, and M. O. Scully (Plenum Press, New York, London, 1978), p. 173; Appl. Phys., 15, 219 (1978)
2. R. L. SHOEMAKER, in Laser and Coherence Spectroscopy, edited by J. L. Steinfeld (Plenum Press, New York, London, 1980)
3. V. P. CHEBOTAYEV, B. Ya. DUBETSKY, Appl. Phys., B31, 45 (1983)
4. E. V. BAKLANOV, B. Ya. DUBETSKY, V. P. CHEBOTAYEV, Appl. Phys., 9, 171 (1976)
5. B. Ya. DUBETSKY, Izv. AN SSSR, Ser. Fiz., 46, 990 (1982)
6. R. L. BARGER, Opt. Lett., 6, 145 (1981)
7. V. P. CHEBOTAYEV, in Laser Spectroscopy IV, edited by H. Walther and K. W. Rothe (Springer, Berlin, Heidelberg, New York, 1979), p. 106.
8. B. Ya. Dubetsky, The XIXth All-Union Symposium on Spectroscopy, Abstracts (Tomsk, 1983), part II, p. 228.
9. A. I. FERGUSON, J. E. M. GOLDSMITH, T. W. HÄNSCH, E. W. WEBER, in Laser Spectroscopy IV, edited by H. Walther and K. W. Rothe (Springer, Berlin, Heidelberg, New York, 1979), p. 31.
10. L. S. VASILENKO, I. D. MATVEYENKO, N. N. RUBTSOVA, Report at the VIII Vavilov Conference on Nonlinear Optics, Novosibirsk, 1984; Preprint 114-84 (Institute of Thermophysics, Novosibirsk, 1984)
11. S. G. RAUTIAN, A. M. SHALAGIN, ZhETF, 58, 962 (1970)
12. E. V. BAKLANOV, B. Ya. DUBETSKY, V. M. SEMIBALAMUT, E. A. TITOV, Kvantovaya Elektronika, 2, 2185 (1975)
13. A. M. SHALAGIN, ZhETF Pisma, 34, 193 (1981)
14. V. A. ALEXEYEV, N. G. BASOV, M. A. GUBIN, V. V. NIKITIN, A. V. NIKULCHIN, V. N. PETROVSKY, E. D. PROTSENKO, D. A. TYURIKOV, Kvantovaya Elektronika, 11, 648 (1984)
15. V. A. ALEXEYEV, N. G. BASOV, M. A. GUBIN, V. V. NIKITIN, N. S. ONISHCHENKO, ZhETF, 84, 1980 (1983)
16. B. Ya. DUBETSKY, A. P. KAZANTSEV, V. P. CHEBOTAYEV, V. P. YAKOVLEV, ZhETF Pisma, 39, 531 (1984)

6.2 Fiber Optic Sensors

H. P. WEBER and Q. MUNIR

Institute of Applied Physics, University of Berne, 3012 Berne, Switzerland

An overview on optical fiber sensors is given with particular emphasis on sensors employing fiber optic interferometers. The optoacoustic fiber sensor is discussed in some details.

1. INTRODUCTION

The use of optical fibers as sensors has gained a lot of interest in the past few years[1]. The reason being that they possess nearly all the advantages that the fibers have in the optical communication techniques, e.g. immunity against electromagnetic radiation, the possibility of using them in explosive and corrosive environments, and as they are made out of glass or plastic they can be used in high electric or magnetic fields.

The fiber sensors are based on the fact, that light may be modulated in intensity, phase or polarization that is transmitted through the optical fibers. If the modulation is introduced onto the initially unmodulated lightfield while the light is carried in the optical, the sensors are classified as "All Fiber Sensors". This is contrary to the "Hibrid Sensors", where the fiber is simply used to carry light to an attached sensing element. Both multimode optical fibers have been used to construct fiber sensors. In section 2 we will discuss sensors constructed out of multimode

528 LASERS IN ATOMIC, MOLECULAR AND NUCLEAR PHYSICS

fibers and in section 3 sensors constructed using singlemode fibers with the emphasis on an optoacoustic fiber sensor.

2. MULTIMODE FIBERSENSORS

The multimode fibersensors are attractive because of their simplicity of construction (large core diameter makes it easier to couple light in a multimode fiber) and compatibility with incoherent broadband light sources. In the following we give examples for the three basically employed principles:

a) The light is coupled out of the fiber and back into another fiber. The displacement of the mutual geometry changes the degree of coupling which is then measured.

b) The critical angle of total internal reflection is used to sense changes along the surface of the fiber.

c) Controlled introduced microbending allows to measure pressure that is applied to a plate.

a) Displacement Sensors

Two fibers are used with their ends polished at such an angle with respect to the fiber axis, that total internal reflection (TIR) is obtained for all modes propagating in the fibers. If the fibers are placed very close ($< 1\,\mu m$) to each other a large fraction of light can be coupled from one fiber to the other (Figure 1) and this can be used as a displacement sensor^2. If one fiber is kept fixed and the other experiences a vertical displacement the inten-

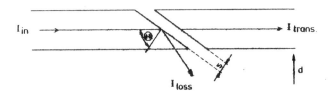

FIGURE 1 Displacement sensor using fibers, cut at an angle Θ beyond the critical total internal reflection.

sity of this coupled light gets modulated. These sensors can be made to detect displacements of ~ 10 nm or pressure of ~ 1 mPa.

Sensors employing moving optical line gratings attached to diaphragms that get displaced, e.g. under acoustic excitation (Figure 2) are good examples of hybrid displacement sensors[3]. Relative displacement between a pair of gratings modulates the transmitted light. Displacements as small as ~ 0.1 nm or prssure ~ 0.1 mPa can still be resolved.

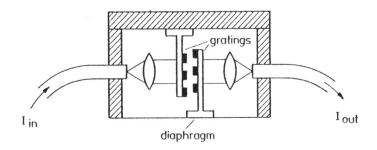

FIGURE 2 Displacement sensor using two line gratings

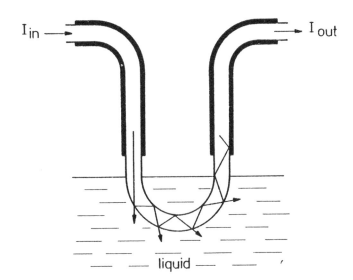

FIGURE 3 Liquid level sensor based on losses at individual internal reflections.

b) 1. Liquid Level Detector (Liquid Refractometer)

An optical fiber can be used as a liquid level detector[4], provided the cladding from a small part of the fiber is removed and then bent into a U-shape (Figure 3). Inserting this "naked" U shape in the liquid will couple out come the light propagating in the core, depending on the refractive index of the liquid. The resulting intensity reduction when the fiber is in contact with the liquid is in the range of 1.5 to 2 for a typical geometry and index difference. Such a sensor cannot just detect the presence of the liquid, rather can also be used to determine the refractive index of the liquid with an accuracy of $\sim 10^{-3}$.

b) 2. Temperature Sensor

The basic principle of the fiberoptic temperature sensor[5] is related to the liquid level sensor. At one fiber end the cladding of the fiber is replaced by a liquid whose refractive index is the same or slightly higher than the core of the optical fiber (Figure 4). The fiber end has also a high reflective coating. Light is coupled over a beamsplitter into the fiber and the reflected light is monitored with a photodiode. With an increase in temperature there is a characteristic change in the intensity of the reflected light, due to the decrease in the refractive index of the liquid.

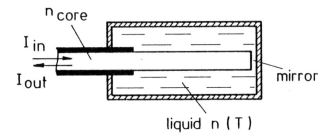

FIGURE 4 Temperature sensor based on losses at the interface dependent on $\Delta n(T) = n_{core} - n(T)$

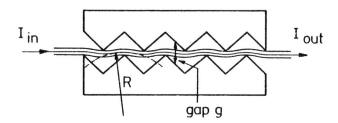

FIGURE 5 Microbend-Sensor. The narrower the introduced bend radii, the higher the transmission losses.

The sensitivity of such a system is ~ 0.1°C.

c) Microbending Pressure Sensor

This sensor uses the well known microbending effect which causes the coupling of light from the core to the cladding of an optical fiber[6]. A fiber is placed between the two grooved plates that are slightly separated (Figure 5). This separation decreases with an increase in pressure and the result is the introduction of microbends that lead to a decrease in the intensity of the light propagating in the core of the fiber. The sensitivity of such a sensor is ~ 1 mPa.

3. SINGLEMODE FIBERSENSORS

The single mode fiber sensors, though technically somewhat more complicated for construction as compares to multimode sensors, offer a sensitivity theoretically orders of magnitude higher then many other existing technologies, if they are used together with interferometric techniques. Interferometry becomes possible as light remains coherent if guided in singlemode waveguides. The three interesting fiberoptic interferometer arrangements are:

   a) The Sagnac interferometer to measure rotation
   b) polarization sensors, and
   c) Mach-Zehnder interferometers.

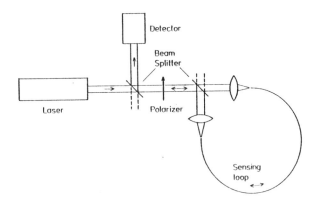

FIGURE 6   Fiber Gyroscope based on Sagnac Interferometer

a) <u>Fiberoptic Rotation Sensor</u> (Fiber Gyro)

The basic principle of a fiber gyro[7] is a Sagnac interferometer (Figure 6). Light is coupled into both directions of a single mode fiberoptic coil. Rotation of the gyro at a rate about an exis parallel to the coil axis results in a nonreciprocal phase difference $\Delta\Phi$ between the couterpropagating light beams as given by $\Delta\Phi = 8\pi A\Omega/(\lambda c)$. A is the total area enclosed by the fiber coil, λ and c are the vacuum wavelength and velocity of the light. When the two beams are recombined the phase shift leads to the rotation dependent intensity of light, which can be measured with a photodiode. Sensitivity approaching 1 deg/h seems easily realizable, the best reported results are 0.1 deg/h.

b) <u>Polarization Sensor</u>

A fiber tightly wound on a compliant cylinder induces large birefringence in the fiber (Figure 7). This birefringence allows two orthogonally polarized modes of the coupled light to propagate independently in the fiber. External perturbations such as acoustic or magnetic fields, change the birefringence of the fiber coil due to the strain induced via the compliant cylinder, which further

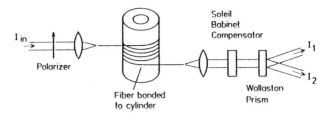

FIGURE 7 Polarization Sensor. By expansion of the compliant cylinder the birefringence of the fiber is modified

changes the phase relation between these two modes. The result is a relative delay between the two orthogonally polarized modes[8]. With the polarizing elements and a photodiode the light emerging from the fiber is analyzed. When used as a magnetic field sensors, polarization sensors have shown a sensitivity ~ 0.8 rad/Oe per meter of fiber.

c) Mach-Zehnder Interferometric Sensors

The Mach-Zehnder interferometer is the most used arrangement when the two light passes should be separated. The basic arrangement (Figure 8) can detect anything that affects the phase of light in the sensing fiber[9]. The most common application is the detection of soundwaves in liquids. Part of a laser beam is sent through the

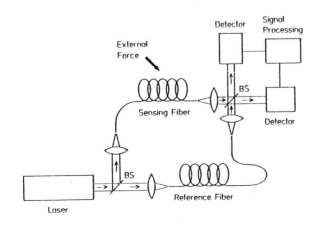

FIGURE 8 Mach-Zehnder arrangement of Fiber Interferometer

sensing fiber, the other part through a reference fiber kept in a stable environment. A disturbance such as acoustic pressure distorts the sensing fiber, changing its effective optical length ($D = n\ell$) and altering the relative phases of light in the two fibers. This phase shows up in the interference pattern produced when the two beams are recombined, and a suitable demodulator is then employed to detect the phase retardation introduced in the signal arm. The sensitivity of an interferometric fiberoptic sensor is determined by the shot noise and the thermal noise limit. Phase changes as small as 10^{-5} to 10^{-6} rad have been measured.

Optoacoustic Fibersensor

For gaseous trace analysis optoacoustic spectroscopy is a very efficient method[10]. In an optoacoustic cell a soundwave is generated and usually detected with an electronic microphone. This microphone can be replaced by an optical fiber interferometer in the Mach-Zehnder arrangement. Such a system was built and experimentally tested[11]. Figure 9 shows the experimental set-up. Two mono-

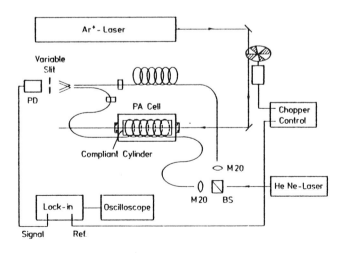

FIGURE 9   Optoacoustic fiber sensor

mode optical fibers of equal length (~10m) were used to construct the fibersensor. One piece of fiber wound on an aluminium foil cylinder (diameter 82 mm, length 50 mm, thickness 20 μm) served as a sensor arm and was placed coaxially in the optoacoustic cell. The optoacoustic cell itself was constructed from an aluminium cylinder of 200 mm length, 87.5 mm inner diameter and having a wall thickness of ~ 8 mm. It was filled with a mixture of 0.25% NO_2 in Ar. NO_2 having a comparatively broad absorption band in the visible spectrum was excited with an intensity modulated Ar-ion-laser (λ = 496.5 nm). The absorption of the modulated Ar-ion-laser leads to the generation of the sound waves due to the photoacoustic effect. The interaction of the sound field with the signal arm results in a phase change, causing the intensity molulation of the interference pattern, which can be detected with a photodiode. A linear dependence between the cell pressure and the optoacoustic signal as measuted with a fiberoptic sensor could be established (Figure 10).

The concentration of NO_2 molecules was kept constant during the experiment and the cell pressure was changed by adding buffer

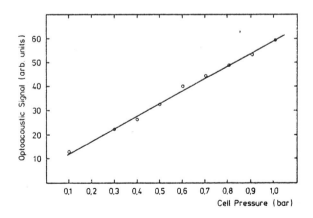

FIGURE 10 Optoacoustic signal dependence on the pressure of the buffer gas. Partial pressure of NO_2 was kept constant

536 LASERS IN ATOMIC, MOLECULAR AND NUCLEAR PHYSICS

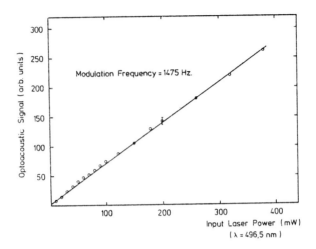

FIGURE 11 Optoacoustic signal dependence on the power of the exciting laser beam

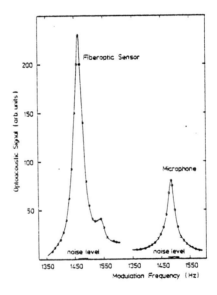

FIGURE 12 Fiberoptic sensor response in the region of an acoustic resonance of the cell. For comparison the response from a microphone, placed in the same cell, is also shown

gas. A linearity between the optoacoustic signal and the intensity of the modulated Ar-ion-laser was also found (Figure 11). A frequency response of the fibersensor was obtained by operating it around the resonance having the highest quality factor Q (Figure 12). The minimum detectable concentration as obtained from the signal to noise ratio of unity is ~0.5 ppm NO_2 in Ar for an input power of 50 mW. Extrapolating this result for an input power of 500 mW renders a minimum detectable concentration ~50 ppb. The corresponding minimum observable absorbed power is $7 \cdot 10^{-8}$ W/(cm$\sqrt{Hz}$). We believe that the sensitivity can be highly impoved by introducing some straightforward modifications such as acoustic isolation of the cell, the use of a longer aluminium foil cylinder and of a longer optical fiber.

REFERENCES

1. T. G. GIALLORENZI et al., IEEE J. Quant. Electron., QE-18, 626(1982)
2. W. B. SPELLMAN, D. H. MCMAHON, Appl. Opt., 19, 113 (1980)
3. W. B. SPELLMAN, D. H. MCMAHON, Appl. Opt. Lett., 37, 145 (1980)
4. K. SPENNER et al., Proceedings 1st Intern. Cong. on Optical Fiber Sensors (London, IEE, 1983), pp. 96-99
5. A. M. SCHEGGI et al., Proceedings 1st Intern. Conf. on Optical Fiber Sensors (London, IEE, 1983), pp. 55-63
6. C. M. DAVIS, T. A. LITOVITZ, D. B. MACEDO, Proceedings 1st Intern. Conf. on Optical Feber Sensonrs (London, IEE, 1983)
7. S. EZEKIEL, H. J. ARDITTY, Fiber-Optic Rotation Sensors and Related Technologies in Springer Series in Optical Sciences (Springer Verlag, 1982), pp. 2-42
8. S. C. RASHLEIGH, Opt. Lett., 6, 19-21 (1981)
9. J. HECHT, High Technology, Julay/Aug. 1982
10. L. B. KREUZER, Optoacoustic Spectroscopy and Detection, edited by Y.-H. Pao (Academic Press, New York, 1977), pp. 1-25
11. Q. MUNIR, R. BÄTTIG, H. P. WEBER, Proceedings 2nd Intern. Conf. on Optical Fiber Sensors (Stuttgart, 1984) to be publ. and
    Q. MUNIR, H. P. WEBER, (appears in Opt. Commun.)

6.3 Horizons of Detection of Gravitational Waves on the Basis of Frequency Stable Lasers

S. N. BAGAYEV and E. V. BAKLANOV

Institute of Thermophysics, Siberian Branch of the USSR Academy of Sciences, 630090 Novosibirsk, USSR

Astrophysical sources of gravitational waves may be divided into two types: pulsed and periodic ones. Pulsed sources emitting gravitational waves are bursts of supernovae, gravitational collapse of star. The energy of emission from pulsed sources is much greater than that from periodic ones. So, practically all gravitational antennae in almost all known research groups in the world (Maryland, Stanford, Lousiana, Rome, Moscow, Rochester, München, Glasgow, Perth) are designed just for pulsed sources[1]. Pulsars are an example of the periodic sources. For instance, the antenna for detecting gravitational radiation from the PSR 0532 pulsar (Crab) has been produced in Tokyo[2]. Periodic sources have two advantages: first, the radiation may be observed for a long time, which allows using high-quality antennae and thereby increasing their sensitivity; second, periodic sources are, as a rule, controllable, i.e. we know at what frequency gravitational waves should occur and prceeding from astrophysical data can estimate their amplitude. In addition, correlation between the repetition rate of radio pulses from the source and the frequency of gravitational waves will enable one to establish reliabley the fact of detection of gravitational waves. The most sensitive element of the antenna is the sensor that detects mechanical displacements of the antenna under the action of gravitational waves. Various types of detectors are now used: piesoceramic, volumetric, magnetic, superconducting ones and others the sencitivity of which is 10^{-12}-10^{-14} cm. Laser methods may also be used to measure small lengths. The researchers at the Institute of Thermophysics in Novosibirsk have measured the periodic displacements with an accuracy of $6 \cdot 10^{-14}$ cm by using the laser detector with a narrow optical resonance in a gas^3, which allows one to speak about a new type of the detector competing with the other types. An analysis shows that the potentialities of this method are not yet exhausted: the instrument sensitivity^3 may be increased up to 10^{-16} cm, estimations for other designs, in principle, give 10^{-21} cm.

In this paper we shall discuus the possibility of detecting gravitational waves from pulsars by using the laser detectors^3.

1. Gravitational Waves from Pulsars

Pulsars are interesting objects that are expected to produce a sufficiently intense gravitational radiation. The most known pulsar is the PSR 0532 (Crab) pulsar that is at the distance $R \approx 2$ kps away from the Earth. The electromagnetic radiation from the pulsar in a form of pulses with the repetition rate of 30 Hz has a wide spectrum: it emits in the radio-frequency and visible ranges and most intensively in the X-ray range. This has enabled one ot establish reliably that the pulsar is a neutron star with a mass that is one and half times as large as that of the Sun and a radius of about 10 km. The pulsar is an almost spherically symmetric body with the moment of inertia $I = 4 \cdot 10^{31}$ kg·km^2. The rotation frequency of the pulsar around the axis is $f = 30$ Hz (period T = 0.33 ms). This frequensy varies very slowly with time, its decrease per year is about 0.01 Hz, which corresponds to the relative variation of the period $\Delta T/T = 3 \cdot 10^{-13}$.

Gravitational waves that should be emitted by the pulsar have a frequency equal to the doubled rotation frequency of the pulsar $\nu = 2f = 60$ Hz. If the pulsar is considered to be an almost symmetric body with the moment of inertia I, the overall power emitted by the pulsar is

$$P = (1/10)(\gamma \omega^6 I^2 \varepsilon^2 / c^5) = 0.3 \chi (f/f_c)^2 (I/I_c)^2 \text{ [erg/s]} \quad (1)$$

wher $\omega = 2\pi\nu$, γ is the gravitational constant, a dimensionless quantity $\chi = (I_1 - I_2)/I$ characterizes the degree of axial asymmetry of the pulsar with respect to the rotation axis, I_1 and I_2 are the moments of inertia along the main axes perpendicular to the rotation axis, $f_c = 30$ kHz and $I_c = 4 \cdot 10^{31}$ kg·km^2 are the rotation frequency and the moment of inertia of the Crab pulsar.

Energy loss due to radiation of gravitational waves results in slowing-down of the pulsar rotation. the relative variation of the period $T = 1/f$ is

$$\Delta T/T = (8\pi/5)(\gamma \omega^3 I \varepsilon^2 / c^5) = 3 \cdot 10^{-7} \chi (f/f_c)^2 (I/I_c) \quad (2)$$

According to (1), the density of energy flux on the Earth $J = P/(4\pi R^2)$ may be written in the form

$$J = (c^3 \omega^2 h^2)/(32\pi\gamma) = 6(f/f_c)^2 h^2 \text{ [erg/cm}^2\text{s]} \quad (3)$$

where

$$h = (2/\sqrt{\pi})(\gamma \omega^2 I \chi)/(c^4 R) = 7 \cdot 10^{-22} \chi (f/f_c)^2 (IR_c/I_c R) \quad (4)$$

is the dimensionless wave amplitude, $R_c = 2$ kps.

Let us discuss the amplitude h for the Crab pulsar. The upper limit of h follows from the requirement that the increase of the rotation period (2) due to radiation of gravitational waves should not exceed the observed value $\Delta T/T = 3 \cdot 10^{-13}$. According to (2-4), this corresponds to $\chi < 10^{-3}$, $h < 7 \cdot 10^{-25}$, $J < 3 \cdot 10^{-7}$ erg/cm^{2}s. In the other words, the maximum achievable amplitude of the gravitational wave from the Crab pulsar is $h_{max} = 7 \cdot 10^{-25}$.

The axial symmetry of the pulsar PSR 0532 is prsently known, the quantity h is therefore calculated from the theoretical models of a pulsar. According to^4, $h = 2 \cdot 10^{-26} - 2 \cdot 10^{-29}$.

Thus, it is possible to speak seriously on an attempt of detecting gravitational waves only when antenna sensitivity is sufficient for measuring $h_{max} = 7 \cdot 10^{-25}$. With $h = 10^{-26} - 10^{-27}$ gravitational waves may be expected to be detected, with $h = 10^{-28} - 10^{-29}$ gravitational waves must be detected. Further astrophysical observations with the PSR 0532 pulsar and their theoretical analysis will give a more definite value of h.

Besides the smallness of h (which is natural), a serious drawback of pulsars is the low frequency of radiation of gravitational waves, which restricts technological potentialities of receivers. So, a great interest has been aroused by the discovery of a new pulsar PSR 1937+21 in November, 1982 in the Fox cluster at the distance of 2.5 kps away from the Earth that has an extremely short period of revolution T = 1.5 ms (f = 670 Hz) and a very high pulse repetition rate $\Delta T/T \approx 10^{-19}$ ^{5}. One may expect that such a small deceleration of the rotation of this pulsar is due to the radiation of gravitational waves at the doubled frequency of pulsar rotation ν = 1.33 kHz. If so, taking the moment of inertia similar to that for the Crab pulsar for estimations in (2) we shall obtain a very high degree of axial symmetry $\varepsilon \sim 5 \cdot 10^{-9}$ and the amplitude of a gravitational wave on the earth $h \sim 10^{-27}$. This is more than two orders as little as those for the Crab pulsar. However, owing to a greater frequency (by 22 times), detection of gravitational waves from this pulsar may turn out to be simpler than that from the Crab pulsar (see Section 3).

2. <u>Resonance Gravitational Antenna</u>

A gravitational wave coming from the pulsar must be detected on the Earth. The gravitational wave receivers may be divided into three

NEW LASER METHODS AND SOURCES

kinds: i) antenna, the mechanical system acted upon by the wave; ii) detector, the system that follows small displacements (and transforms them into electric oscillations); iii) computer, the system of treatment of the signal. The three systems should satisfy extremely high requirements.

Let us first consider the antenna. On the Earth a gravitational wave may be detected through its action on test bodies. Two free bodies are a simple antenna. If they are placed at a distance l along the x axis, they would acquire a relative acceleration

$a = (\hbar \omega^2 l/2) \cos \omega t$.

The antenna sensitivity may be improved by making the antenna as an oscillatory system. A real resonance antenna is a high-quality mechanical resonator that is an elastic body of a certain form. The frequency of the principal mode ω_o is close to the frequency of the gravitational wave ω.

An elastic rod can be the simple antenna. The stationary amplitude of displacement between the ends is

$$\Delta l = 4QlH/\pi^2 \qquad (5)$$

where l is the rod length, Q is the quality of the principal mode. For the principal mode rod oscillations are $l = \lambda/2$, where $\lambda = v_s/\nu$ is the sound wave length in the rod, v_s is the sound speed. For aluminium $v_s \approx 5000$ m/s, so with $\nu = 60.2$ Hz we have $l \approx 40$ m. Such an antenna is, due to its awkwardness, impracticable, some other antennae should therefore be used.

The researchers from the University of Tokyo[2] have designed the antenna of a special form that is a quadratic plate of mass M with four slots (see Figure 1). The square side is l. The mode of interest if a quadrupolar one, when two opposite squares oscillate in phase with each other and out of phase with respect ot the other squares, as is shown by the arrows. By varying the length of four slots the resonance part of the mode may be made equal to 60.2 Hz.

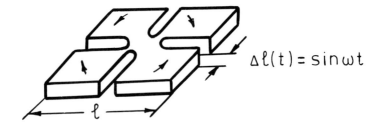

FIGURE 1

The stationary amplitude of variation of the slot size is

$$\Delta l = (3/8)Qhl \qquad (6)$$

the mode energy

$$\mathcal{E} \approx 0.5\ h^2 Q^2 M\ l^2 \nu^2 \qquad (7)$$

The quadratic antenna produced in Tokyo had l = 1.65m, M = 1400 kg, the resonance frequency was 60,2 Hz. The linear sizes of the antenna could be considerably decreased by choosing the resonance shape.

3. Thermal Noise of the Antenna

One of the pricipal factors determining an antenna sensitiviey is the termal noise that is related to thermal oscillations on a resonance frequency. When estimating the effect of this noise we shell proceed from the fact that each degree of freedom of the oscillator, including the principal operating mode, has an energy kT. This means that

$$\int_0^\infty d\omega\, \mathcal{E}_n(\omega) = kT \qquad (8)$$

where $\mathcal{E}_n(\omega)$ is the spectral density of the thermal noise energy that is of the form

$$\mathcal{E}_n(\omega) = kT \frac{1}{\pi}\ (\gamma/2)/[(\omega - \omega_o)^2 + (\gamma/2)^2]$$

where $\gamma = \omega/Q$. At the resonance frequency ω_o

NEW LASER METHODS AND SOURCES 545

$$\mathcal{E}_n(\omega_0) = kT \, 2Q/\hbar\omega$$

For definiteness we shall consider a quadratic antenna. Let t be a measurement time, i.e. the time for which a monochromatic wave acts upon the antenna. Then $\Delta\omega = 2\hbar/t$ is the spectral width of an effective signal, $\mathcal{E}(\omega) = \mathcal{E}/\Delta\omega$ is its spectral density ($\mathcal{E}$ is an effective signal energy of which is given by (7) for the quadratic antenna). Then

$$\mathcal{E}(\omega) \approx 0,5 \, h^2 Q^2 M l^2 \nu^2 t$$

We shall assume that detection of a gravitational wave requires $\mathcal{E}(\omega)/\mathcal{E}_n(\omega_0) = 1$. Hence

$$h_{min} \approx kT/(QMl^2\nu^3 t) \tag{9}$$

One can see from the formula that obtaining small h requires as low as possible temperature and high quality. With a given frequency it is advisable to have an antenna with great mass and length. A measurement time is mostly determined by the systems of recording and treatment of information.

The researchers from Tokyo[2] have made an attempt to observe gravitational waves from the Crab pulsar. The antenna in the form of a quadratic plate had the following parameters: ν =60.2 Hz, Q = 4500, M = 400 kg, l = 1.1 m, T = 300 K, t =420 hours. The experimental result was negative. Substitution of the parameters of the antenna into (9) gives an upper experimental limit for the amplitude of gravitational wave from the Crab pulsar

$$h < 1.6 \cdot 10^{-19}.$$

This result is natural. As we have already noted, one may speak seriously about an attempt to detect gravitational waves only with the antenna sensitivity $h \approx 7 \cdot 10^{-25}$.

In future this group is planning to do an experiment with much better parameters of the antenna: ν = 60.2 Hz, Q = $2 \cdot 10^8$,

M = 1400 kg, l = 1.65 m, T = $3 \cdot 10^{-3}$ K, t = 6 months. According to (9), the antenna sensitivity will be h $\approx 10^{-25}$. From (6) it follows that the displacement amplitude the detector must measure is $\Delta l \approx 10^{-15}$ cm.

Using (9) we shall estimate the antenna sensitivity for the rapid PSR 1937+21 pulsar. As its rotation frequency is 22 times as high as that of the Crab pulsar, with the other approximately identical parameters its antenna sensitivity will be two orders as high as that for the Crab pulsar, i.e. may be at the level of h $\approx 10^{-27}$. It is this value of h_{max} that may be oriented at when carrying out an experiment on detection of gravitational waves from the pulsar. According to (6), the detector sensitivity should be $\Delta l \sim 10^{-17}$ cm. Thus, as compared with the Crab pulsar our pulsar loses only in detector sensitivity, however, a higher frequency of gravitational radiation can largely facilitate the construction of antenna and detector.

4. <u>Laser Detectors Using Narrow Optical Resonances</u>

Under the detector we mean a system that follows small dispacements of an antenna. The detector is the most sensitive part of the antenna and requires most attention. Different types of detectors are presently used: piesoelectrical, volumetric, magnetic, superconducting ones and others. All presently available detectors of different types have a sensitivity $\Delta l \sim 10^{-13} - 10^{-14}$ cm. In the systems under development this value is supposed to be improved up to $10^{-14} - 10^{-15}$ cm.

Detectors using optical methods allow recording small displacemets of the phase of light beams arising when an optical path varies under the action of gravitational wave. We shall consider the laser detector using narrow optical resonances and high-stable lasers[3] that has allowed obtaining the sensitivity of length mea-

NEW LASER METHODS AND SOURCES 547

surement of $6 \cdot 10^{-14}$ cm on the base of 5 m.

Over the last fifteen years new physical methods were developed to obtain very narrow optical resonances of sufficiently high intensity on the background of a Doppler brodened line based on the nonlinear interaction of optical fields with gas. With complete elimination of Doppler broadening, the limiting spectral linewidth is determined by a linewidth of one particle γ that is refered to as a homogeneous width and can be many orders as small as a Doppler width. The narrowest resonances were obtained by the method of saturated absorption. A saturated absorption resonance can be observed, when a resonantly absorbing medium with gas is placed under a low pressure into the laser resonator. As a result, the effective amplification of such a two-component medium acquires a narrow peak in the absorption line centre, the output power on the laser frequency near the resonance may be written in the form

$$P = P_o[1 + k\Gamma^2/((\omega - \omega_o)^2 + \Gamma^2)]$$

were ω is the lasing frequency, ω_o is the resonance centre, $\Gamma = \gamma/2$ is the resonance half-width, k is its contrast, P_o is the laser power at $\Gamma \ll \omega - \omega_o$.

The narrowest resonaces of less than 1 kHz wide have been obtained in the output power of a 3.39 µm He-Ne laser using methane as an absorber ($F^{(2)}$ component of the P(7) vibrational-rotational transition of the ν_3 band)[6]. This resonance was used to stabilize a He-Ne/CH_4-laser. A relative frequency stability was 10^{-15}[7], a radiation linewidth 0.05 Hz[8].

Let us consider the schematic diagram of measurement of small displacements by using narrow resonances (see Figure 2). An external perturbation acting upon one of the mirrors of the laser resonator produces a variation of the resonator length l to Δl and, consequently, of the generation frequency ω to $\Delta \omega = \omega \Delta l/l$. With

548 LASERS IN ATOMIC, MOLECULAR AND NUCLEAR PHYSICS

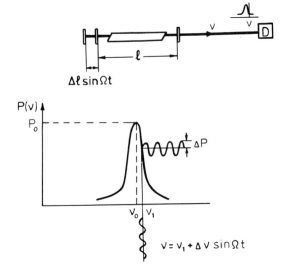

FIGURE 2

the narrow resonances of saturated absorption the variation of the generation frequency transforms into the variation of the laser radiation power ΔP that is recorded by a photodetector. With periodic perturbation of the resonator length, when $\Delta l(t) = \Delta l \sin\Omega t$, $\omega - \omega_o = (\omega_1 - \omega_o) + \Delta\omega\sin\Omega t$, where $(\omega_1 - \omega_o)$ is the detuning of the generation frequency ω_1 with respect to the resonance center ω_o, $\Delta\omega = \omega \Delta l/l$. With $\Delta\omega/\Gamma \ll 1$ from (11) we have

$$P = P_o - \Delta P \sin\Omega t$$

$$\Delta P = 2P_o k(\omega_1 - \omega_o)\Delta\omega\Gamma^2/[(\omega_1 - \omega_o)^2 + \Gamma^2]^2.$$

To achive the maximum sensitivity requires to tune the laser frequency to the region of the largest steepness of the resonance $(\omega_1 - \omega_o = \Gamma/\sqrt{3})$. In doing so

$$\Delta P_{max} = (3\sqrt{3}/8)P_o k\Delta\omega/\Gamma.$$

Transformation sensitivity grows with increase in the resonance intensity and decrease in its width. The minimum recorded variation of the resonator length is determined by noises of laser radiation and of a measuring system. An important limiting factor is the photon noise conditioned by the quantum character of fluctuations of laser radiation.

We shall explain the nature of the noise. Let us consier a laser to be a monochromatic source of radiation. If τ is a time constant of detector ($\Delta f = 1/\tau$ is its bandwidth), one may assume that the detector would take the laser radiation as a train of τ duration. The number of photons in the train is $N = P_o\tau/\hbar\omega$, their fluctuation is $\Delta N = \sqrt{N} = \sqrt{P_o\tau/\hbar\omega}$ (distribution of photons is considered to be a Poisson one). The noise power appears to be $P_n = \Delta N \hbar\omega/\tau = \sqrt{\hbar\omega P_o/\tau}$.

The minimum recorded frequency variation is found from the condition of signal-to-noise equality ($P_{max}/P_n = 1$)

$$\Delta\omega_{min} = (8/3\sqrt{3})(\Gamma/k)\sqrt{(\hbar\omega/P_o\tau)}$$

The minimum recorded length variation is $\Delta l_{min} = \Delta\omega_{min} l/\omega$:

$$\Delta l_{min} = (8/3\sqrt{3})(l\Gamma/\omega)\sqrt{\hbar\omega\Delta f/P_o} \qquad (12)$$

An estimate for real parameters of the laser and nonlinear resonance ($\omega = 10^{14}$ Hz, $P_o = 10^{-2}$ W, $\Gamma = 10^4$ Hz, $l = 1$ m, $K = 1$, $\Delta f = 1$ Hz) gives $\Delta l_{min} \approx 10^{-16}$ cm ($\Delta l/l \approx 10^{-18}$).

Under realistic experimental conditions the measurement sensitivity falls down due to additional noises conditioned by fluctuations of the radiation power because of laser frequency instability and discharge plasma, by noises of a photodetector and so on. The amount of these noises depends on real experimental conditions and can largely exceed the photon noise.

Experiments have been made with a 3.39 μm He-Ne laser with a methane absorption cell. The nonlinear resonance of $5 \cdot 10^4$ Hz wide in methane was used, $K \approx 0.7$, $P_o \approx 1$ mW. The resonator length was 5 m. Periodic perturbation of the resonator length was provided by

550 LASERS IN ATOMIC, MOLECULAR AND NUCLEAR PHYSICS

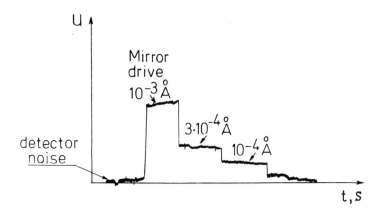

FIGURE 3

supplying a sinusoidal signal with frequency Ω = 15 kHz to a preliminarily calibrated piesoelectrical element on which one of the resonator mirror was mounted.

The Figure 3 shows the typical record of the signal at the output of the photodetector with different amounts of periodic perturbation. The averaging time was 10 s. The minimum detectable quantity of displacement of the mirror determined by the noise level was $6 \cdot 10^{-6}$ Å. The relative sensitivity was $\approx 10^{-16}$. Measurements have showed that the observed noise was entirely determined by the photodetector noise which was approximately three orders as large as the photon noise.

Use of narrower saturated absorption resonances for improving sensitivity is limited due to decrease in their intensity. Promising may be the use of Doppler shift free two-photon resonances[9] in the short-wavelength spectral region. In these resonances intensity saturation at power of ~ 1 W doesn't yet occur. Use of two-photon resonances requires development of high stable tunable lasers. An estimation over (12) for the following parameters of the system: λ = 500 nm, $P_o \sim 1$ W, $l \sim 1$ cm, $K \approx 1$, $\Gamma \approx 1$ kHz, Δf = 1 Hz gives $\Delta l \sim 10^{-21}$ cm ($\Delta l/l \sim 10^{-21}$).

5. Conclusion

Thus, the laser detector using a saturated absorption resonance allowed measuring the displacement of $6 \cdot 10^{-14}$ cm. Estimations over the photon noise indicate that a sensitivity of 10^{-16} cm can be achieved in this system. Use of the two-photon absorption resonance in a standing-wave field opens up horizons for achieving a better sensitivity. The laser detector may be used in constructing an antenna for the Crab pulsar and the rapid PSR 1937 pulsar. As has been already noted, this requires a sensitivity of $10^{-15} - 10^{-17}$ cm.

The problem of detection of gravitational waves is a serious scientific and technological problem. So, naturally, we could not consider here all the aspects. We didn't discuss the methods of obtaining high-quality mechanical resonator, the noises of detector that affect the antenna high quality, intersting problems of use of superconducting technique were also omitted, we didn't consider the system of singal treatment which could be used in some cases to increase sensitivity by two or three orders. All these problems are treated in detail in[1] and in the literature cited there.

The authors are indebted to Prof. V. P. Chebotayev for valuable discussion.

REFERENCES

1. E. AMALDI, G. PIZZELLA, in Astrofisica e Cosmologia Gravitazione, Quanti e Relatività (Guinti Barbera, Firenze, 1979), p. 241.
2. H. HIRAKAWA, Detection of Gravitational Radiation from Pulsars, in Gravitational Radiation, Collapsed Objects and Exact Solutions, edited by C. Edwards (Springer, 1980), p. 331.
3. S. N. BAGAYEV, A. S. DYCHKOV, V. P. CHEBOTAYEV, ZhETF Pis'ma, 33, 85 (1981)
4. M. ZIMMERMANN, Nature, 271, 524 (1978)
5. D. BACKER, S. KULKARNI, C. HEILES, M. DAVIS, M. GROSS, Nature, 300, 615 (1982)

6. S. N. BAGAYEV, V. P. CHEBOTAYEV, A. K. DMITRIYEV,
   A. S. DYCHKOV, V. G. GOLDORT, L. S. VASILENKO,
   Appl. Phys., 13, 291 (1977)
7. S. N. BAGAYEV, A. S. DYCHKOV, V. P. CHEBOTAYEV,
   ZhETF Pis'ma, 5, 590 (1979)
8. S. N. BAGAYEV, V. P. CHEBOTAYEV, A. S. DYCHKOV, S. V. MALTSEV,
   J. de Physique, Coll. C8, 42, 21 (1981)
9. L. S. VASILENKO, V. P. SHEBOTAYEV, A. B. SHISHAYEV,
   ZhETF Pis'ma, 12, 161 (1970)

6.4 Coherent Vacuum UV Sources Using the Methods of Nonlinear Optics

C. R. VIDAL

Max Plank Institut fuer Extraterrestrische Physik, 8046 Garching, Federal Republic of Germany

Coherent vacuum ultraviolet souces using the methods of four wave frequency mixing in gases are now understood in a quantitative manner and have reached a degree of perfection where they are able to extend high resolution laser spectroscopy into the vacuum ultraviolet spectral range. A brief review of the fundamental processes is given.

1. INTRODUCTION

In recent years four wave parametric processes have been successfully used to extend the tuning range of lasers into the vacuum ultraviolet as well as into the infrared. In the visible and the infrared spectral region this has generally been done by sum or difference frequency mixing in suitable nonlinear crystals. In the vacuum ultraviolet spectral region, however, most solids become opaque and can therefore no longer be used. In this case gaseous nonlinear media can be used which offer several advantages:

1. Extended spectral regions of high transparency for the ultraviolet radiation can be provided.
2. The anomalous dispersion can be used for phase mathing of gaseous two-component systems [1,2] and for providing large column densities as required for efficient nonlinear media.

3. Large nonlinear susceptibilities can be achieved by a suitable resonant enhancement$^{3-5}$.

4. The material properties such as the linear and nonlinear susceptibilities can be accurately calculated from first principles allowing a reliable test of the theoretical model.

In this contribution the two-photon resonant and the nonresonant sum frequency mixing in gaseous nonlinear media$^{6-8}$ is briefly reviewed. This technique has become increasingly interesting as a means of extending high resolution laser spectroscopy into the vacuum UV where many atoms and a number of the most important small molecules such as H_2, CO, NO and others have some of their most prominent absorption features.

In presenting the fundamental physical processes of four wave sum frequency mixing it is useful to distinguish essentially three regimes which differ with respect to the electric field amplitudes to be considered:

1. small signal limit
2. onset of saturation
3. high intensity saturation

With growing electric field amplitudes an increasing number of nonlinear processes are requied for a quantitative description of four wave sum frequency mixing in gases.

2. SMALL SIGNAL LIMIT

In the small signal limit the intensity of the generated wave is proportional to the product of the intensities of the incident waves and for the simplest case of the third harmonic generation the well-known cubic dependece for the harmonic intensity is obtained. In the small signal limit it basically suffices to know the phase matching curve which primarily depends on the complex linear susceptibilities, and to know the conversion efficiency which depends on the absolute value of the corresponding nonlinear susceptibility.

The phase matching curve is furthemore affected by the properties of the incident beams and of the nonlinear medium. For focused beams the intensity distribution has to be considered explicitly which is typically characterized by the confocal parameter of a Gaussian beam. For practical applications the confocal parameter should be comparable or larger than the length of the nonlinear medium[9-11]. Any excessive focusing does not improve the power conversion efficiency any more and leads to additional perturbing nonlinear processes which are associated with the increased intensity. A further modification of the phase matching curve is due to the line profiles involved[12,13] and the mode structure of the incident beams[14].

Finally, one has to consider the fact that nonlinear medium which in many cases is provided by a concentric heat pipe oven[15,16], does not have a rectangular density profile as assumed in most theoretical models. Instead, density gradients at the end of the metal vapor zone which give rise to an asymmetry of the phase matching curve, have to be taken into account for a quantitative analysis[11,17,18].

In the small signal limit the conversion efficiency reaches its optimum value if one of the optical depths for the different waves approaches a value of the order of unity[8,17]. The large column density associated with this experimental condition, generally requires a very careful phase matching and hence a highly homogeneous nonlinear medium. Excessively large column densities, on the other hand, do not gain any further improvement because eventually the power generated per unit length is balanced by the power absorbed and the conversion efficiency then becomes independent of the column density.

3. ONSET OF SATURATION

In the small signal limit the intensity of the harmonic wave follows a simple power law over a wide range of input intesities until eventually the harmonic intensity starts to level off long before energy conservation leads to a depletion of the fundamental waves. This onset of saturation can be explained by additional nonlinear polarizations which give rise to field dependent changes of the refractive index and which are responsible for a destruction of the phase matching condition. For gasous media this was shown in nonresonant systems[11] as well as in two-photon resonant systems[19]. For the nonresonant case it is the Kerr effect which causes an intensity dependent change of the refractive index. For the two-photon resonant situation it is the two-photon absorption which induces an intensity dependent redistribution of the population densities[19] and modifies the effective index of the nonlinear medium.

Furthermore, the intensity dependence of the incident beams in space and time complicates calculations of the conversion efficiency for a particular nonlinear medium. Since the destruction of the phase matching condition is highly intensity dependent, the intensity distribution in space and time which generally is assumed to be Gaussian, has to be considered explicitly. For nonresonant cases the field dependent modification of the refractive index occurs instantaneously, whereas for resonant situations the population densities inside the nonlinear medium have to be determined which have been accumulated at some point in space and time and which lead not only to an intensity dependent, but also to a time dependent change of the refractive index. In the small signal limit the intensity distribution can simply

NEW LASER METHODS AND SOURCES 557

be taken care of by a constant factor.

4. HIGH INTENSITY SATURATION

Due to the large number of different nonlinear processes the high intensity saturation regime is the most difficult one to handle in a quantitative analysis. In the high intensity saturation regime one is to a first approximation still dealing with the same nonlinear polarizations which have already dominated the onset of saturation. They are, however, strongly enhanced and give rise to additional effects. Furthermore, a variety of higher nonlinearities become important which in part may be viewed at as field dependent corrections of the preceding third order nonlinearities, whereas other nonlinearities give rise to a new class of higher order parametric processes.

All of these processes show up most clearly in two-photon resonant situations where they are resonantly enhanced. They can be detected, for example, by measuring the conversion efficiency around a particular two-photon resonance as a function of the input intensity. These conversion profiles show several very characteristic features[20]. One observes a pronounced broadening due to power broadening and a narrow dip which causes a minimum of the conversion efficiency right on the two-photon resonance. In this case the highest conversion efficiency occurs near a two-photon resonance. The latter effect was first predicted theoretically[21] and is due to a bleaching of the two-photon transition.

As a further phenomenon self-(de)focusing of the incident beams can be observed which originates from the same field dependent changes of the refractive index which already destroyed the phase matching condition near the onset of the saturation. A theoretical treatment of this effect is exceedingly difficult because also the radial derivatives of the Laplacian operator inside the basic equations of nonlinear optics have to be considered which otherwise can be neglected.

As a consequence of the strong two-photon absorption a severe redistribution of the population densities occurs inside the nonlinear medium and gives rise to population inversions. The resulting stimulated emission is collinear with the incident beams and leads to additional parametric processes which may be as intense as the original sum frequency wave[20].

In the high intensity saturation regime additional effects will generally occur such as higher harmonics of the incident waves, ionization processes and level shifts due to the AC Stark effect. All of these phenomena give rise to a highly intricate behaviour of the conversion efficiency. For practical applications, however, they are generally of little interest if a monochromatic vacuum UV source of high conversion efficiency has to be designed which can be applied, for example, to high resolution laser spectroscopy in the vacuum ultraviolet spectral region. The most favorable condition for such an application is achieved with a phase matched system of optimum column density at an input intensity just below the high intensity saturation regime.

5. SPECTROSCOPIC APPLICATIONS

In our laboratory we have built several tunable and narrow band coherent vacuum UV sources. They consist of two dye lasers which are pumped either by a nitrogen laser or by an excimer laser or by flash lamps. The first dye laser is tuned to a particular two-photon resonance of the nonlinear medium, whereas the second dye laser provides the tunability of the resulting sum frequency wave in the vacuum ultraviolet spectral region. The laser systems were in most cases pressure tuned to synchronize the various dispersive elements over the entire tuning range. The nonlinear medium consists of a modified concentric heat pipe oven[16] containing a magnesium krypton or a strontium xenon mixture.

NEW LASER METHODS AND SOURCES 559

The vacuum UV systems have so far been designed to cover a spectral region from about 100 to 200 nm depending on the dyes used. The line width is typically 0.1 to 0.3 cm^{-1} in the vacuum UV. The number of photons are about 10^{11} to 10^{13} per shot at a repetition rate of up to 20 sec^{-1}. The spectral brightness of these systems exceeds the one of any synchrotron and of other light sources by several orders of magnitude[22]. Also the spectral resolution $\lambda/\Delta\lambda$ exceeds the one of conventional instruments such as 10 m vacuum UV spectrometers by almost an order of magnitude and further improvements are expected once the dye lasers are designed to approach the Fourier transform limited linewidth.

Third harmonic generation and sum frequency mixing offer also the great advantage of allowing a wavelength calibration of the vacuum UV radiation by calibrating the wavelength of the fundamental beams in the visible part of the spectrum where the secondary length and frequency standards are significantly more accurate. So far we used the iodine spectrum as measured by Gerstenkorn and Luc[23] as a calibration spectrum. The wavelengths of the fundamental waves were determined from a least squares fit of the iodine spectrum with a typical standard error of 0.003 cm^{-1} corresponding to an accuracy of 0.01 cm^{-1} or 2 parts in 10^7 in the vacuum UV.

In our laboratory the first experiments of high resolution laser spectroscopy in the vacuum ultraviolet region were carried out on the NO and the CO molecule. The following methods have already been demonstrated:

1. Frequency selective excitation spectroscopy on the CO intercombination bands[24] and on the electronic transitions from the ground state to the A-, B-, C- and D-states of the NO molecule[25].

2. Laser induced fluorescence spectroscopy on the first excited states of the NO molecule[25].

3. Two step excitation spectroscopy on the CO molecule using a vacuum UV system and an additional laser in the visible part of the spectrum. With this technique we circumvented the lithium fluoride cutoff limit in a cell condined by magnesium fluoride windows. These experiments showed collision induced transitions in the A-state of the CO molecule, predissociation, isotope shifts, perturbations and fine structure splittings of excited states in CO[26].

4. Life time measurements on selected levels of the excited triplet states of the CO molecule.

REFERENCES

1. S. E. HARRIS, R. B. MILES, Appl. Phys. Lett.,19, 385 (1971).
2. J. F. YOUNG, G. C. BJORKLUND, A. H. KUNG, R. B. MILES, S. E. HARRIS, Phys. Rev. Lett., 27, 1551 (1971).
3. D. M. BLOOM, J. T. YARDLEY, J. F. YOUNG, S. E. HARRIS, Appl. Phys. Lett., 24, 427 (1974).
4. R. T. HODGSON, P. P. SOROKIN, J. J. WYNNE, Phys. Rev. Lett., 32, 343 (1974).
5. K. M. LEUNG, J. F. WARD, B. J. ORR, Phys. Rev., A 9, 2440 (1974).
6. C. R. VIDAL, Appl. Opt., 19, 3897 (1980).
7. W. JAMROZ, B. P. STOICHEFF, Generation of Tunable Coherent Vacuum Ultraviolet Radiation, in Progress in Optics, edited by E. Wolf (North Holland, Amsterdam, 1983), Vol. 20, pp. 326-380.
8. C. R. VIDAL, Four Wave Frequency Mixing in Gases, to be published in Tunable Lasers, edited by L. F. Mollenauer and J. C. White in Topics in Applied Physics (Springer, Heidelberg, 1985).
9. J. F. WARD, G. H. C. NEW, Phys. Rev., 185, 57 (1969).
10. G. C. BJORKLUND, IEEE J. Quant. Electr., QE-11, 287 (1975).
11. H. PUELL, K. SPANNER, W. FALKENSTEIN, W. KAISER, C.R. VIDAL, Phys. Rev., A 14, 2240 (1976).
12. A. STAPPAERTS, G. W. BEKKER, J. F. YOUNG, S. E. HARRIS, IEEE J. Quant. Electr. QE-12, 330 (1976).

13. C. LEUBNER, H. SCHEINGRABER, C. R. VIDAL, Opt. Commun., 36, 205 (1981).
14. Y. M. YIU, T. J. McILRATH, R. MAHON, Phys. Rev., A 20, 2470 (1979).
15. C. R. VIDAL, F. B. HALLER, Rev. Scient. Instr., 40, 3370 (1969).
16. H. SCHEINGRABER, C. R. VIDAL, Rev. Scient. Instr., 52, 1010 (1981).
17. H. SCHEINGRABER, H. PUELL, C. R. VIDAL, Phys. Rev., A 18, 2585 (1978).
18. H. JUNGINGER, H. PUELL, H. SCHEINGRABER, C. R. VIDAL, IEEE J. Quant. Electr., QE-16, 1132 (1980).
19. H. PUELL, H. SCHEINGRABER, C. R. VIDAL, Ptys. Rev., A 22, 1165 (1980).
20. H. SCHEINGRABER, C. R. VIDAL, IEEE J. Quant. Electr., QE-19, 1747 (1983).
21. H. SCHEINGRABER, C. R. VIDAL, Opt. Commun., 38, 75 (1981).
22. K. RADLER, J. BERKOWITZ, J. Opt. Soc. Am., 68, 1181 (1978).
23. S. GERSTENKORN, P. LUC, Atlas du Spectre d'Absorption de la Molecule d'Iode(editions du CNRS, 15, quai Anatole France, Paris, 1978).
24. P. KLOPOTEK, C. R. VIDAL, Can. J. Phys., to be published.
25. H. SCHEINGRABER, C. R. VIDAL, J. Opt. Soc. Am., to be published.
26. C. R. VIDAL, P. KLOPOTEK, H. SCHEINGRABER, AIP Conf. Proc., to be published.

6.5 Recent Advances in Tunable Solid-State Lasers

G. S. KRUGLIK, G. A. SCRIPKO and A. P. SHKADAREVICH

For years the possible sources of tunable coherent emission were considered to be realized by dye lasers and nonlinear crystals. Still, a new direction in laser physics started to develop - phonon terminated solid-state lasers. These lasers are based on the use of impurity ions, as well as crystal lattice defects, the first being of the greatest practical interest because of its high photo and thermal stability. The physical studies of such lasers were initiated in the 60th, but these lasers became competitive as a result of room tempereature laser action[1,2] and cw gas laser pump[3].

The present paper presents the results obtained in this direction during the recent two years at the chair "Applied Optics" in Byelorussian polytechnical institute.

1. CHROMIUM-DOPED CRYSTALS

Trivalent chromium ions in crystalline matrix can have a luminescence spectrum in the form of narrow R-lines (ruby), R-lines and phonon wing combination (alexandrite) or of a developed phonon band. The spectrum shape depends upon the mutual position of the ^2E metastable state and the $^4T_2$ vibronic band. The crystalline matrices, whose energy gap $\Delta E(E_{4T_2} - E_{2E} = \Delta E)$ is small or negative are promising for obtaining tunable coherent emission. This condition will be met in case of chromium ion in weak crystal field. The field is considered weak under the condition

$$D_q/B < 2.3 \qquad (1)$$

where D_q is the crystal field parameter, B is the Racah parameter for d-electrons interaction.

It is known[4], that

$$D_q = \text{const } Q\langle\overline{r}^4\rangle/ R^5 \qquad (2)$$

where Q is the effective ligand charge, r is the average 3d-orbital radius, R is the ion-ligand spacing.

We studied two crystal classes: the silicates, based on the SiO_2 group and also the complex oxides of the $A^+B^{3+}(M^{6+}O_4)_2$ type, where A^+ - Li^+, Na^+, K^+, ...; B^{3+} - Y^{3+}, Lu^{3+}, Gd^{3+}, ...; M^{6+} - W^{6+}, Mo^{6+}, ... The presence of Si^{4+} and M^{6+} in both classes reduces the effective charge Q and increases the average R, and it is therefore may be hoped that they have weak crystal field. Besides, it is known[4], the crystal field strength decreases in the spectrochemical series

$$C^- > N^- > S^- > O^- > F^- > Cl^- > Br^- \ldots \qquad (3)$$

Therefore, the transition from the oxygen environment to fluorine seems perspective. Chromium-doped perovskite crystals (e.g. $KMgF_3$, $KZnF_3$) have been also studied.

Preliminary investigations of spectral and luminescence characteristics were carried out on the polycrystalline samples, obtained with a coprecipitation method. Subsequent annealing according to definite regimes made it possible to obtain prescribed crystal structures, they were identified by X-ray phase analysis and IR spectroscopy. At present we have developed methods, which may be used to study the spectral and kinetic luminescence characteristics of the polycrystalline samples as a function of concentration and temperature, luminescence, absorption and excitation spectra, to estimate excited-state absorption.

Spectral and kinetic measurements were carried out over the range of 4.2 K to room temperature. The Table presents some spec-

TABLE

No	Material	λ^{max} [nm] $^4A_2-^4T_1$	λ^{max} [nm] $^4A_2-^4T_2$	λ_R^{max} [nm]	λ_l^{max} [nm]	$\Delta\lambda_l$ [nm]*	τ_l^{max} [μs]	$\Delta\tau_l$ [μs]**
1.	$3MO \cdot Al_2O_3 \cdot 3SiO_2$	455	615	686	765	670-1020	80	180-30
2.	$3K_2O \cdot 3CaO \cdot 3ZnO_2 \cdot 12SiO_2$	460	588	–	830	785-995	28	40-16
3.	$MgO \cdot 2CaO \cdot 2SiO_2$	438	614	–	795	710-1055	54	78-10
4.	$2MgO \cdot 2Al_2O_3 \cdot 5SiO_2$	452	630	–	760	685-920	72	150-36
5.	$3MgO \cdot 2SiO_2$	–	–	690	888	685-1100	14	67-12
6.	$Na_2O \cdot Al_2O_3 \cdot 2SiO_2$	445	628	–	785	680-940	4	–
7.	$Na_2O \cdot BeO \cdot 2SiO_2$	440	630	–	890	750-900	20	–
8.	$3Al_2O_3 \cdot 2SiO_2$	432	602	701	775	670-1055	78	130-34
9.	$CaMg[Si_2O_6]$	450	655	685	775	685-965	23	–

* Luminescence bandwidth is indicated to level 0.1

** Variation limits of τ_l are indicated to luminescence intensity decrease from maximum to level 0.1

troscopic characteristics of chromium-doped silicates. As can be seen, there is a wide class of materials with a broadband luminescence in the range of 600-1100 nm with the room-temperature lifetime 1-180 μs. Some of them may be perspective for tunable laser action in the near IR range under lamp excitation, others as tunable laser converters.

The silicate growth of the proposed compositions is a complicated technological problem, because, in particular, of their tendency to vitrification. At present the production process for crystals of the required size and quality is in progress. We have developed a method, which may be of interest, to obtain optically homogeneous vitreous compositions with spectral and kinetic characteristics close to single crystals.

From the results for polycrystal series obtained with the mentioned technique some of the media have been chosen perspective for near IR tunable generation. High efficiency laser action was obtained from one of them $KZnF_3:Cr^{3+}$ with coherent and lamp excitation. (The generation is obtained together with the research group of scienticts directed by N. I. Silkin, University in Kazan.)

For the first time, $KZnF_3:Cr^{3+}$ laser action was reported in[5] under krypton laser pump and dye laser pump with the conversion slope efficiency ~ 1%.

Independet of[5] we at first obtained generation under free-running ruby laser pump. (The results are reported at the Annual Conference of Lecturers of Byelorussinan Polytechnical Institute, Minsk, March, 1-3, 1984 and in[6].) We present some of its characteristics: the conversion efficiency is 45% (overall), 73% (slope), (Figure 1), the maximum pulse energy up to 8 J, the tuning range is 780-860 nm. We ralized pulse laser action under rube laser pump with the pulse repetition rate 10 Hz. The average output power is 8.2 W, conversion efficiency ~ 31%. Overall efficiency with

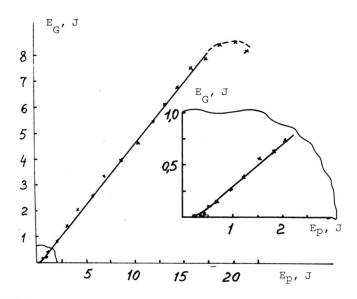

FIGURE 1

reference to the power source reached 0.18%.

The Q-switched laser operation from $KZnF_3:Cr^{3+}$ was studied with tree-running laser pump. Q-switched pulses with the energy 270 mJ and pulse duration 120 ns were obtained. Here we report for the first time the generation with lamp excitation.

The $KZnF_3:Cr^{3+}$ element ∅5 × 40 mm with 1 at.% chromium in charge was used in the experiment. The generation characteristics were studied both in a nonselective and selective cavities. The nonselective cavity was formed by confocally placed dielectric mirrors with curvature radius ∼ 1 m and 1% transmittance in the range of 725-950 nm. The threshold pump energy under these conditions was 12 J and the laser emission near the threshold was observed in the range with a center at 827 nm and FWHM of ∼ 8 nm. The measured slope efficiency was 0.3% with the output mirror reflectivity 85%.

The tuning characteristics were measured in the cavity with 1% transmittance mirrors within the whole tuning range. The Abbe

568 LASERS IN ATOMIC, MOLECULAR AND NUCLEAR PHYSICS

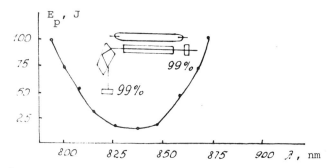

FIGURE 2

dense flint prism, placed into the 50 cm-cavity was used as a despersive element. Figure 2 shows $KZnF_3:Cr^{3+}$ generation threshold versus lasing wavelength for laser tuning range. The tuning range is 787-873 nm at the pump level not more than 100 J.

The given $KZnF_3:Cr^{3+}$ laser characteristics, obtained without optical scheme optimization suggests that this laser will successfully compete with alexandrite and $GSGG:Cr^{3+}$ lasers.

Like alexandrite, $KZnF_3:Cr^{3+}$ luminescence band center does not coincide with spectral minomum of the threshold curve, this shows the presence of the chromium excited-state absorbtion. This is the main factor limiting the tuning range and lasing efficiency in chromium containing crystals.

2. TITANIUM-DOPED SAPPHIRE LASER

Trivalent titanium ions are advantageous over chromium ions because d^1 configuration gives only one ^2D term of interest which splits in the cubic field into two levels $^2T_{2g}$ and ^2E_g, the next excited configuration has the energy of 80 379 cm^{-1} and this excludes induced absorption in Ti^{3+} lasing band.

The generation from $Al_2O_3:Ti^{3+}$ with lamp excitation was originally reported in[7]. Still, the short lifetime (3.7 μs) of the Ti^{3+} excited state and the high crystal solarization hampers obtaining high efficiencies with lamp excitation. That is why we

NEW LASER METHODS AND SOURCES 569

studied $Al_2O_3:Ti^{3+}$ lasing characteristics with doubled frequency neodimium laser excitation. In this case $Al_2O_3:Ti^{3+}$ is used as a tunable converter. Excellent matrix properties and high photo and thermal stability make it incompetitive in the analogous series - organic dyes and alkali-halides with colour centers. The maximum conversion efficiency of 40% and slope efficiency of 50% were obtained in a nonselective cavity. The generation spectrum FWHM was 40 nm with the center at $\lambda = 775$ nm and had a form of a set of equidistant bands. The periodic spectral structure depended upon the specific crystal and its orientation and was independent on cavity length, mirrir wedgegorm, crystal position in the cavity. The spectral structure is constantly reproducable. Such peculiarities of the generation spectrum suggest the presence of periodic structures in $Al_2O_3:Ti^{3+}$ crystal with specific spatial period ~ 100 μm. This agrees with weak dependence of the threshold and output energy upon the output mirror transmittances.

The $Al_2O_3:Ti^{3+}$ laser tuning characteristics were measured in a selective savity. The dense flint prism set was used as a selector. The maximum tunable range 680-930 nm was obtained with the mirrors $R_1 = R_2 = 99$%. Tuning in 706- 910 nm range with maximum efficiency of 25% was realized using the interfernce reflector as a coupling mirror.

An effective frequency doubling of the obtained emission in $LiIO_3$ is realized in the 353-455 nm range. Figure 3 shows tuning curves of $Al_2O_3:Ti^{3+}$ laser fundamental frequency (a) and its second harmonic generation (b). An interesting feature of the laser is a time delay between excitation and generation pulses. It varied versus the excitation excess over the threshold in the range of 70-4 ns. The generation pulse duration strongly depended on cavity length (it could vary from 2 to 14 ns after cavity length modifications from 2 to 18 cm with a constant excitation pulse duration of

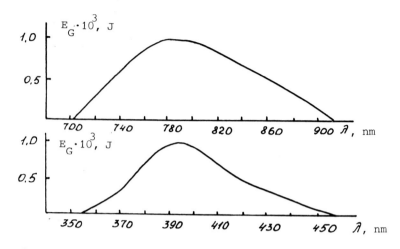

FIGURE 3

10 ns).

The studied crystal is of great practical interest. The initial investigations of the generation characteristics resulted in a laser system with the use of GSGG:Nd^{3+}, Cr^{3+} pump laser and frequency doubler on CDA crystal, with overall efficiency 0.25%, with reference to power source over the range 700-910 nm.

To enlarge the Ti^{3+} concentration in a grown single crystal we have developed a method of thermochemical colouring which substantially increases the concentration of active centers. Simultaneously, an absorption band appears in the generation range, caused by the colour centers, which have a high photo and thermal stability. The subnanosecond pulse generation is obtained from such crystals with the appropriate cavity characteristics and pump condition.

The sapphire single crystal is not the only matrix of interest for Ti^{3+} doping. Figure 4 is a plot of ΔE of the ground state ^2D ($\Delta E = {}^2E_g - {}^2T_{2g}$) versus Dq. It is seen from the figure that tunable generation up to 1.5 μm is possible in the media with weaker crystal fields. Other ions, such as V^{4+}, Cu^{2+}, have a simi-

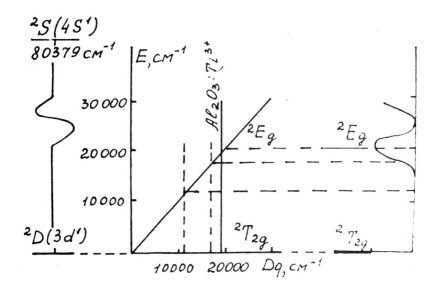

FIGURE 4

lar to Ti^{3+} configuration. Therefore, the use of these ions in the appropriate matrices seems of interest. In this case one of the main limiting factors for generation energy and tuning range of such lasers - an excited-state absorption - will be actually eliminated. For example, the nearest Cu^{2+} excited ^4F $(3d^84s^1)$ configuration is $60\,804$ cm^{-1} apart the ground state and cannot cause significant excited-state absorption. The transitions themselves are in a more longwave range and therefore it seems possible realize generation in 1.0- 1.8 μm region. On the contrary, V^{4+} will probably remove tunable generation region to shortwave side to 600 nm.

3. MONOVALENT COPPER-DOPED GLASS

It is interesting to find dopants which unlike Ti^{3+} and Cr^{3+} make it possible to obtain coherent emission in the visible and even in the UV ranges. The monovalent copper ions are the most promissing in this field. Figure 5 shows absorption (curve 1) and

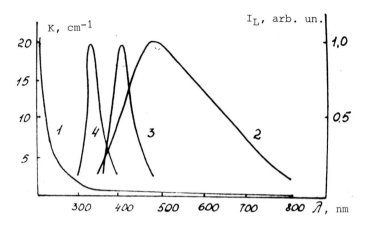

FIGURE 5

luminescence (curve 2) spectra of alumoborosilicate(ABS) glass, doped with monovalent copper (copper luminescence spectra in $KMgF_3$ and Al_2O_3 (curves 3,4 - respectively) are also presented). As is seen from Figure 5, copper luminescence spectrum in ABS-glass has a wide featureless band at 340-700 nm and copper absorption band overlapped by matrix fundamental absorption. Luminescence decay constant is 30 µs. To clear up the possibility of obtaining generation in this medium the gain is measuted with intracavity spectroscopy methods. The copper-doped glass plate excited by the YAG fourth harmonic generation was placed in the dye laser cavity, pumped by YAG doubled frequency radiation. The measured gains are: $\alpha \geq 2.4 \cdot 10^{-4}$ cm^{-1} for $\lambda = 560$ nm and $\alpha \leq 5.5 \cdot 10^{-4}$ cm^{-1} for 585 nm. the gain frequency dependence indicates the excted-state absorption. Estimation of the induced luminescence transition cross-section gives $\sigma = 2.1 \cdot 10^{-21}$ cm^2. That means that the active elements with nonactive loss factor $< 5 \cdot 10^{-4}$ cm^{-1} are necessary for the generation. Such active element characteristics with present day tichnology are practicable.

An important feature of the proposed medium is a possibility of Cu^+ excitation not only by UV radiation but also by X-ray,

NEW LASER METHODS AND SOURCES   573

electron beam and other types of radiation. The excitation is
realized through matrix, which effectively transfers energy to the
impurity ions. Such system is supposed to have not only high efficiency, but also a variety of exctation energy sources.

To summarize, as a result of the study we proposed the copper-doped ABS-glass as an active medium amplifying emission in
560-585 nm with excitation into the fundamental absorption band;
this may be a precondition for development of a tunable glass
laser over visible range.

REFERENCES

1. J. C. WALLING, O. G. PETERSON, H. P. JENSSEN, R. C. MORRIS,
   E. W. O'DELL, J. Quant. Electr., 16, 1302 (1980)
2. Yu. D. GUSEV, S. I. MARENNIKOV, V. P. CHEBOTAREV,
   JETP Letters (USSR), 3, 305 (1977)
3. L. MOLLENAUER, D. J. OLSON, J. Appl. Phys., 46, 3109 (1975)
4. A. S. MARPHUNIN, Introduction into Physics of Minerals
   (Moscow, 1974)
5. U. BRAUCH, U. DÜRR, Optics Communications, 49, 61 (1984)
6. G. S. KRUGLIK, G. A. SCRIPKO, A. P. SHKADAREVICH,
   Doped-Crystals Tunable Lasers (Minsk, 1984)
7. P. F. MOULTON, Optics News, No. 6, 9 (1982)

6.6 Line Competition of Optically Pumped Lasers

RAINER SALOMAA
Helsinki University of Technology, Department of Technical Physics, SF-02150 Espoo, Finland

and

M. A. DUPERTUIS and M. R. SIEGRIST
Centre de Recherches en Physique des Plasmas, Ecole Polytechnique Federale de Lausanne, CH-1007 Lausanne, Switzerland

> To investigate line competition phenomena and interactions between coherently coupled transitions in laser pumped multilevel systems we have calculated the atomic polarization in the presence of an external pump laser beam and several competing radiation modes. The results are applied to discuss the preferred Raman-line generation as also to describe interference effects between filling and refilling transitions in FIR-lasers.

1. INTRODUCTION

In studies of nonlinear light-matter interactions one often has to calculate the material susceptibility in the presence of an intense, external monochromatic laser beam coupling one pair of atomic levels. The polarization considered is usually induced by other coherent fields (of varying intensity) coupled either to the same or to adjacent transitions of the pumped one. Examples of this situation include stability analysis of lasers, operation of multicolor lasers (one common level shared by lasing transitions), laser pumped lasers, stimulated scattering processes, four-wave mixing and other parametric generation mechanisms. Our interest has partly its origin in the optimization of far infrared (FIR) laswes pumped by CO_2-lasers and used for fusion diagnostics. Despite the extensive studies on single mode operation of FIR lasers only a very limited amount of work has been published on their multimode behavior.

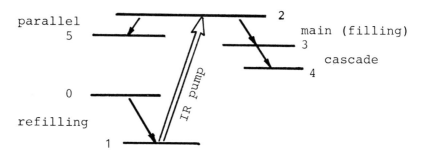

FIGURE 1

We consider the levelscheme of Fig. 1. This system contains all the main ingredients of a FIR laser[1]: the externally pumped transition 1 ↔ 2, the "filling" and "refilling" transitions 2 ↔ 3 and 0 ↔ 1, respectively, the cascade 2 → 3 → 4 and the pair (2 ↔ 3, 2 ↔ 5) describing color competition. The pump field and the FIR fields may contain several modes. A general solution of the model is not available - instead we have solved all the important subcases of which we will discuss firstly the competition between Raman and line center modes in the system 1 ↔ 2 ↔ 3 and secondly describe interferense phenomena occuring when both the filling and refilling transitions lase in the system consisting of levels 0, 1, 2 and 3.

2. THEORETICAL MACHINERY, NOTATION

We have based our calculations on semiclassical laser theory employing density matrix formalism[2]. The fields are described classically by Maxwell equations. The response of the "few-level" atoms coupled to the fields and thermal reservoirs (collisional effects and radiative damping) is obtained by solving the master

equation of the density matrix. Several simplifying approximations are made to avoid unnecessary compications:
- rotating wave approximation
- single mode pump (for multimode pump see ref. 3)
- homogeneous broadening
- level degeneracy neglected
- collisions accounted with the aid of simple relaxation rates etc.

On the other hand we generally do not restrict the field amplitudes by using ordinary perturbation treatments. The case where only one of the transitions is coupled by a multimode field is solvable with the aid of matrix continued fractions[3]. For a detuned pump field a convenient solution is a perturvation expansion in terms of the mode-mode interaction strength (self-effects accounted exactly).

The resonance frequencies of the model atom of Fig. 1 are denoted by ω_{ij}. For simplicity all levels are assumed to relax at a rate γ outside of the relevant states. In the absence of EM fields the low-lying levels have a population density $n_0 = n_1$ and the excited states are empty i.e. $n_i = 0$ for $i \geq 2$. The pump fild $\frac{1}{2}E_0 \exp[i(Kz - \Omega t)]$ + c.c. gives rise to two relevant parameters: the pump detuning $\Delta = \omega_{21} - \Omega$ and the flipping rate $\alpha = \mu_{21} E_0/2\hbar$ (μ_{mn} is the dipole matrix element between states m and n). The FIR field coupling to the level pair 2 and 3 may contain several modes which are described by the flipping parameters $\beta_0, \beta_1, \ldots$ ($\beta_0 = E_{FIR,0} \mu_{23}/2h$ analogously to the pump case) and whose frequencies are $\nu_0, \nu_1, \ldots$. The corresponding quantities appearing in the $0 \leftrightarrow 1$ transition are denoted by $\epsilon_0, \epsilon_1, \ldots$ and $\delta_0, \delta_1, \ldots$. When performing the rotating wave approximation one of the frequencies in each of the transition is chosen as a central one and only the beat frequencies remain. Then for example in a

578 LASERS IN ATOMIC, MOLECULAR AND NUCLEAR PHYSICS

two- mode FIR-field the total flipping rate will retain time dependence i.e.

$$\beta_{eff} = \beta_0 + \beta_1 \exp[i(\nu_1 - \nu_0)t] \qquad (1)$$

which greatly complicates the solveng of the density matrix equations. In the following we only discuss the results leaving out the details of the lengthy calculations.

3. COMPETITION BETWEEN RAMAN- AND LINE CENTRE OSCILLATION

Many of the characteristics of laser pumped lasers are describable by a model consisting of a three-level atom interacting with two laser beams - one pump beam and the second one a signal beam [1,4]. The small signal gain for a detuned pump shows two maxima - one at the line center $\nu = \omega_{23}$ and the second one at the Raman resonance resonance $\Omega - \nu = \omega_{31} = \omega_{21} - \omega_{23}$. (Note that the Raman emission is tunable in contrast to the line center radiation.) Despite the fact that the small signal gain at both resonances is usually nearly the same, FIR-lasers almost exclusively operate on the Raman line.

To explain the preferred Raman-line operation of FIR-lasers we have to calculate the gain spectrum assuming two cotravelling FIR modes. For two modes (or several equidistant modes) the problem can be exactly solved with matrix continued fractions. Very simple analytical expressions are obtained if the pump laser is is detuned enough to allow a perturbation expansion in terms of Δ^{-1}. Assuming that the FIR-mode indexed by 0 in Eq.(1) oscillates at the Raman resonance ($\nu_0 = \Omega - \omega_{31}$) and the mode 1 is at the line center ($\nu_1 = \omega_{23}$) we get for the Raman mode gain

$$G_0 = C \frac{\alpha^2}{\Delta^2} \frac{1}{1 + \beta_1^2/\gamma^2} \qquad (2)$$

and for the line center gain

$$G_1 = C \frac{\alpha^2}{\Delta^2} \frac{1}{1+\beta_1^2/\gamma^2} \frac{1 - 3\beta_0^2/\gamma^2 + \beta_1^2/\gamma^2}{1 + 4\beta_1^2/\gamma^2} \quad (3)$$

where $C = \nu\mu_{23}^2 n_0/c\hbar\gamma\epsilon_0$. From Eq.(3) we see that Raman-oscillation is able to suppress the line center gain (G_1 turns negative) for $\beta_0^2 > \gamma^2/3$ (the actual critical value of β_0 is more complicated when the relaxation rates are not taken equal). The Raman gain remains positive even though it may be considerably saturated by oscillation at line center.

To study the transient behavior we have to numerically integrate the equations

$$\frac{d}{dz}\beta_k^2 = G_k(\beta_0^2, \beta_1^2)\beta_k^2 \quad (k = 0.1) \quad (4)$$

with appropriate boundary conditions at the entrance plane $z = 0$. One example is shown in Fig. 2.

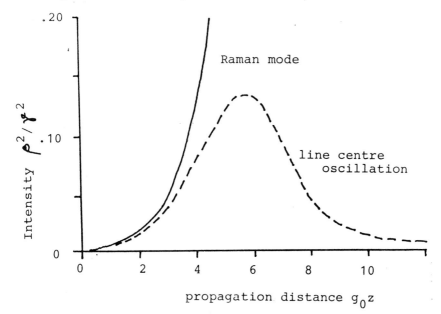

FIGURE 2

Both modes initially grow at the same rate, but once they reach saturating values ($\beta \sim \gamma$) the Raman oscillation suppresses the line center oscillation in a short distance. To obtain some insight into the underlying physical effects we study the equation of motion of the density matrix element $\rho_{23}(1)$ (responsible of the line center polarization)

$$\dot{\rho}_{23}(1) + (\gamma + i\Delta)\rho_{23}(1) = i\beta_1 [\rho_{22}(0) - \rho_{33}(0)]$$
$$+ i\beta_0 [\rho_{22}(1) - \rho_{33}(1)] - i\alpha\rho_{31}^*(-1) \quad (5)$$

On the right hand side the source terms can be classified as a DC-population term, a population pulsation term and an off-diagonal ρ_{31}-term presenting the ordinary two-photon coherence. The three terms contribute a factor $2 - 2\beta_0^2/\gamma^2$, $-\beta_0^2/\gamma^2$, and -1, respectively, to the factor $1 - 3\beta_0^2/\gamma^2$ appearing in Eq.(3). Thus the main mode suppression mechanism is due to population changes (a simple view is obtained by noting that the two-photon transition transfers population from 1 to the final level 3 and efficiently reduces the population inversion between levels 2 and 3 created by off-resonant single photon pumping $1 \rightarrow 3$ and required for lasing at the line center).

The result that the FIR gain disappears above a critical Raman intensity remains valid in a wide range around the line center as we have verified by calculating the gain spectrum. This task requires the calculation of the weak probe (β_1, ν_1) gain in the presence of a strong pump (α_1, Ω) and an arbitrarily intense FIR mode (β_0, ν_0). As long as the FIR modes are at different resonances the single mode probe approach suffices. When one, however, wants to study the gain behavior under the Raman resonance housing the strong mode, too, a bichromatic probe[6] is needed. This is due to the fact that the beating of β_1 and β_0 automatically drives a polarization term at $\nu_0 + (\nu_0 - \nu_1)$ creating a field mode

NEW LASER METHODS AND SOURCES   581

symmetrically located with ν_1 with respect to ν_0.

4. INTERFERENCE BETWEEN FILLING AND REFILLING TRANSITIONS

In FIR-lasers both the filling (2 ↔ 3) and refilling (0 ↔ 1) transitions may lase simultaneously[7]. Theoretically a homogeneously broadened lambda-configuration is exactly equivalent to a V-configuration and, therefore, one would expect - according to the discussion above - that at both transitions the Raman oscillation would survive, if there were no interference between the FIR-emissions. It has, however, been shown experimentally that the refilling oscillation may also occur at the line center[8]. To explain this we have calculated the gain spectrum assuming a detuned pump which allows a perturbation expansion in terms of Δ^{-1} (exact calculation can be performed with continued fraction techniques).

Assuming again that two of the FIR modes are at Raman resonances, i.e. $\nu_0 = \Omega - \omega_{31}$ and $\delta_0 = \Omega - \omega_{20}$ for β_0 and ε_0, respectively, and two of them at line centre, i.e. $\nu_1 = \omega_{23}$ and $\nu_1 = \omega_{01}$ for β_1 and ε_1, respectively, we obtain for the field equations (amplitudes in unites of γ)

$$\frac{d}{dz}|\varepsilon_0| = C_\varepsilon \frac{\alpha^2}{\Delta^2} \frac{\gamma^2}{D} [N|\varepsilon_0| - 2\cos\Phi|\varepsilon_1||\beta_1||\beta_0|] \qquad (6)$$

$$\frac{d}{dz}|\beta_0| = C_\beta \frac{\alpha^2}{\Delta^2} \frac{\gamma^2}{D} [N|\beta_0| - 2\cos\Phi|\varepsilon_1||\varepsilon_0||\beta_1|] \qquad (7)$$

$$\frac{d}{dz}|\varepsilon_1| = C_\varepsilon \frac{\alpha^2}{\Delta^2} \frac{\gamma^2|\varepsilon_1|}{1+4\varepsilon_1^2}[1 - (\varepsilon_0^2 - \beta_0^2)(3 + 3\varepsilon_1^2 + \beta_1^2)/D] \qquad (8)$$

$$\frac{d}{dz}|\beta_1| = C_\beta \frac{\alpha^2}{\Delta^2} \frac{\gamma^2|\beta_1|}{1+4\beta_1^2}[1 + (\varepsilon_0^2 - \beta_0^2)(3 + 3\beta_1^2 + \varepsilon_1^2)/D] \qquad (9)$$

where $N = 1 + |\varepsilon_1|^2 + |\beta_1|^2$, $D = N^2 - 4|\varepsilon_1|^2|\beta_1|^2 > 0$ and the global phase $\Phi = \Psi(\beta_0) - \Psi(\beta_1) + \Psi(\varepsilon_0) + \Psi(\varepsilon_1)$ obeys the equation

$$\frac{d}{dz}\Phi = \frac{\alpha^2}{\Delta^2}\frac{\gamma^2}{D}\;(C_\varepsilon/\mathcal{E}_0^2 + C_\beta/\beta_0^2)\;|\mathcal{E}_0||\mathcal{E}_1||\beta_0||\beta_1|\sin\Phi \qquad (10)$$

The small signal gain coefficients are denoted by C_β and C_ε. Equations (6)-(10) reduce to Eqs. (2)-(3) when we let $\mathcal{E}_0, \mathcal{E}_1 \to 0$ (or $\beta_0, \beta_1 \to 0$; note the symmetry). A novel feature is the phase dependence typical for interference phenomena. According to (10) Φ tends to lock to π; the other possible solution $\Phi = 0$ is unstable. For $\Phi = \pi$, $\cos\Phi = -1$ and, therefore, the Raman modes assist mutually each other. The evolution of the line center oscillations is independent of Φ but sensitively depends on the relative magnitudes of the Raman-intensities. An example based on the numerical integration of Eqs. (6)-(10) is shown in Figure 3.

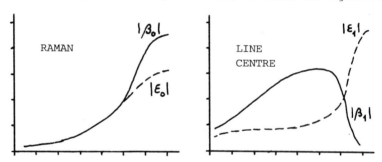

FIGURE 3

The phase (not shown) locks rather abruptly to π. Both Raman modes grow, but only one of the line center modes survive. The system behavior turns out to depend on the initial conditions, too, in addition to e.g. on C_β and C_ε and therefore there does not exist a simple general prediction of which of the modes survive and which are suppressed.

5. CONCLUDING REMARKS

In addition to the practical needs met e.g. in laser design, the unexpected theoretically interesting features obtained from solutions of multimode n-level models provide good motivation for the study of these systems. Astonishingly, the apparently complicated formulas lead, after cumbersome algebraic manipulations, to rather simple final results. Computationally the equations are at present tractable - simple theoretical explanations are still required. Quite often the fields used are not intense enough to render e.g. dressed atom interpretation superior to a bare atom picture.

In the previous examples we have illustrated some mode competition aspects applying stationary solutions. As expected the system may also exhibit interesting dynamical behavior which in part aids in interpreting experimental results. In practical FIR lasers these considerations are of importance since the relaxation, pulse propagation and pumping times are all of the same order of magnitude.

REFERENCES

1. T. A. DETEMPLE, Pulsed Optically Pumped Far Infrared Lasers, in Infrared and millimeter waves, edited by K. J. Button (Academic, New York, 1979), pp. 129-184.
2. See e.g. M. Sargent III, M. O. Scully, W. E. LAMB, Jr., Laser Physics (Addison-Wesley, Reading, 1974).
3. M. A. DUPERTUIS, R. R. E. SALOMAA, M. R. SIEGRIST, IEEE J. Quantum El., QE-20, 440 (1984).
4. B. WELLEGEHAUSEN, IEEE J. Quantum El., QE-15, 1108 (1979).
5. M. A. DUPERTUIS, M. R. SIEGRIST, R. R. E. SALOMAA, to be published in Phys. Rev. A Rapid Communications; Proc. of 8th Internat. Conf. on Infrared and Millimeter Waves, Dec. 12-17, 1983, Miami, Fla, paper TH2.6.
6. L. W. HILLMAN, R. W. BOYD, C. R. STROUD, Jr., Opt. Lett., 7, 426 (1982); S. T. HENDOW, M. SARGENT III, Opt. Commun., 40, 385 (1982) and 43, 59 (1982).
7. G. DODEL , N. G. DOUGLAS, IEEE J. Quantum El., QE-18, 1294 (1982); G. D. WILLENBERG, J. HEPPNER, F. B. FOOTE, J. Quantum El., QE-18, 2060 (1982).
8. J. S. MACHUZAK, P. WOSKOBOINIKOV, W. J. MULLIGAN, Proc. of The 8th IRMMW (see/5/), paper W5.1.
9. M. A. DUPERTUIS, R. R. E. SALOMAA, M. R. SIEGRIST, Proc. of

The 3rd International Conf. in IR Physics, Zürich, July, 23-27, 1984, p. 433.

6.7 Optical Pumping and Coherence Effects in Doppler-Free Laser Spectroscopy

W. GAWLIK

Instytut Fizyki, Uniwersytet Jagielloński, 30-059 Kraków, Reymonta 4, Poland

Velocity selective optical pumping effects are discussed and illustrated with examples of polarization spectroscopy experiments on sodium. Reduction of effective saturation parameters, changes of ratios of signal amplitudes and stability of induced population anisotropies are considered as the most important consequences of optical pumping. Also discussed are coherence and nonstationary effects of optical pumping which are shown to be responsible for substantial deformations of the experimantal signals.

1. INTRODUCTION

Optical pumping is the method of altering populations of atomic (molecular) energy levels with the help of light of given frequency and polarization. In some cases one may also create a coherence of atomic sublevels[1,2]. Optical pumping and coherence effects show up also in laser spectroscopy. In particular, they are substantial for polarization spectroscopy, so we will mainly illustrate this lecture with the results of polarization spectroscopy experiments on sodium vapours. From this results we will, however, derive general conslusions valid also for other spectroscopic Doppler-free methods based on velocity selection. More detailed discussion may be found in [3]. Other related works are [4-6].

Polarization spectroscopy (PS) introduced by Wieman and

586 LASERS IN ATOMIC, MOLECULAR AND NUCLEAR PHYSICS

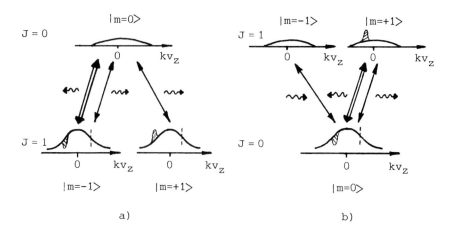

FIGURE 1 Three-level model for discussion of polarization spectroscopy consisting of a pair of degenerate Zeeman sublevels $|J=1,m=\pm 1\rangle$ and a single state $|J=0,m=0\rangle$ as the third level arranged in the Λ or the V-system.
a) For the Λ-system: population difference to be sustained against relaxation from the ground states;
b) For the V-system: population difference to be sustained against relaxation from the excited state.

Hänsch[7] is a very sensitive Doppler-free technique of spectroscopy in which opposing beams of different intensity and polarization traverse the sample. In this way the probe beam monitors the dichroism and birefringence of the medium rather than absorption changes (as it was in the case of saturated absorption spectroscopy (SAS)).

While a two-level system provides an adequate model for explanation of the salient features of SAS one needs at least a three-level structure to discuss PS[8]. The simplest such structure is a three-level system where we have a pair of degenerate Zeeman sublevels $|J=1,m_J=\pm\rangle$ and a single, zero-angular momentum state $|J=0,m_J=0\rangle$ as the third level (Figure 1).

The anisotropy responsible for the signal induced by the circularly polarized pump beam depends on the difference $N = N_+ - N_-$ of populations of the Zeeman sublevels. For small optical density, weak light intensity and steady-state conditions the amplitude E_{scat} of the scattered field of the probe beam which passes the analyser is proportional to

$$\int_{-\infty}^{\infty} W(v_z) \Delta N(\delta + k v_z) \alpha(\delta - k v_z) \, dv_z \qquad (1)$$

where $W(v_z)$ is the distribution of the velocity components along the beams, $\delta = \omega_L - \omega_o$ and $\alpha(x) = \gamma_{01}^2/(x^2 + \gamma_{01}^2)$ is a Lorentzian factor resulting from optical resonance between the probe beam and the investigated transition 0 - 1 having the homogeneous linewidth γ_{01}.

In our simple model two cases shown in Figure 1 need to be considered:

-) the Λ system with

$$E_{scat} = CN_o \beta' \frac{\beta^2}{\gamma} /(\delta + i\gamma_{01}) = CN_o \beta' \ G_\gamma^2 \gamma_{01}/(\delta + i\gamma_{01}) \qquad (2)$$

and

-) the V system where we have

$$E_{scat} = CN_o \beta' \frac{\beta^2}{\Gamma} /(\delta + i\gamma_{01}) = CN_o \beta' \ G_\Gamma^2 \gamma_{01}/(\delta + i\gamma_{01}) \qquad (3)$$

Here: N_o is an initial population of lower states; β and β' are Rabi frequencies for the pump and the probe respectively; γ, Γ are relaxation constants for the lower and upper states respectively; C denotes a constant depending on the length of the medium, and G_γ and G_Γ are saturation parameters defined as:

$$G_\gamma^2 = \beta^2/\gamma \gamma_{01} \ , \quad \text{and} \quad G_\Gamma^2 = \beta^2/\Gamma \gamma_{01} \qquad (4)$$

In addition to the scattered field given by (2) and (3) we allow for a small background $E_b = E_o(\theta + i\varphi)$ transmitted by the polarizer-analyser system, where θ represents a deviation from exact crossing of the polarizers of their possible finite extinction at crossed position (or both) and φ represents possible phase difference due to, e.g. a strain-induced birefringence of the cell windows. The resulting intensity measured by the detector is now given by

$$I = I_p[\Theta^2 + \varphi^2 + 2C'N_oG_a^2(\Theta\gamma_{01}\delta/(\delta^2 + \gamma_{01}^2) - \varphi\gamma_{01}^2/(\delta^2 + \gamma_{01}^2)) +$$
$$+ (C'N_oG_a^2)^2\gamma_{01}^2/(\delta^2 + \gamma_{01}^2)] \qquad (5)$$

where G_a becomes G_γ for the Λ-system and G_Γ for the V-system and C' is a constant.

When $\Theta = \varphi = 0$, signals are Lorentzian resonances on zero background with the halfwidth γ_{01} and the amplitude proportional to $N_o^2I_pI_s^2$ (I_p, I_s are intensities of the probe and pump beam respectively). When $E_b \neq 0$, signal (5) becames a complex mixture of Lorentzian and dispersion resonances of various amplitudes together with a constant background $\Theta^2 + \varphi^2$. Appropriately large background heterodynes a weak scattered field, which makes interference terms (proportional to N_o) larger than the intensity of the scattered light (proportional to N_o^2). When additionally $\varphi = 0$, the signal change is despersive and its amplitude is proportional to $N_oI_pI_s$. (With the pump beam linearly polarized at 45° with respect to the probe, signals remain always symmetric, even when heterodyned[9].)

Since usually the rates γ and Γ are very different, the saturation parameters G_γ and G_Γ differ very much and the signal intensity expressed by (5) depends very strongly on the level configuration; usually signals are very much stronger for the Λ-system than for the V one. In the Λ-system the main effect of the pumping beam is to redistribute the populations between the sublevels of the lower state (optical pumping). The dynamics of the redistribution is determined by G_γ, i.e. the ratio between the perturbation due to the pumping beam and the relaxation of the lower states. In the V-system, on the other hand, the population difference $\Delta N = N_+ - N_-$ on which the signal depends is between the upper state sublevels. To sustain this population difference the

NEW LASER METHODS AND SOURCES 589

pump beam must work against the relaxation of the upper states. If this is faster than that of the lower state, the intensity of the pumping beam needs to be correspondingly higher. Formation of the ΔN difference of the upper state populations is associated with the saturation of the optical transition.

We may, therefore, distinguish between Ps as a saturation phenomenon and as a pumping phenomenon - the latter being realizable when the lower state is not single.

A detailed discussion of experimental aspects of PS is presented in[3]. Here we wish only to stress the importance of compensation of the Earth's and stray magnetic fields (or placing the cell in a magnetic shield) as they are another source of polarization anisotropy.

3. CONSEQUENCES OF OPTICAL PUMPING:

3.1 Change of Saturation Conditions

An example of Doppler-free polarization spectrum of Na D_1 (589 nm) line is shown in Figure 2 (from [8]). It was obtained with the in-

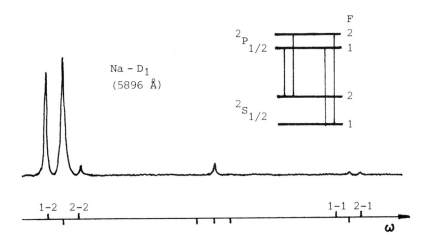

FIGURE 2 High resolution polarization spectrum of Na D_1 line. Frequency scale hfs components are marked F_e-F_g. Crossovers are marked below the line.

tensities 2.5 mW/cm^2 and 0.05 mW/cm^2 for the pump and the probing beam respectively. This yields G_Γ^2 of the order of 0.1, which means that the stimulated emission is far weaker than the spontaneous one and the saturation of the optical transition (i.e. equalization of the upper and lower level populations) is negligible. On the other hand $G_\gamma^2 \approx 10$, which indicates that optical pumping processes may be very important even when there is no saturation of the optical transition. They allow observation of Doppler-free signals with beam intensities far weaker than would be necessary to achieve the saturation. Indeed, according to formulae (2,3) and definitions (4), amplitude E_{scat} and therefore also the amplitude of Doppler-free resonance in PS changes drastically with the configuration of the levels. Since in general we have $\gamma \ll \Gamma$, with the same intensity of laser beams the signals obtained for the Λ-system are $(\Gamma/\gamma)^4$ times higher than those fo the V-system in the case of crossed polarizers (and $(\Gamma/\gamma)^2$ times higher than the signals for the V-system in the case to heterodyned signals).

This property is clearly visible in the polarization spectrum of the $H_\alpha D_1$ $(2S_{1/2} - 3P_{1/2})$ line of hydrogen[19] with two hfs components of very different amplitudes: the component $2S_{1/2}(F=1) - 3P_{1/2}$ for which optical pumping in the ground state is possible (Λ-system) is strong, whereas the component $2S_{1/2}(F=0) - 3P_{1/2}$ for which there is no optical pumping in the ground state (V-system) is hardly visible. This very strong increase of the signal amplitude seems to be the most signaficant consequence of optical pumping effects from the practical point of view in the systems where optical pumping is possible.

Reduction of the light intensity necessary for the observation of PS signals also reduces the influence of the dynamic Stark effect on the shape and position of Doppler-free resonances.

Efficient optical pumping decreases the number of atoms which interact resonantly with the pumping beam, which results in a non-linear dependence of the signal on the beam intensity (saturation of the signal amplitude). It is convenient to introduce effective saturation parameter G_{ef} which reproduces G_γ and G_Γ in limiting cases and is defined as $G_{ef}^2 = I/I_{sat}$, where the saturation intensity is given by

$$I_{sat}^{-1} = \frac{4\pi}{\hbar c} \frac{|D_{01}|^2}{\gamma_{01}} (\bar{\gamma}^{-1} + \bar{\Gamma}^{-1} - A_{eg} \bar{\gamma}^{-1} \bar{\Gamma}^{-1}) \qquad (6)$$

and A_{eg} is the probability of spontaneous emission via the same transition which was perturbed by the pump. Without optical pumping $A_{eg} = \Gamma$, which yields $G = G_\Gamma$, whereas in the case where optical pumping is efficient (i.e. we have a strong branching in the upper state) $A_{eg} < \Gamma$, which increases G_{ef} even to the value G_γ when optical pumping is maximal.

3.2 Changes of Signal Amplitudes

According to (5), the signal amplitude in PS with crossed polarizers ($\Theta = \varphi = 0$) is proportional to G_α^4, i.e. to $|D_{01}|^4$, or to the square of the line strength (and transition probability). So, in contrast to SAS, amplitude ratios of various components of polarization spectra are not equal to the coresponding ratios of the transition probabilities.

This property is illustrated in Figure 2 where, e.g. hfs components $F_e = 1 - F_g = 2$ and $F_e = 2 - F_g = 2$ have their equal line strength. With parallel polarizers (SAS case), however, we have components of equal amplitudes (see Figure 12 in [8]).

In many approaches to the theory of PS[7,9,11,12] repopulation of lowers sublevels via spontaneous emission is neglected. Such a simplification, tolerable in the case of very strong branching to the other levels (as e.g. in molecular transitions), fails for the atomic transitions where the repopulation pumping is as important as the depopulation one. Consequently, the signal amplitudes estimated in the papers[7,9,11,12] differ significantly from the observed ones (Figure 2). On the other hand a correct analysis of a full optical pumping cycle in two-, three-, and four-level system[13,14] gives a very good agreement with the experiment.

3.3 Stability of Induced Population Anisotropies

The PS signal decays as fast as the population asymmetry $N_+ - N_-$ relaxes to zero value. If the V-configuration is considered and collisions are neglected then spontaneous emission is the dominant relaxation mechanism. Relaxation times of the lower levels are in general longer than those of the upper ones, particularly in the case of nonradiating ground states. If the lower level has a structure allowing for optical pumping (Λ-system), then the population imbalance $N_+ - N_-$ persists for a much longer time than the lifetime of the upper level or the decay time of optical coherences (i.e. coherence between the upper and the lower states). This property is discussed in[8,15] where PS signals were observed even when the pump and probe beams were so separated that the optically pumped (oriented) atoms had to diffuse within some µs (to be compared with 16 ns lifetime of the excited state) to the probing region.

The high stability of the anisotropy induced by optical pumping in the lower level offers a possibility of studying many phenomena unperturbed by a strong light field. For example, a free precession of optically oriented atoms in a magnetic field[15] and velocity-changing collisions could be studied in this way without additional perturbation by the light field. The separation of the beams extends, moreover, the time scale of the observation, which may be helpful with low frequencies of the collisions.

3.4 Coherence Effects

The light beam interacting with an atomic system does not only change its populations but is also capable of establishing a coherence between perturbed atomic states, in particular, between Zeeman sublevels within the upper and/or lower state[16,17].

In our situation of PS the Zeeman coherences may be introduced between sublevels with m = ±2, ±4, ... by an appropriately strong probe beam (linearly polarized) propagating across the cell also without the pumping beam[19]. As long as there is no magnetic field the coherence contributions have no direct influence of the signal with crossed polarizers (they depend in a dispersive way on the Zeeman splitting). A necessary condition for these contributions to be revealed in PS is a partial removing of the degeneracy of the Zeenam sublevels. When the degeneracy is removed, so called "stimulated Hanle effect"[18] can be observed in forward scattering of the probe beam which simultaneously induces and detects the coherences. Figure 3e presents an example (from reference[19]) of the coherence signals observed at Na D_1 ($3^2S_{1/2} - 3^2P_{1/2}$) transition as a function of a magnetic field. The narrow structure around zero field (resembling a squared dispersion curve) is a superposition of the contributions due to the Zeeman and the optical coherences. Their width is associated with the width of corresponding levels; hence, it may be very small with the long-lived states.

Such coherences can have interesting effect on the Doppler-free signals obtained with two beams as in PS. Figures 3a-d present various possible cases described in[8], of interference between the coherence contributions due to a strong probe beam and the population ones associated with the pump beam. Curve (a) is a zero signal obtained when the frequency of a single probe beam was scanned across the Doppler contour with zero magnetic field. All conributions to the signal are zero under these conditions. Curve (b) was recorded with a counter-propagating pump beam. It is a part of a standard polarization spectrum of the Na D_1 line in zero magnetic field including the $^2S_{1/2}(F=2) - ^2P_{1/2}(F=1,2)$

594 LASERS IN ATOMIC, MOLECULAR AND NUCLEAR PHYSICS

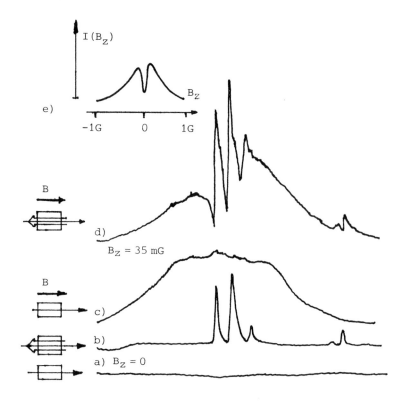

FIGURE 3 Interference between the coherence contributions and the population ones (from[6]).
a) No magnetic field. No pumping beam. Null signal recorded on probe beam.
b) No magnetic field. Pumping beam. Standard polarization spectrum (population contributions due to the pumping beam).
c) Magnetic field. No pumping beam. Doppler-broadened coherence contribution due to the strong probe beam.
d) Magnetic field and pumping beam simultaneously applied.
e) An example of the coherence signal observed at the Na D_1 transition (from[5]).

hfs components and cross-over resonances. No coherence contributions due to the probe beam are seen in the spectrum. Curve (c) is the spectrum obtained when the frequency of the single probe beam was scanned but, differently from the previous cases, Na atoms were in a weak (35 mG) longitudinal magnetic field. This time

NEW LASER METHODS AND SOURCES 595

a large coherence contribution shows up. Unlike the magnetic dependence from Figure 3e which was narrow, the frequency dependence is Doppler-broadened since the coherence generation by the single beam is not velocity-selective. Finally, curve (d) represents the signal when both the pump beam and the magnetic field are acting simultaneously. A drastic deformation of the polarization spectrum occurs due to the interference of the population contributions and the coherence ones (revealed by the degeneracy removal). Nearly Lorentzian Doppler-free signals from curve (b) are changed into dispersion resonances similar to those obtained with heterodyning by the coherent background when polarizers are uncrossed. Described effect demonstrates the very high sensitivity of PS (and other methods where optical pumping and coherence effects may show up) to minute external magnetic fields (of the order of some tens of mG) and indicates that a careful shielding or compensation is necessary to eliminate serious deformations of the signal and possible systematic errors.

4. NONSTATIONARY EFFECTS OF OPTICAL PUMPING

The effects discussed above could be described in terms of the density matrix equations (see e.g.[5,13,14]. There are, however, some effects that may not be explained with standard stationary solutions of the equations.

One such effect is the perturbation of the polarization spectrum of the Na D_1 line when the intensity of the probe beam is so much increased that it perturbs atoms about as strongly as the pump beam[8,20,21]. Figure 4a shows the spectrum recorded with the pump beam intensity 5.6 mW/cm^2 and the probe beam attenuated to 0.9 mW/cm^2. Subsequent curves (Figures 4b, 4c, and 4d) illustrate the deformation associated with an increase of the pump intensity to 30 mW/cm^2 and the probe beam one to 23 mW/cm^2. Besides power

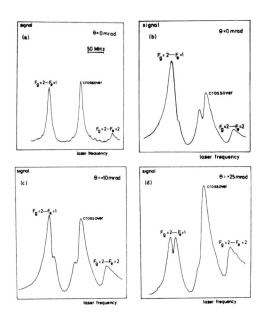

FIGURE 4 Sections of polarization spectra in the Na D_1 line: a) weak probe beam, crossed polarizers ($I_P = 0.9$ mW/cm^2, $I_S = 5.6$ mW/cm^2); b)-d) deformation if the polarization spectra associated with an increase of the beam intensities ($I_P = 23$ mW/cm^2, $I_S = 30$ mW/cm^2) and various rotations of the polarizations from crossed position.

broadening of the resonances a new narrow structure appears close to the resonance centers. It consists of dispersion forms and symmetrical dips (Figure 4b) with relative proportions depending on particular conditions of the interference with the background. By an appropriate choice of the uncrossing angle θ the dispersion contributions can be recorded (Figure 4c) or the symmetrical dip may be obtained in the center either of the "genuine" resonances or of the cross-over one. The width and the depth of the dip grow with the increase of the beam intensities. It is possible to obtain dips much narrower than the natural width of the investigated transition. For example, Figure 5 shows the dip with FWHM equal to 2.6 MHz, while the natural width of the Na D_1 line is 10 MHz.

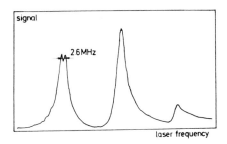

FIGURE 5 Example of the narrowest dips in the Na D_1 line with a linewidth of 2.6 MHz.

Two mechanisms have been considered as a possible physical ground for the appearance of the described dips: Firstly, they could be due to Zeeman coherences introduced by a sufficiently strong probe beam and revealed by velocity-selective differential light-shifts, removing the degeneracy of the magnetic sublevels while the laser is tuned across the resonance[20]. A second possible ground could be nonstationary effects of optical pumping to some other levels that could not be further excited by any of the beams. Nonstationarity arises from the fact that the atoms move across the light beam, hence the interaction time is finite and steady-state conditions may not be reached during typical transit times. Such effects have recently been shown to be responsible for some pecularities of two-photon resonances[22].

To distinguish between the two possible mechanisms several experimental tests have been performed[21]. Their results favour the interpretation that the nonstationary effects of optical pumping are responsible for the narrow "subnatural" structures. To discuss the effects within our simple model of Figure 1 we have to extend it by adding the fourth "trap" level to which the excited atom may spontaneously decay. In the simplest possible case with the lower state $J = 1$ it is the sublevel $|J=1, m_J=0\rangle$ which acts as the trap level.

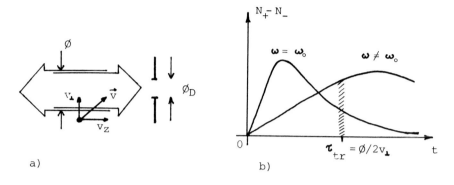

FIGURE 6 a) Schematic diagram showing an atomic movement across the pumping and probing beams of a diameter $\emptyset$ to the central derection region with a diameter $\emptyset$ ($\emptyset_D > \emptyset$).
b) Time evolution of the population asymmetry $N_+ - N_-$ for given intensities and various detunings of the beams.

For a simultaneous pumping by the pump and probe beams of comparable intensities and diameters (Figure 6a) the rate equations predict time evolution of the population asymmetry $N_+ - N_-$ such as shown in Figure 6b. The evolutions for $\omega_L = \omega_o$ and $\omega_L \neq \omega_o$ have similar characters but differ in their time scales and magnitudes of the $N_+ - N_-$ asymmetry. The time scale is determined by the transit time $\tau_{tr} = \emptyset/2v_\perp$ of an atome flying across the beam to its central detection region and by the pumping rates, which depend on the intensities and detunings of the beams (the pumping is faster for $\omega_L = \omega_o$). An initial increase of the $N_+ - N_-$ curves in Figure 6b reflects a build-up of this ground state orientation due to the pump beam pumping, while its later decrease is associated with the probe beam depopulation of both N_+ and N_- to the trap level.

If the pumping is strong (high intensities, long interaction times, and $\omega_L = \omega_o$), nearly all atoms which enter the detection region (see Figure 6a) are in the trap state. Consequently, after τ_{tr} the asymmetry $N_+ - N_-$ may be very low (for a sufficiently long interaction time steady-state conditions, i.e. an equilibrium between pumping and relaxation, might be achieved). If, however, the pumping is weaker (e.g. $\omega_L \neq \omega_o$), the evolution of $N_+ - N_-$ is slower and the steady-state conditions may not yet been reached within a given τ_{tr}. Then the probe beam pumping to the trap level would not completely reduce the orientation $\Delta N = N_+ - N_-$ built by the pump beam. Consequently, if the detuning $\omega_L - \omega_o$ is not too large (say $\omega_L - \omega_o \approx \gamma_{01}$), we may have for given beams intensities and τ_{tr}

$$\Delta N(\omega_L = \omega_o, t = \tau_{tr}) < \Delta N(\omega_L \neq \omega_o, t = \tau_{tr})$$

Since ΔN determines the amplitude of the PS signal a dip may occur in the very center of the Doppler-free resonance (for $\omega_L = \omega_o$) when the above effects take place.

To obtain ensemble averaged expressions for the signal affected by the velocity selective optical pumping two integrations have to be performed

$$I = I_p |\langle \overline{\Delta N}(v_z) \alpha(\delta - kv_z) \rangle_{v_z}|^2$$

Firstly, the solutions of the time dependent equations yielding $\Delta N[\omega_L, v_z, t(v_\perp)]$ are to be averaged over various transit times arising from a spread of the transversal velocities $v_\perp$ which yields $\overline{\Delta N}$, and, secondly, such $\overline{\Delta N}$ multiplied by the probe beam resonance factor α (also frequency- and v_z- dependent) has to be integrated over the $W(v_z)$ distribution like in formula (1). As usual, the latter intergration yields Doppler-free signals (as expression (5) obtained with the steady-state calculations) can be given when nonstationary optical pumping affects the measurements.

It should be noted that although the width of the considered dips may be significantly smaller than the natural width it is not possible to use them for measurements exceeding the resolution limits imposed by the natural width. The main reason is a complicated nonlorentzian shape of the dips with a relatively flat center which can not be described with an analytical expression. Nevertheless, a full understanding of the nonstationary effects of optical pumping is necessary for a correct interpretation of experimental results.

5. CONCLUSIONS

We have discussed, and illustrated with numerous examples of the polarization spectroscopy in sodium, the most important consequences of optical pumping for Doppler-free laser spectroscopy. Many of them are desirable from the point of view of experimental convenience, as for instance the facility of obtaining Doppler-free signals; some other consequences, however, might present serious complications. Optical pumping may drastically affect the results obtained with very well known, as it is generally admitted, techniques of laser spectroscopy. That is why a thorough understanding of all aspects of optical pumping is necessary and further studies are desirable.

REFERENCES

1. A. KASTLER, C. COHEN-TANNOUDJI, Progress in Optics, 5, 3 (1966)
2. W. HAPPER, Rev. Mod. Phys., 44, 169 (1972)
3. W. GAWLIK, Acta Phys. Polon., A, in print (1984)
4. M. PINARD, C. G. AMINOFF, F. LALOË, Phys. Rev., A19, 2366 (1979)
5. P. G. PAPPAS, M. M. BURNS, D. D. HINSHELWOOD, M. S. FELD, D. E. MURNICH, Phys. Rev., A21, 1955 (1980)
6. H. RINNEBERG, T. HUHLE, E. MATTHIAS, A. TIMMERMAN, Z. Phys., A295, 17 (1980)
7. C. WIEMAN, T. W. HANSCH, Phys. Rev. Lett., 36, 1170 (1976)
8. W. GAWLIK, G. W. SERIES, in Laser Spectroscopy IV, edited by H. Walther and K. W. Rothe (Springer-Verlag, Berlin, 1979), vol. 21, pp. 210-222.
9. R. E. TEETS, F. V. KOWALSKI, W. T. HILL, N. CARLSON, T. W. HANSCH, Proc. Soc. Photoopt. Instr. Eng., 113, 80 (1977)
10. J. E. M. GOLDSMITH, E. W. WEBER, T. W. HÄNSCH, Phys. Rev. Lett., 41, 1525 (1978)
11. M. Sargent III, Phys. Rev., A14, 524 (1976)

12. S. SAIKAN, J. Opt. Soc. Am., 68, 1184 (1978)
13. S. NAKAYAMA, G. W. SERIES, W. GAWLIK, Opt. Commun., 34, 382 (1980)
14. S. NAKAYAMA, J. Phys. Soc. Jap., 50, 606 (1981)
15. S. NAKAYAMA, G. W. SERIES, W. GAWLIK, Opt. Commun., 34, 389 (1980)
16. C. COHEN-TANNOUDJI, in Frontiers in Laser Spectroscopy, edited by R. Balian, S. Haroche, and S. Liberman, Proceedings of Les Houches Summer School (session XXVII) (North-Holland, Amsterdam, 1976), pp. 3-104.
17. B. DECOMPS, M. DUMONT, M. DUCLOY, in Laser Spectroscopy of Atoms and Molecules, Topics in Applied Physics, edited by H. Walther (Springer-Verlag, Berlin, 1976) Vol. 2, pp. 283-347.
18. M. FELD, A. SANCHEZ, A. JAVAN, B. J. FELDMAN, in Méthodes de Spectroscopie sans Largeur Doppler de Niveaux Exctés de Systèmes Moléculaires Simples, Colloques Internationaux du C.N.R.S. (C.N.R.S., Paris, 1973), No. 217, pp. 87-104.
19. W. GAWLIK, J. KOWALSKI, R. NEUMANN, F. TRÄGER, Opt. Commun, 12, 400 (1974) (see also S. GIRAUD-COTTON, V. P. KAFTANDJIAN, L. KLEIN, Phys. Lett., 88A, 453 (1982); and W. GAWLIK, Phys. Lett., 89A, 278 (1982))
20. W. GAWLIK, J. KOWALSKI, F. TRÄGER, M. VOLLMER, Phys. Rev. Lett., 48, 871 (1982)
21. W. GAWLIK, J. KOWALSKI, F. TRÄGER, M. VOLLMER, in Laser Spectroscopy VI, Springer Series in Optical Sciences, edited by H. P. Weber and W. Lüthy (Springer-Verlag, Berlin, 1983), Vol. 40, pp. 136-137.
22. J. E. BJORKHOLM, P.F. LIAO, A. WOKAUN, Phys. Rev., A26, 2643 (1982)

Index of Contributors

Arakelian S. M. 229
Aussenegg F. R. 357

Bagayev S. N. 539
Baklanov E. V. 539
Bamford D. J. 125
Bonch-Bruevich A. M. 111
Borisov A. Yu. 397
Brückner V. 213

Chebotayev V. P. 509
Chilingarian Ju. S. 229

Danielius R. V. 397
Dubetsky B. Ya. 509
Dupertuis M. A. 575

Freiberg A. 321

Gawlik W. 585

Hodel W. 457
Huizer A. H. 487

Kaarli R. 197
Kaniauskas J. M. 77
Karaian A. S. 229
Kchromov V. V. 111
Kharlamov B. M. 259
Kikas J. 197
Koroteev N. I. 291
Korppi-Tommola J. 503
Kruglik G. S. 563

Laczkó G. 307
Letokhov V. S. 1
Lippitsch M. E. 357

Maróti P. 307
Minogin V. G. 61
Mishin V. I. 47
Moore C. B. 125

Munir Q. 527

Neugart R. 23

Papageorgiou G. C. 379
Personov R. I. 259
Petnikova V. M. 245
Plenshanov S. A. 245
Prjibelsky S.G. 111
Puretzky A. A. 135

Razjivin A. P. 397
Rebane A. 197
Rentsch S. K. 473
Riegler M. 357
Rotomskis R. J. 397
Rudzikas Z. B. 77

Saari P. 197
Salomaa R. 575
Scripko G. A. 563
Shkadarevich A. P. 563
Shuvalov V. V. 245
Siegrist M. R. 575
Suijker J. L. 487
Szalay L. 307

Valk B. 457
Valkunas L. 341
Van Grondelle R. 413
Varma G. A. G. O. 487
Vetter R. 181
Vidal C. R. 553
Vidolova-Angelova E. 93

Weber H. P. 457, 527
Wilhelmi B. 425
Wolfrum J. 157

Zheludev N. 281

Subject Index

Aggregated dyes 473, 482

^{26}Al 2

Alkali-halide crystals 288

Al_2O_3:Ti^{3+} 569

Atomic
- collisions 111
- trap 62

Anacystic nidulans 387, 388

Antenna pigments 357

Autoionization states 93, 105

Ba 90, 183

Bacteriochlorophyll 322, 401

^{10}Be 2, 4

Biharmonic pumping 253

Br 183

^{14}C 2, 3, 11, 12, 19

Ca 183

^{41}Ca 2

CARS 291
- spectroscopy 131, 157
- thermometry 295

C_6F_6 487

CH_4 160

Charge separation process 397

C_2H_3CN 143, 145

Chirp
- compression 451
- generation 451

1,1,1-chlorodifluoroethane 157, 167

Chlorophyll 308, 322, 341, 358

Chloroplast 332

Chromatium minutissimum 404

Chromatophore 333

Chromium doped crystals 563

^{36}Cl 2

Classical echo 509

CO 126, 132

$^{14}C^{16}O$ 11, 12

CO_2 293, 294

Coherence effects 585, 592

Coherent radiation 522

Collinear
- beam technique 25
- laser fast beam spectroscopy 23

Combustion 157

Compressed pulse 445

Condenced media 291

Cooling of atoms 12

Copper-doped glass 571

Correlation function 217

CrO_2Cl_2 143, 144, 147

Cs 117, 121, 181, 184, 190, 191

Cs_2 116

CsH 116

Cu-laser 51

Cyanobacteria 357, 379, 386

Cyanobacterial thylakoids 380, 385

605

SUBJECT INDEX

Dewar isomer 487, 490
1,2-dichloroethane 157, 167
Dimethyl pyrromethenone 365
Direct electronic relaxation 136
Dispersive
- delay line 440
- linear optical samples 432
- self-phase modulation 465
Displacement sensors 528
Distributed-feedback systems 233
Doppler-free laser spectroscopy 585
Dy 29, 30, 32

Energy
- migration 341
- transfer 328
Ensemble of oscillators 511
Er 29, 32
Eu 28, 34, 36, 49, 54, 55

F_2CO 143, 144
Femtosecond light pulses 425
Fiber optic sensors 527
Fluorescent bursts 8, 9
Formaldehyde 125, 126
Frequency stable lasers 539

GaAs 283, 285
- surface 304
GaP 304
GaSe 251

Gravitational
- antenna 542
- waves 539
Group-velocity dispersion 458

H_2 126, 131, 132, 165, 166, 181
HCl 183
H_2CO 19, 125, 126, 130
Hexafluorobenzene 487, 499
Highly-excited
- molecules 291
- states 93
HO_2 162, 163
Hole burning
- optical band shape 265, 266
- spectroscopy 260
Holography 198
Homogeneous line width 266, 268
- temperature dependence 269
Hyperfine structure 47, 50

I_2 183
Induced
- paramagnetism 273
- population anisotropies 592
Inhomogeneous broadening 259, 5
Inverse electronic
relaxation 135, 139
IR-multiple-photon
excitation 135
Isospin formalism 84
Isotope shift 47, 51

SUBJECT INDEX 607

K 183

Laser
- cooling 61
- induced chemical reactions 157
- induced fluorescence 125, 126
- plasma 297
- radiation pressure 61

Light
- harvesting antenna 328, 341, 397, 413, 422
- pressure force 65

$LiIO_3$ 200, 209

Line competition 575

Linear frequency sweep 442

Liquid crystals 229

Lu 94, 102, 107

Lu^+ 94, 102

Lu^{2+} 94

Lysozyme 387, 388

Mach-Zehnder Interferometric Sensors 533

Magnetic trap 71

Many-electron
- atoms 78, 82, 95, 96
- ions 78

Mean-square charge radius 57

Mg 183

Microbending pressure sensor 531

Molecular beam 182, 187

Monocrystalline silicon 219

Monomeric dyes 473, 480

Multimode fibersensors 528

Multistep
- laser ionization 93
- photoionization 14, 25
- resonance excitation 8, 18

N_2 297

Na 13, 63, 64, 73

Narrow optical resonances 546

Nd 28
- phosphate glass laser 361

Ne 117

N-methyl etiobiliverdin 369

Nonhomogeneous media 234

Nonlinear
- Fabry-Perot resonators 230
- optical activity 281, 282, 288
- optical diagnostics 301
- optical samples 437 441
- optics 553
- oscillator 232, 235
- spectroscopy 245, 291
- susceptibility 246, 287

OH 161

Optical
- bistability 229, 232
- fibers 457
- Kerr-effect 457
- pumping 585, 595
- storage devices 276

Optically pumped laser 575
Optoacoustic fibersensor 534
Optoelectronic switching 214, 226
OsO_4 143, 144, 145

Parametric picosecond light oscillator 399
Passively modelocked lasers 451
Phase
- fluorometer method 398
- modulation 438
Pheophytin-a 308
Photochemical holes 207, 262, 263
Photochromic media 197
Photodissociation dynamics 125
Photoinduced isomerization 487
Photoreceptor molecules 357, 358, 360, 363
Photosynthesis 307, 321, 328, 341, 357, 397
Photosynthetic
- electron 379
- reaction center 308
- unit 313, 322
Phycobiliproteins 359
Phytochrome 358, 363
Picosecond
- absorption spectroscopy 399
- laser system 503
- processes 213, 397, 473
- pulses 197
- spectrochronography 325
- spectroscopy 375, 473, 479

Pigment molecules 329, 397
Plastoquinone 309
Polarization sensor 532
Polymethine dyes 475
Propagation of light pulses 430, 448
Prosthesochloris aestuarii 342
Purple bacteria 413

Quantum echo 509, 523
Quasispin formalism 87

Ra 36, 40
Radiationless dissipation 322
Radical
- chain reactions 167
- reactions 161
Radioactive
- nuclides 47, 48
- isotopes 47, 53
Radioisotope dating 3
Rare
- earth elements 93
- isotope 1, 13
Rb 119, 121
Rb_2 116
Reaction center 328, 357, 397
Reactive collisions 181
Reflected optical harmonics 302
Retinal 373

SUBJECT INDEX 609

Rhodopseudomonas
- capsulata 413
- sphaeroides 333, 342, 403, 413, 422
- viridis 401

Rhodopsin 358

Rhodospirillum rubrum 402

Ring laser 451

Rotation sensor 532

Rotational state distribution 161

Rydberg states 50, 93, 101, 102, 105

$S_2C_2F_4$ 143

Selective photoreactions 260

Selectivity 5, 6, 15, 17, 18

Self-Phase Modulation 458

Semiclassical echo 509

Semiconductors 245, 250
- processes 213
- surface 301

SF_6 295

Short-lived isotopes 47

Silicon-on-Sapphire 224

Single
- mode fibers 457
- fibersensors 531
- photon counting 473

Site selection spectroscopy 259

Sm 28

SO_2 143

Solid-State lasers 563

Soliton 471

Soret band 354

Space-time holography 197

Sr 183

Stark
- effect 270
- experiments: "field curves" 272

State selective 125, 127
- excitation 182

Step photoionization 48

Stepwise excitation 11

Sum-frequency generation 305

Superhigh resolution spectroscopy 520

Temperature sensor 530

Third harmonic generation 559

Thylakoid membranes 379, 384

Time
- resolved photoconductivity 213, 217
- resolved reflectivity 215
- resolved transparency 215
- space holography 203, 204, 207

Titanium doped sapphire lasers 568

Tm 94, 104´, 105

Tm^+ 102

Two-photon Raman excitation 292

Ultrashort light pulses 197, 426

SUBJECT INDEX

Unimolecular
- chemical reaction 125
- dissociation 142
Unstable nuclides 23
UV-sources 553

Velocity selectivity 61
Vibrational excitation 165
Vibronic relaxation 275
Vinylchloride 158, 173, 177
$VOCl_3$ 143

Wave front 208, 210

Xe 119
Xe^{47+} 85

Yb 29, 32, 94, 102, 105, 107
Yb^+ 94, 102

Zeeman effect 273
Zero phonon lines 259, 264